Miguel A. Sierra
María C. de la Torre
Dead Ends and Detours

Further Reading from WILEY-VCH

K. C. Nicolaou, S. A. Snyder

Classics in Total Synthesis II
More Targets, Strategies, Methods

2003
ISBN 3-527-30685-4 (Hardcover)
ISBN 3-527-30684-6 (Softcover)

C. Bittner, A. S. Busemann, U. Griesbach, F. Haunert,
W.-R. Krahnert, A. Modi, J. Olschimke, P. L. Steck

Organic Synthesis Workbook II

2001
ISBN 3-527-30415-0

J.-H. Fuhrhop, G. Li

Organic Synthesis
Concepts and Methods

2003
ISBN 3-527-30272-7 (Hardcover)
ISBN 3-527-30273-5 (Softcover)

H.-G. Schmalz, T. Wirth (Eds.)

Organic Synthesis Highlights V

2004
ISBN 3-527-30611-0

Miguel A. Sierra
María C. de la Torre

Dead Ends and Detours

Direct Ways to Successful Total Synthesis

With a Foreword by K. C. Nicolaou

WILEY-VCH

WILEY-VCH Verlag GmbH & Co. KGaA

Authors:

Prof. Miguel A. Sierra
Departamento de Química Orgánica
Facultad de Química
Universidad Complutense
28040 Madrid
Spain

Dr. María C. de la Torre
Instituto de Química Orgánica General
Consejo Superior de Investigaciones Científicas (CSIC)
C/Juan de la Cierva 3
28006 Madrid
Spain

All books published by Wiley-VCH are carefully produced. Nevertheless, authors, editors, and publisher do not warrant the information contained in these books, including this book, to be free of errors. Readers are advised to keep in mind that statements, data, illustrations, procedural details or other items may inadvertently be inaccurate.

Library of Congress Card No.: Applied for
British Library Cataloging-in-Publication Data: A catalogue record for this book is available from the British Library.

Bibliographic information published by
Die Deutsche Bibliothek
Die Deutsche Bibliothek lists this publication in the Deutsche Nationalbibliografie;
detailed bibliographic data is available in the internet at http://dnb.ddb.de.

© 2004 Wiley-VCH Verlag GmbH & Co. KGaA, Weinheim

Printed in the Federal Republic of Germany.
Printed on acid-free paper.

Composition   Manuela Treindl, Laaber
Printing   Strauss GmbH, Mörlenbach
Bookbinding   Litges & Dopf Buchbinderei GmbH, Heppenheim

ISBN 3-527-30644-7

A Virginia y a Javier,
veros crecer es nuestra alegría

Foreword

Despite the fact that the art of total synthesis and the rationale for its pursuit as a scientific endeavor have changed over the last century, its fundamental nature as an opportunity for discovery and invention has not. Arguably, no avenue within the realm of the chemical sciences demands from, and provides to, its practitioners so much. The reason for the never-ending wealth of knowledge derived from such endeavors lies within the beauty and diversity of Nature's molecules, each of which places its own demands, challenging the suitor in terms of three dimensional and, often, sensitive molecular complexity. For instance, stunningly unprecedented molecular motifs are certain to inspire, and often require, the development of new synthetic methods and creative strategies and tactics for their laboratory construction. Equally likely, such structural domains can serve as stringent testing grounds revealing weaknesses in the power of existing synthetic technologies to effectively fashion such complexity, and thus provide impetus for further sharpening of the tools of the art. To be sure and despite our present sophistication as synthetic artisans, however, we still cannot predict with absolute certainty the outcome of our ideas regarding a particular synthetic scheme.

It is within this arena where *Dead Ends and Detours – Direct Ways to Successful Total Synthesis*, for it lucidly presents many of the unexpected pitfalls, challenges, and failures that accompany every original research plan in total synthesis. Indeed, it is in the name of originality that we face the most unknowns and the highest risks. There is no denying that the science of chemical synthesis, despite its present power, leaves much to be desired as compared to the reach of Nature. Empowered by a clear and highly appealing pedagogical format, the authors beautifully illustrate the creative nature of synthesis by carefully explaining a number of synthetic downfalls followed by a series of inventive solutions developed by some of its leading practitioners. As such, this text effectively illustrates the dynamic quality of synthesis, and more specifically, it demonstrates how strategies seeking to access a given target molecule are often in a constant state of evolution as obstacles are encountered and practitioners exercise their creativity in efforts to overcome them. This feature of constant struggle to surmount obstacles, coupled with serendipity, is often under-appreciated even though it is pivotal for invention, discovery, and, dare we say, adventure and excitement in chemical synthesis.

Thanks are certainly due to *Sierra* and *de la Torre* for bringing this key element of the endeavor into stark relief in a new format. Students, teachers, and researchers alike should benefit from this text, which should also serve to inspire the next generation of synthetic chemists in their quest to develop even more powerful and predictable strategies to solve the puzzles discussed within this volume and others like them. Ultimately, such endeavors should turn dreams into reality much to the benefit of chemical synthesis and society at large.

K. C. Nicolaou

Dead Ends and Detours: Direct Ways to Successful Total Synthesis
Miguel A. Sierra and María C. de la Torre
Copyright © 2004 WILEY-VCH Verlag GmbH & Co. KGaA, Weinheim
ISBN: 3-527-30644-7

Preface

Is organic synthesis dead? Are we relegated to serve as simple technicians for other magnificent disciplines such as bioorganic chemistry, nano technology, material chemistry, genetics, and proteomics? Does this mean that we can produce any molecule *à la carte*? Apparently, this is the impression that others seem to have of us – (the proud people working in organic chemistry) – in the 21st century: namely that we are able to produce any molecule they require at any time. But, the happy truth is that "no", a lot of thinking and man-hours are necessary to prepare even the simplest molecules.

Organic synthesis is not only a living discipline, still more it is the key science for the development of the world in this, rather young, century. Who else than us know how to produce molecules? We are not the architects of the nano-world but we are the humble masons that designed and built these structures that are the nano-world. We are not the magic doctors able to cure diseases from all around the world, but we are the people able to prepare the molecules that produce the miracle. Organic synthesis is the central discipline in modern science and technology. However, in spite of our respect for our work, our knowledge in preparing molecules is still very limited. This is why we are still here, trying to understand and, like the phoenix, arising from our mistakes and failures and more than anything else having a need for knowledge.

The aim of the book is to present and learn from failures, to achieve a synthetic target by a well-designed route. The knowledge, effort, enormous amount of work and aptitude, which reside behind the magnificent achievements of modern organic synthesis are contained in each single synthetic step of the rather many schemes in the book.

We tried to keep the references to a minimum. It is not the goal of this book for the reader to become lost in a wast collection of chemical literature. Other books and primary literature contain the necessary information for understanding the chemistry in this work. Please read this book as a novel about a journey that is just beginning. Is organic synthesis dead? The answer is up to you.

September 2004

Miguel A. Sierra
María C. de la Torre

Dead Ends and Detours: Direct Ways to Successful Total Synthesis
Miguel A. Sierra and María C. de la Torre
Copyright © 2004 WILEY-VCH Verlag GmbH & Co. KGaA, Weinheim
ISBN: 3-527-30644-7

Contents

Dead Ends and Detours: Direct Ways to Successful Total Synthesis
Miguel A. Sierra and María C. de la Torre
Copyright © 2004 WILEY-VCH Verlag GmbH & Co. KGaA, Weinheim
ISBN: 3-527-30644-7

Abbreviations

Ac	Acetyl
AIBN	2,2'-Azobisisobutyronitrile
All	Allyl
Bn	Benzyl
BOC	*tert*-Butoxycarbonyl
BOM	Benzyloxymethyl
Bz	Benzoyl
Cbz	Benzyloxycarbonyl
CSA	10-Camphorsulfonic acid
dba	*trans,trans*-dibenzylideneacetone
DBU	1,8-diazabicyclo[5.4.0]undec-7-ene
DCC	1,3-dicyclohexylcarbodiimide
DCM	Dichloromethane
DDQ	2,3-dichloro-5,6-dicyano-1,4-benzoquinone
DEAD	Diethyl azodicarboxylate
DEIPS	Diethylisopropylsilyl
DET	Diethyl tartrate
DHP	3,4-Dihydro-2*H*-pyran
DIAD	Diisopropyl azodicarboxylate
DIBALH	Diisobutylaluminium hydride
DIPT	Diisopropyl tartrate
DMAP	4-Dimethylaminopyridine
DMPI	Dess–Martin periodinane
DMPS	2,3-dimercapto-1-propanesulfonic acid, sodium salt
DMPU	1,3-dimethyl-3,5,6-tetrahydro-2-(1*H*)-pyrimidinone
DMS	Dimethyl sulfide
DMSO	Dimethyl sulfoxide
ETSA	Ethyl trimethylsilylacetate
FDPP	Pentafluorphenyl(diphenyl)phosphinate
HMPA	Hexamethylphosphoramide
Im	Imidazole
KHMDS	Potassium bis(trimethylsilyl)amide
LDA	Lithium diisopropylamide
LHMDS	Lithium bis(trimethylsily)amide
MCPA	3-Choroperbenzoic acid
MEM	2-Methoxyethoxymethyl
MMP	Magnesium monoperoxyphthalate
MOM	Methoxymethyl
Ms	Methanesulfonyl
Ms (preceded by nÅ)	Molecular sieves

Dead Ends and Detours: Direct Ways to Successful Total Synthesis
Miguel A. Sierra and María C. de la Torre
Copyright © 2004 WILEY-VCH Verlag GmbH & Co. KGaA, Weinheim
ISBN: 3-527-30644-7

NBS	*N*-bromosuccinimide
NIS	*N*-Iodosuccinimide
NMO	*N*-Methylmorpholine-*N*-oxide
NMP	*N*-Methyl-2-pyrrolidinone
PDC	Pyridinium dichromate
PIFA	Phenyliodine(III)bis(trifluoroacetate)
Piv	Pivaloyl
PMB	*p*-Methoxybenzyl
PPTS	Pyridinium *p*-toluenesulfonate
py	Pyridine
SEM	2-(Trimethylsilyl)ethoxymethyl
Sia	Siamyl
TBAF	Tetra-*n*-butylammonium fluoride
TBDPS	*tert*-Butyldiphenylsilyl
TBHP	*Tert*-Butyl hydroperoxide
TBS	*tert*-Butyldimethylsilyl
TBTH	Tributyltin hydride
TEA	Triethylamine
TEOC	2-(trimethylsilyl)ethoxycarbonyl
TES	Triethylsilyl
TFA	Trifluoroacetic acid
Th	2-Thienyl
THF	Tetrahydrofuran
TIPS	Triisopropylsilyl
TMS	Trimethylsilyl
TPAP	Tetra-*N*-propylammonium perruthenate
TPS	Triphenylsilyl
Tr	Trityl

Chapter 1
Introduction: From the Paper to the Laboratory

Seven years ago we began to write a long article for *Angewandte Chemie International Edition* [1] pointing to an aspect of modern organic synthesis that, in our opinion, had been neglected: the difficulties in reaching a synthetic target. The introduction of that article said *much of the current chemical literature deals with the synthesis of organic molecules or describes the development of methodology for organic synthesis. The achievements of synthetic methodology are impressive and the most complex molecules are, in principle, accessible. From the beginning of the 90's a feeling has been spread throughout the chemical community about the maturity of this branch of chemistry, and today terms such as atom economy, highly efficient homogeneous and heterogeneous catalytic transformations, combinatorial chemistry, and so on, are frequently used when talking about organic synthesis.*

Great optimism arises from the chemical literature when describing how the molecules are synthesized. It seems that our ability to devise synthetic routes has become infallible, and that even the most complex target molecules are prepared without apparent effort. Is this always true? or is a lot of effort still necessary to make every step in a multi-step synthesis possible? The readers of papers that describe the preparation of organic molecules are very familiar with sentences such as "after extensive experimentation it was found, to our delight, the reaction worked nicely", and other examples thereof. Apparently, to go into detail about the failures or to discuss all the unfruitful approaches, decreases the beauty of the synthetic route reported. Therefore, it is not often easy to recognize when the synthesis has developed as planned. In the meantime a lot of useful information may be lost. The aim of this work is to look at the total synthesis from a different point of view. We will focus on the detours from the original synthetic plan, the dead-ends that may arise at specific steps of a total synthesis, as well as the thinking involved in solving the problems en route to the synthetic goal. Often, the final route is at least of equal beauty to that originally planned. The selection of problems discussed below has been extracted from papers that explicitly expressed the failure of the original plan, and the evolution to the final successful (or in some cases unsuccessful) solution of the problem.

Dead Ends and Detours: Direct Ways to Successful Total Synthesis
Miguel A. Sierra and María C. de la Torre
Copyright © 2004 WILEY-VCH Verlag GmbH & Co. KGaA, Weinheim
ISBN: 3-527-30644-7

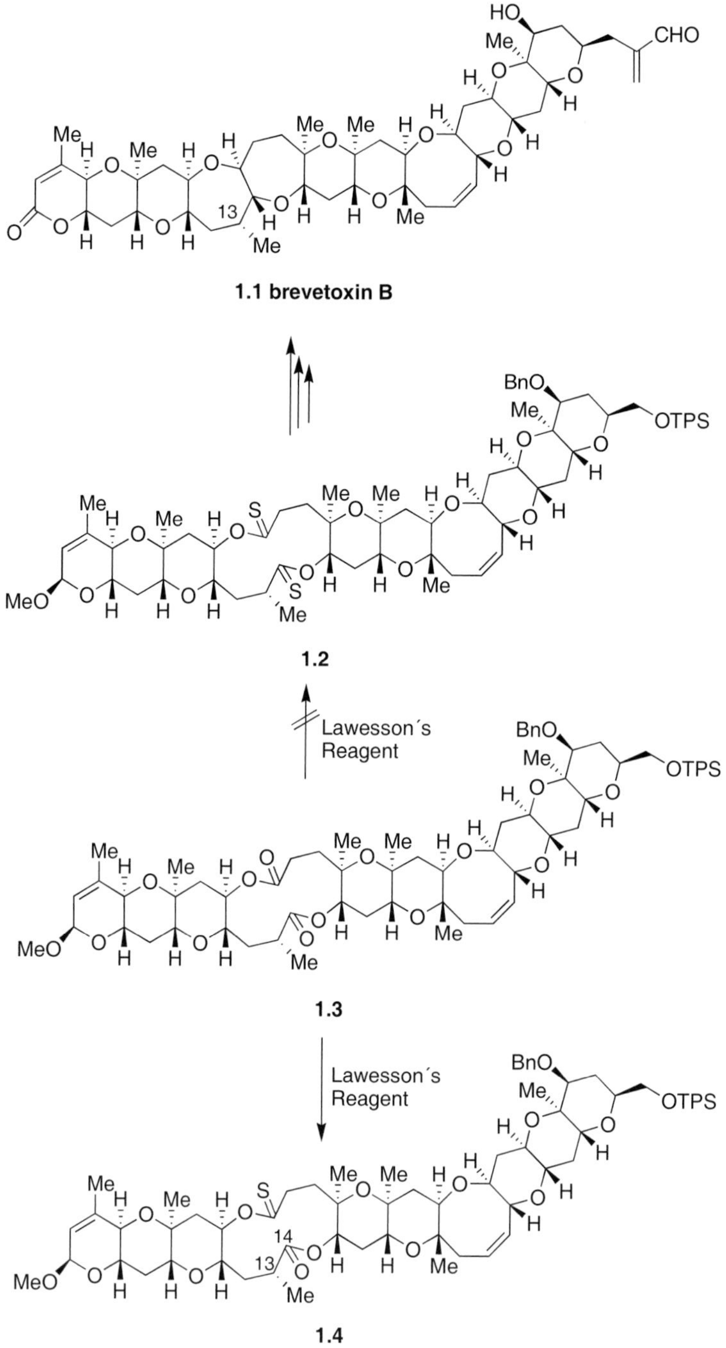

1.1 brevetoxin B

1.2

Lawesson's
Reagent

1.3

Lawesson's
Reagent

1.4

Scheme 1.1

Nicolaou's brevetoxin B synthesis [2] was used as the first example to show how the failure of a well-tested transformation can truncate a total synthesis in an advanced step. The transannular bridging of the 12-membered bis-(thiolactone) **1.2** was designed to build the remaining two rings of brevetoxin B **1.1** (Scheme 1.1).

The process had been thoroughly tested in model systems like **1.6** [3] (Scheme 1.2) and seemed to be a very attractive strategy to prepare the desired final product. However, all the attempts to prepare compound **1.2** by reaction of the macrodilactone precursor **1.3** with Lawesson's reagent produced very low yields of the desired bis(thiolactone) **1.2**. The lactone carbonyl at C14 remained unaltered, and the reaction product was monothiolactone **1.4**. Other Lawesson-like reagents were also unable to form **1.2**, and this route was eventually abandoned. In spite of this apparent failure, a full body of methodology to build fused seven-membered rings was developed in the process. The successful synthesis of brevetoxin B by Nicolaou and co-workers is now a landmark in organic synthesis [4].

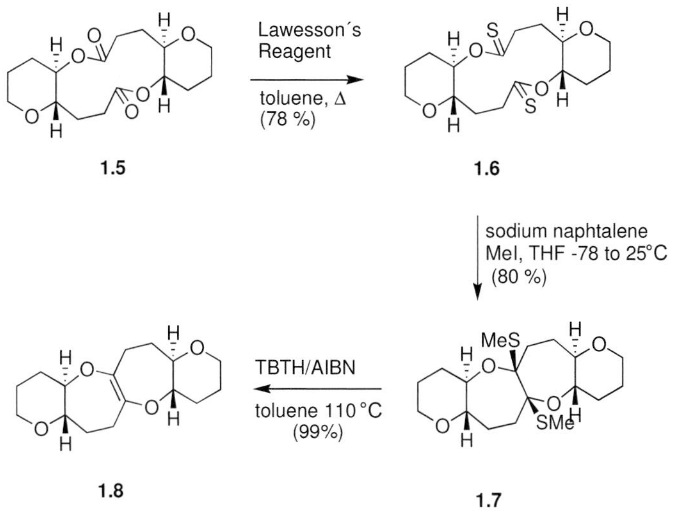

Scheme 1.2

We have searched the literature to find another example that better illustrates the ideas to be discussed in this book but have failed. This example exceptionally shows how the failure of a single transformation can thwart a beautiful idea making the planned synthesis unfeasible. However, the advantage is the acquisition of an exceptional amount of knowledge, which is, in the end, the object of chemical synthesis.

During the long process of publishing our review, interesting discussions with the reviewers demonstrated that our opinions about how to present total organic synthesis were not generally shared by the scientific community. We are going to include two of these points of view; the first one was written by one anonymous referee and stated:

"failed" reactions often lead to results which are not clean. These are usually described in papers as giving "decomposition of starting materials" or complex mixtures of products". For those involved in Organic Synthesis such reactions seem to occur far too often! From such experiments conclusive and/or publishable information is difficult to glean. Reporting of results in this fashion, while being honest, does not provide much information from which the reader can learn and is often avoided. Second: *often Organic Synthesis relies on luck! It doesn't matter how well planned a synthesis is, the route is nearly always going to have stages which are risky and predictably so, in advance. Such risks are assessed in the planning stages before undertaking synthetic work. It may sometimes be judged that a very high-risk strategy may offer significant potential benefits, which warrant investigation. Failures of high-risk work may go unreported as with the benefit of hindsight they may look to have strayed across the border between courageous and foolish!*

The second different point of view against reporting excessive details about failures was written by Danishefsky: *Before relating a few of these episodes* (referring to the synthesis of some of the many molecules synthesized in his group), *some important cautionary notes and attributions are in order. For those schooled in the art* (of Organic Synthesis), *there will be little need for either. The experienced practitioner is well aware that the pathways of synthesis are circuitous, bumpy, and even treacherous. Seldom do straight lines suffice to connect points in a synthesis of real consequence. Hence, the seasoned chemist will appreciate that along with these "magic moments" of success, one could have reported a litany of setbacks and reversals. However, for younger and more optimistic enthusiasts, it is appropriate to underscore the uncertainties, the detours and, yes, the frustrations associated with Organic Synthesis. Success is often a prize reserved for those who temper noble ideas with appropriate measures of realism and skepticism. Given the episodic nature of our science, wisdom may well be more valuable than cleverness. The ability to plumb the implications of each experiment, positive and negative, is central to the process of learning as we go along. Our quest to reach the promised land should not render us insensitive to opportunities for discovery, even as we find our way through the desert* [5].

We are going to try, through this book, to show how the difficulties encountered in a total synthesis, lead to an increase in knowledge. Much has been learnt during the seven years after we commenced the work from which this book is derived. We still believe that the difficulties encountered, while implementing many synthetic routes in the laboratory, are not due to any intrinsic unpredictability of this science. This is an idea that may arise from this book, but is far from reality. We are going to discuss superb synthetic achievements and the idea is not unpredictability but necessity for knowledge. Organic synthesis is still very far from becoming a closed science. The words written by Corey [6]: *In many respects, the development of the synthesis of okaramine N was similar to finding a way up a vertical cliff that offers just a limited number of small cracks and handholds,* clearly demonstrate our necessity for knowledge to increase the number of cracks and handholds available to climb any vertical cliff, present or future.

References

1. Sierra, M. A.; de la Torre, M. C. *Angew. Chem. Int. Ed.* **2000**, *39*, 1538.
2. a) Nicolaou, K. C.; Hwang, C.-K.; Duggan, M. E.; Nugiel, D. A.; Abe, Y.; Reddy, K. B.; DeFrees, S. A.; Reddy, D. R.; Awartani, R. A.; Conley, S. R.; Rutjes, F. P. J. T.; Theodorakis, E. A. *J. Am. Chem. Soc.* **1995**, *117*, 10227; b) Nicolaou, K. C.; Sorensen, E. J. *Classics in Total Synthesis*, VCH, Weinheim, **1996**.
3. a) Nicolaou, K. C.; Hwang, C.-K.; Duggan, M. E.; Reddy, K. B.; Marron, B. E.; McGarry, D. G. *J. Am. Chem. Soc.* **1986**, *108*, 6800; b) Nicolaou, K. C.; Hwang, C.-K.; Marron, B. E.; DeFrees, S. A.; Couladouros, E.; Abe, Y.; Carroll, P. J.; Snyder, J. P. *J. Am. Chem. Soc.* **1990**, *112*, 3040.
4. a) Nicolaou, K. C.; Theodorakis, E. A.; Rutjes, F. P. J. T.; Sato, M.; Tiebes, J.; Xiao, X.-Y.; Hwang, C.-K.; Duggan, M. E.; Yang, Z.; Couladouros, E. A.; Sato, F.; Shin, J.; He, H.-M.; Bleckman, T. *J. Am. Chem. Soc.* **1995**, *117*, 10239; b) Nicolaou, K. C.; Rutjes, F. P. J. T.; Theodorakis, E. A.; Tiebes, J.; Sato, M.; Untersteller, E. *J. Am. Chem. Soc.* **1995**, *117*, 10252.
5. Danishefsky, S. J. *Tetrahedron* **1997**, *53*, 8689.
6. Baran, P. S.; Guerrero, C. A.; Corey, E. J. *J. Am. Chem. Soc.* **2003**, *125*, 5628.

Chapter 2
Tuning-up, Tactical and Strategic Changes

The experimental realization of a synthetic route may encounter different problems leading to changes in the original scheme of synthesis. These changes may be included in the three following cases:

- The optimization of the reaction conditions required to effect a synthetic step.
- A tactical change.
- A strategic change.

The search for the appropriate conditions required to effect a specific transformation is routinely done in any chemistry laboratory. A tactical change assumes in many cases a significant detour from the plan, but without dropping the overall synthetic scheme. A strategic change means that the planned synthetic route is not viable; subsequently it has to be redesigned. On many occasions this implies starting the synthesis again from the very beginning.

2.1   Tuning-up Reaction Conditions

The nature of chemical work obliges one to tune up the reaction conditions of a single transformation in most cases. However, the exception is the situation in which adequate reaction conditions are not found, and the optimization of a pre-determined synthetic transformation does not cause deviations from the original plan. Nevertheless, it is time consuming. The requirement to carry out several attempts in an advanced intermediate in the synthesis of a complex molecule, often involves going back and preparing more material from an earlier intermediate. The following example illustrates how difficult it was to tune up the, apparently simple, oxidation of a primary alcohol to the corresponding carboxylic acid in the presence of a labile dithiane moiety.

Dead Ends and Detours: Direct Ways to Successful Total Synthesis
Miguel A. Sierra and María C. de la Torre
Copyright © 2004 WILEY-VCH Verlag GmbH & Co. KGaA, Weinheim
ISBN: 3-527-30644-7

2.1.1   (+)-13-Deoxytedanolide [1]

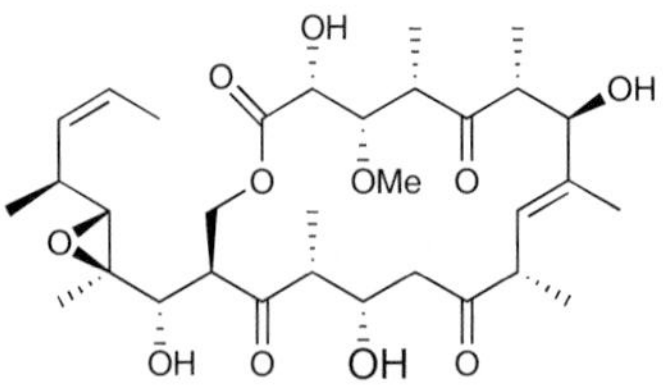

2.1 (+)-tedanolide 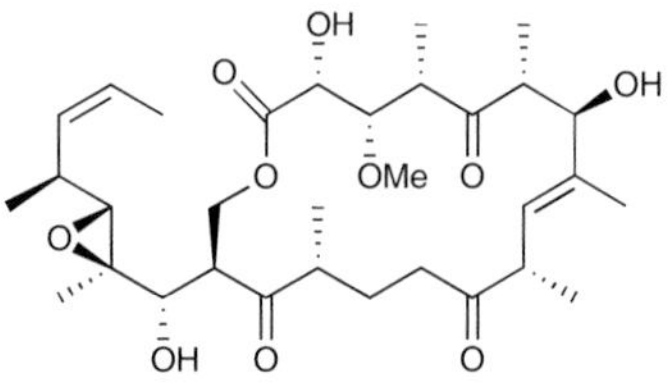   **2.2** (+)-13-deoxytedanolide

Target relevance
The 18-membered macrocycles (+)-tedanolide **2.1** isolated [2] from the Caribbean sponge *Tedania ignis*, and (+)-13-deoxytedanolide **2.2** isolated [3] from the Japanese sea sponge *Mycale adhaerens* exhibit significant antitumor activity.

Synthetic plan for the final stages of the synthesis
(+)-13-deoxytedanolide **2.2** would arise from seco-acid **2.3** that would be available via the union of dithiane **2.4** with iodide **2.5**. The synthetic planning calls for a high-risk epoxidation of C18-C19 olefin at a late stage of the synthesis. Furthermore, the dithiane moiety of **2.3** should be kept during the macrocyclization event, to prevent the potential ketalizations with the C15 and C29 hydroxyls (Scheme 2.1).

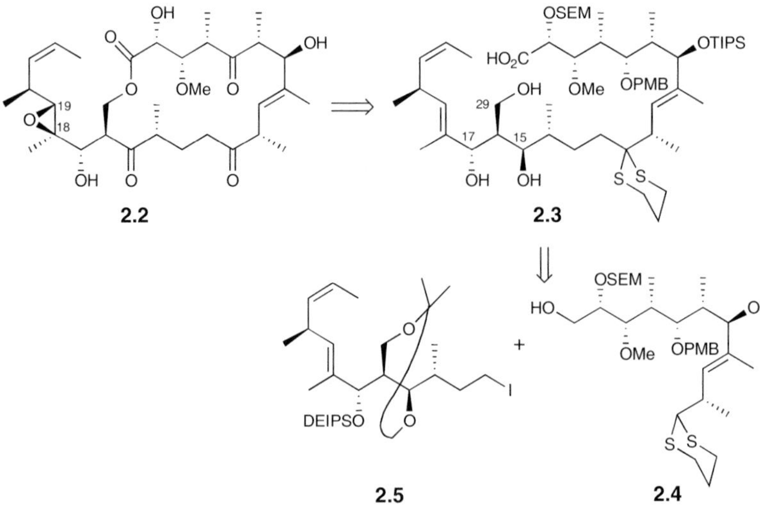

2.2   **2.3**

2.5   **2.4**

Scheme 2.1

Predictable problems

- Stereoselectivity of the building of the epoxide moiety of **2.2**. It was expected to solve this potential problem by using the director effect of the C17-hydroxyl group.
- (+)-13-Deoxytedanolide **2.2** has six epimerizable centers that may be affected by the reaction conditions during the latter steps of the synthesis.
- Seco-acid **2.3** will derive from alcohol **2.4**. The oxidation of the alcohol group should be performed in the presence of the oxidation-sensitive dithiane moiety.

Synthesis

*Step 1. Joining the alcohol **2.4** and iodide **2.5** (Scheme 2.2):* This step assembles the fully elaborated backbone of (+)-13-deoxytedanolide **2.2**. Compound **2.6** was obtained in a 75% yield by generation of the dianion derived from **2.4** by treatment with *t*BuLi, followed by addition of a slight excess of iodide **2.5**.

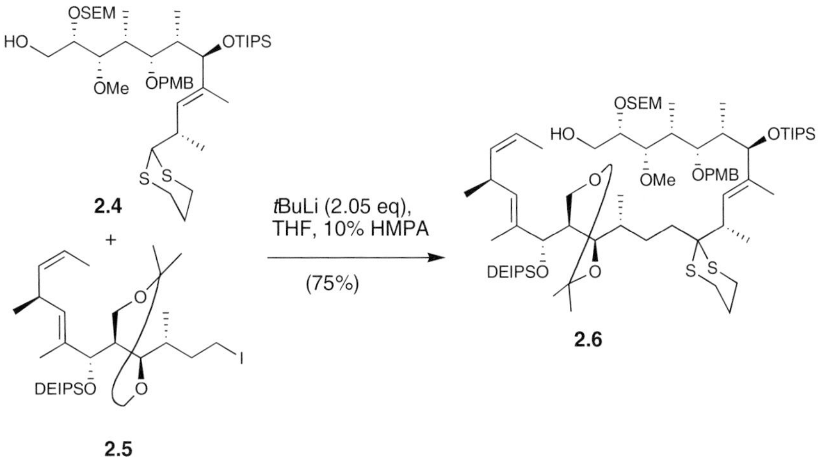

Scheme 2.2

*Step 2. Oxidation of the C1 primary alcohol of **2.6** to the corresponding carboxylic acid **2.7** (Scheme 2.3):* This is a critical step that will be effected in **2.6** and it should maintain the oxidatively labile dithiane moiety intact. According to the authors the exhaustive screening of the standard arsenal of oxidants proved unrewarding [4].

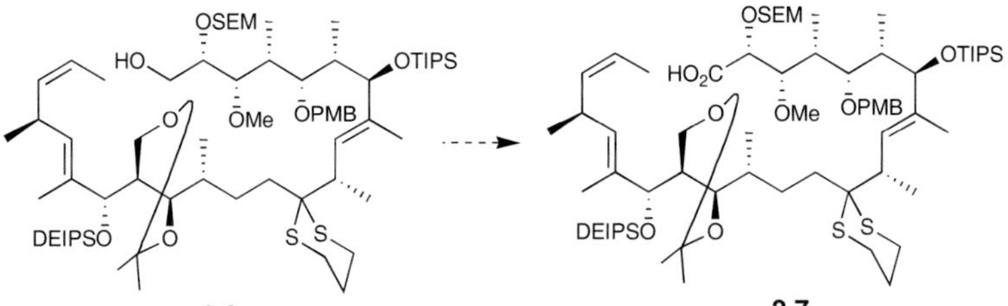

Scheme 2.3

Therefore, a variant of the SmI$_2$ promoted Evans–Tischenko *reduction* specifically developed for this step, was used. The Parikh–Doering oxidation of alcohol **2.6** yielded the corresponding aldehyde **2.8** (Scheme 2.4). Treatment of aldehyde **2.8** with 35 mol % SmI$_2$ in the presence of β-hydroxy ketone **2.9** afforded the ester **2.10** as a diasteromeric mixture. These specifically developed conditions achieve the *oxidation* of the aldehyde at C1 by *reducing* the carbonyl ketone group of **2.9** (this is an internal oxidation–reduction couple like the alcohol–ketone pair used in the Oppenhauer oxidation [5]).

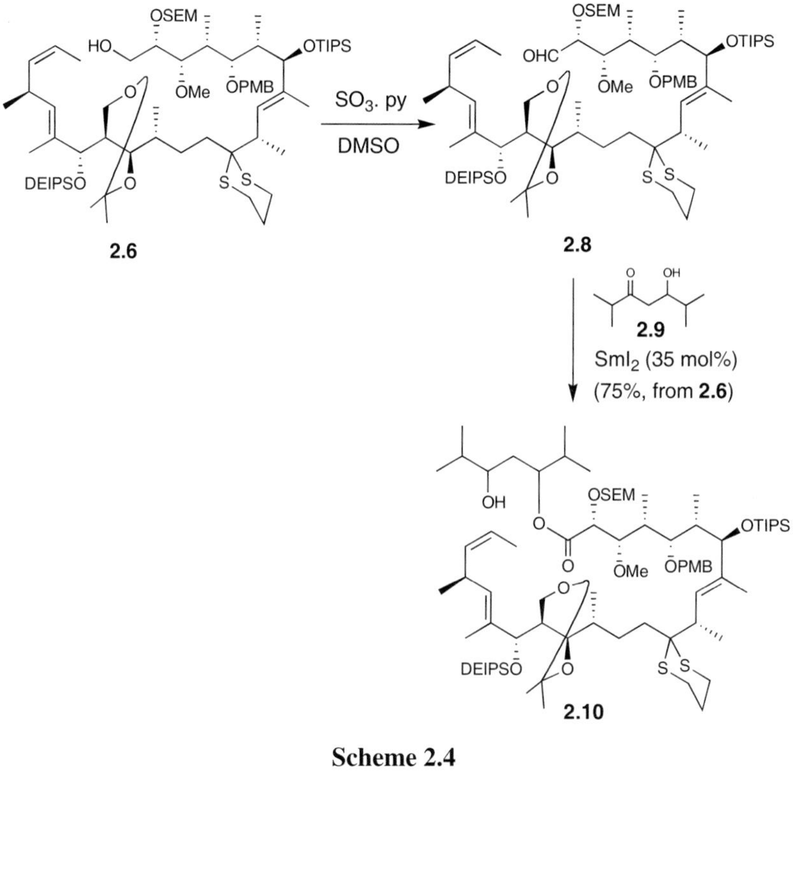

Scheme 2.4

The synthesis of (+)-13-deoxytedanolide **2.2** was completed (Scheme 2.5) by acetonide removal on **2.10**, which resulted in the concomitant loss of the DEIPS-group, followed by Yamaguchi macrolactonization [6]. Now, with three free hydroxyl groups C15, C17 and C29, it was anticipated that the bias of the macrolactonization would be highly favorable for the C29 alcohol. In fact, macrolactone **(+)-2.11** was obtained. Functional group manipulation led to **2.12** in which critical epoxidation had to be undertaken. Treatment of **2.12** with MCPBA/NaHCO$_3$ yielded the desired epoxide with high stereoselectivity (> 15:1 favoring the required α-epoxide). (+)-13-Deoxytedanolide **2.2** was obtained by removal of the protecting group of the silylated alcohol (TBAF/wet DMPU).

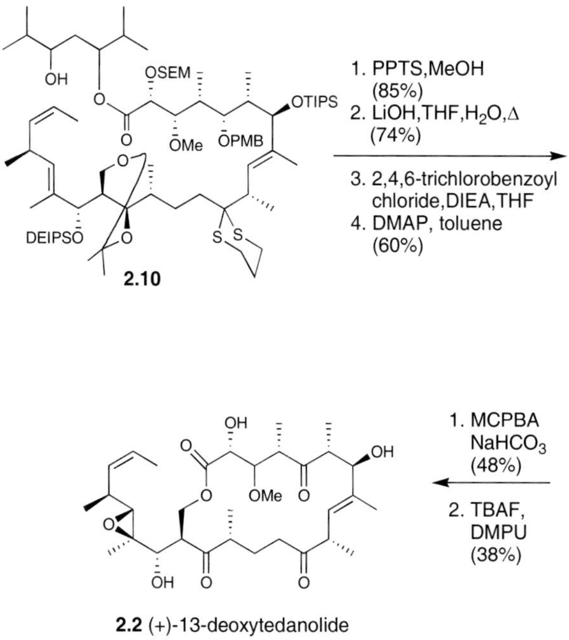

Scheme 2.5

Evaluation

The difficulty in achieving the *a priori* compromising oxidation step was not documented. However, the oxidation of an alcohol to a carboxylic group required developing a set of specific conditions to be used in a single intermediate **2.8**. No changes in the synthetic planning derived from this fine-tuning of an oxidation (**2.8→2.10**), although undoubtedly a lot of effort was required. Nevertheless, no further comments about this extremely important topic will be made in this book, since it falls in the area of routine synthetic work.

Key synthetic reaction

Yamaguchi macrolactonization protocol: Activation of the carboxyl group of an hydroxy acid using 2,4,6-trichlorobenzoyl chloride to form a mixed anhydride followed by lactonization in the presence of DMAP [6]. Usually, high dilution conditions are required.

2.2   Tactical Changes

A planned transformation may occur in an undesired sense or may not take place at all. Therefore, a detour should be taken to avoid the problematic step. The detour usually increases the number of synthetic steps. Alternatively, it may force the recommencement of the synthesis from an earlier intermediate to introduce the necessary changes in order to succeed in the stubborn transformation. A tactical change is then a detour or a drawback to the original synthetic plan, but not a change in its essence. The following examples illustrate this point.

2.2.1   (+)-**Taxusin** [7]

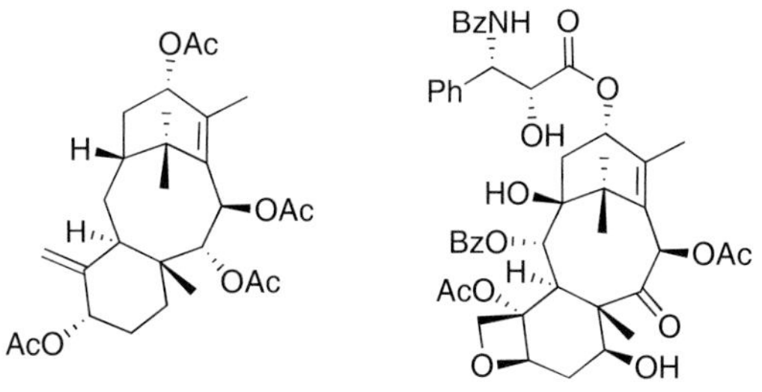

2.15 (+)-taxusin   **2.14 taxol**

Target relevance
The taxane diterpenes isolated from yew trees [8] exhibit promising activity against a number of human cancers (taxol, **2.14**). Other natural taxanes exhibit multidrug resistance reversing activity. The similitude of taxusin **2.15** with these compounds makes it an attractive target for synthesis.

Synthetic plan for the final stages of the synthesis
The access to (+)-taxusin **2.15** was planned from the tricyclic intermediate **2.16** by introduction of the C19-Me group via the selective cleavage of cyclopropane **2.17**. After accomplishing the task of placing the C19-Me group, the exo-double bond will be introduced on **2.18** to yield (+)-taxusin **2.15** (Scheme 2.6).

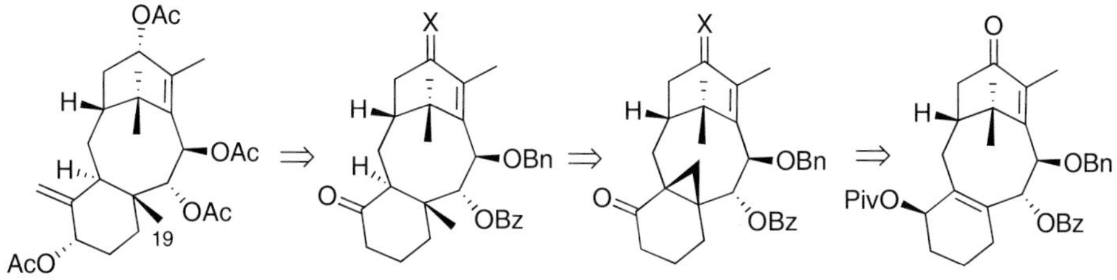

2.15 (+)-taxusin   **2.18**   **2.17**   **2.16**

Scheme 2.6

Predictable problems
Tricyclic ketone **2.16** is designed to allow the introduction of the C19-Me group from the sterically less demanding convex β-face. The methylenation of ketone **2.18** (X = OAc) had been already reported by Holton [9]. Therefore there are not, in principle, predictable problems at the late stages of the synthesis of (+)-taxusin.

Synthesis

Step 1. Introduction of C19 Me-group (Scheme 2.7):

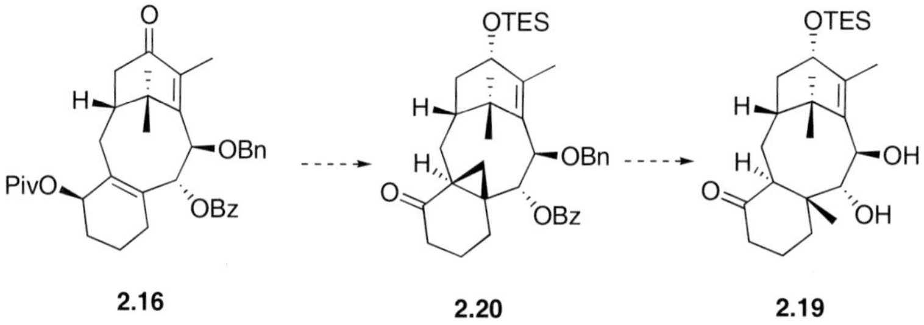

Scheme 2.7

The synthetic manipulations needed to place the C19-Me of the tricyclic skeleton of taxusin started with the reduction of the C13 keto group of **2.16**. This transformation was achieved by using the bulky hydride Li(t-BuO)$_3$AlH. The reduction produces the α-alcohol, as expected from the attack of the hydride from the less hindered convex β-face.

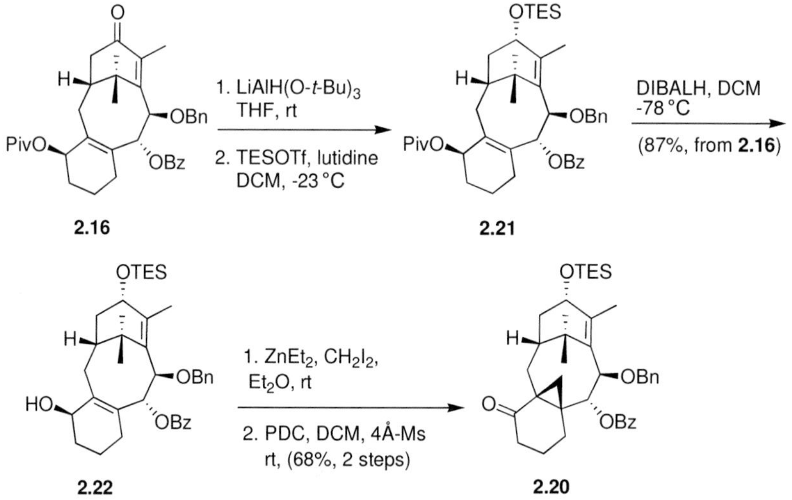

Scheme 2.8

The resulting alcohol was protected as its TES-derivative to yield **2.21** and the pivaloyl ester was removed using DIBALH to form the substrate for the cyclopropanation, the alcohol **2.22**. We should remember that the cyclopropane will become the C19-Me after ring cleavage. The cyclopropanation was effected by reaction of **2.22** with Et_2Zn/CH_2I_2 and the free alcohol oxidized with PDC to yield **2.20**. The cyclopropanation reaction produced as predicted, exclusively, the β-isomer (Scheme 2.8).

Dauben's protocol [10] to install an angular Me-group involves the ring cleavage of a bridging cyclopropyl ketone. Exposure of tetracylic ketone **2.20** to the standard Birch conditions ($Li/tBuOH/NH_{3(liq)}/THF$, −78°C) followed by acid quenching (NH_4Cl) produced the smooth cleavage of the cyclopropane ring. However, the desired ketone **2.19** was not obtained. Instead a very unstable product was detected, for which a hydroperoxide structure **2.23** was assigned. Treatment of the reaction mixture with Me_2S gave the stable triol **2.24**. It is clear that the transannular cyclization of the intermediate enol **2.25**, generated during the Birch cleavage of the cyclopropane **2.20**, occurred in the presence of traces of air. The further reduction of the hydroperoxide **2.23** by the Me_2S formed the triol **2.24** (Scheme 2.9).

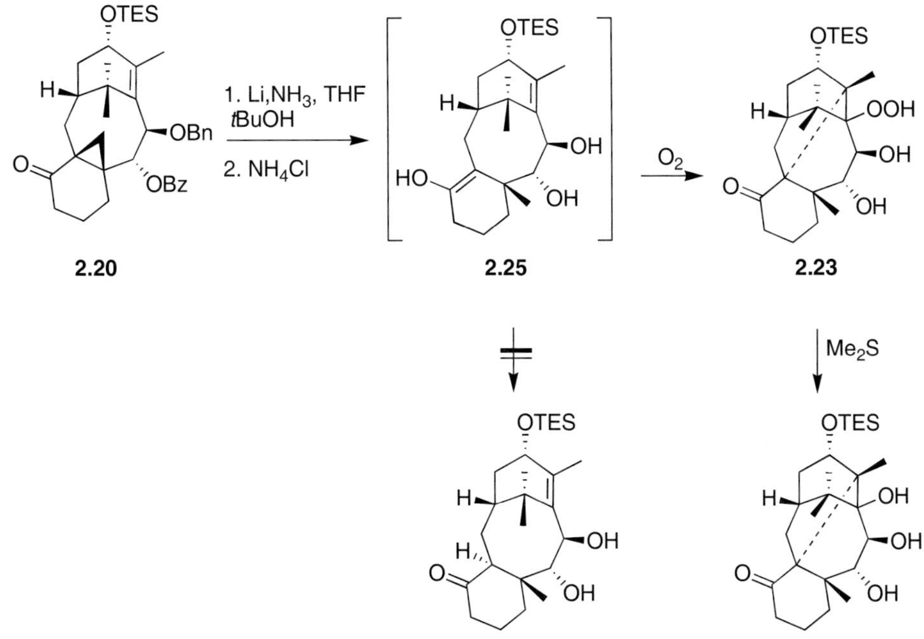

Scheme 2.9

Two additive factors may account for this unexpected failure of tetracyclic ketone **2.20** to produce the needed ketone **2.19**. First, the protonation of the enol **2.25** by the concave α-face, which would lead to **2.19**, has to be highly disfavored due to strong steric hindrance. Second, the protonation by the convex β-face of **2.25** should form ketone **2.26** having the bulky TESO-group proximate to the six-membered C-ring. Subsequently, the enol evolves through an alternate reaction pathway or simply decomposes (Scheme 2.10).

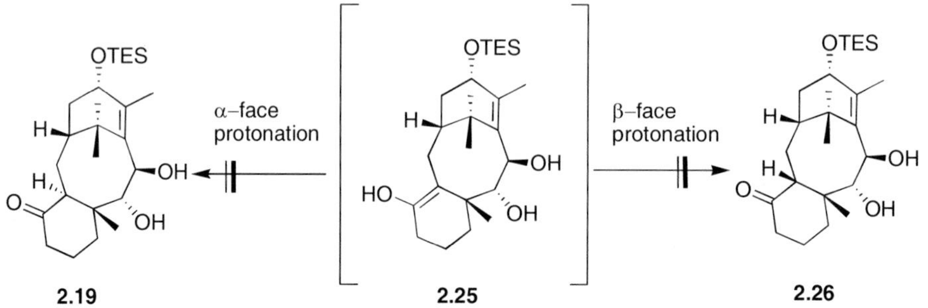

2.19   **2.25**   **2.26**

Scheme 2.10

The simplest solution to the above problem is to remove the bulky C13 protecting group. Thereafter, the directing ability of the alcohol group positioned between the A and C rings will effect the needed α-protonation. In fact, when **2.20** was treated with TBAF and the free alcohol **2.27** was submitted to Birch conditions the ketone **2.28** was obtained in quantitative yield upon quenching with MeOH (Scheme 2.11).

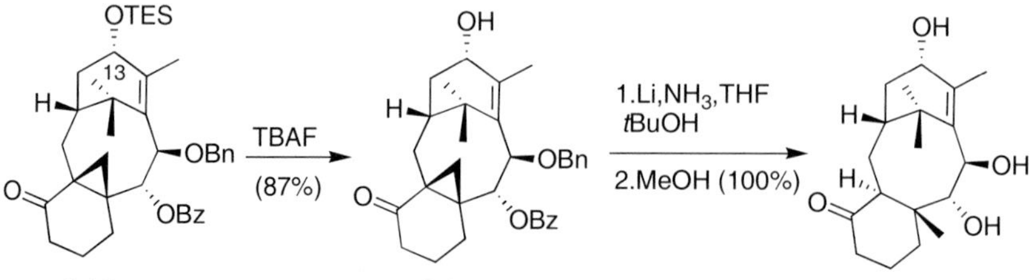

2.20   **2.27**   **2.28**

Scheme 2.11

Step 2. Completion of the synthesis: introduction of the exo-double bond (Scheme 2.12):

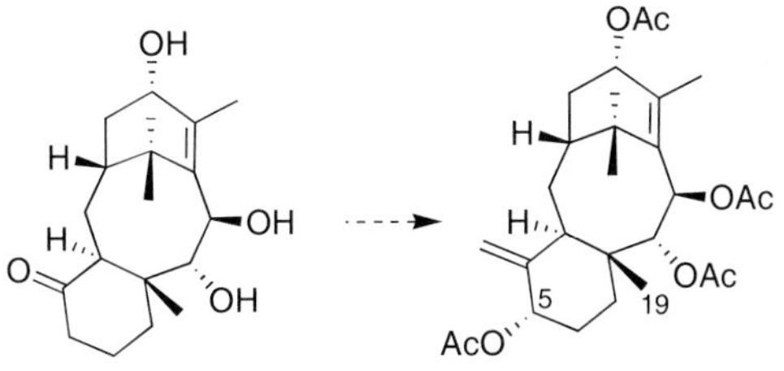

Scheme 2.12

The completion of the synthesis of (+)-taxusin **2.15** from ketone **2.28** required introducing the exocyclic methylene group and the hydroxyl group at C5. This task was accomplished by conversion of the carbonyl group to the enol-triflate **2.29**, after extensive manipulation of **2.28** to selectively protect the alcohols at C9, C10, and C13. Cross-coupling of **2.29** with TMSCH₂MgCl in the presence of Pd(PPh₃)₄ yielded allylsilane **2.30**, that was oxidized exclusively to the α-isomer of alcohol **2.31**. Removal of the silyl protecting groups and peracetylation yielded finaly (+)-taxusin (Scheme 2.13).

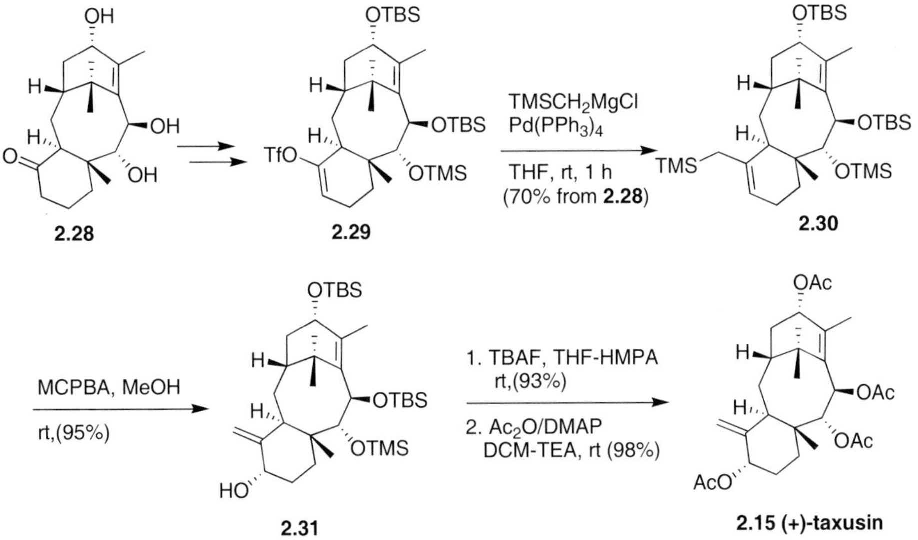

Scheme 2.13

Evaluation

The unexpected reactivity of ketone **2.20**, due to its special topology, which inhibits the protonation of the enol **2.25** by any of the two faces, results in a minor tactical change during the synthesis of (+)-taxusin **2.15**. The problem, due to the presence of an inadequate protecting group was solved by removing it. The desired transformation of ketone **2.27** to ketone **2.28** was then effected and the C19-Me placed.

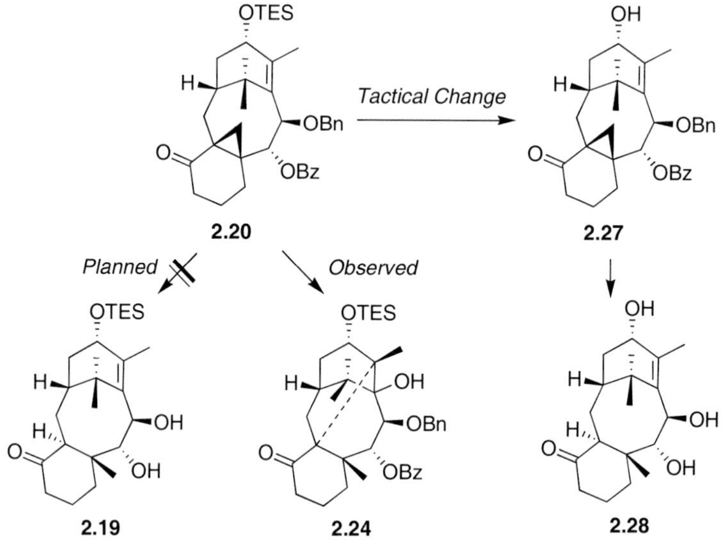

2.20   **2.27**

2.19   **2.24**   **2.28**

Key synthetic transformation

Dauben's protocol: Installation of an angular Me group on a cyclopropyl-ketone by Birch reduction of the cyclopropyl moiety [10].

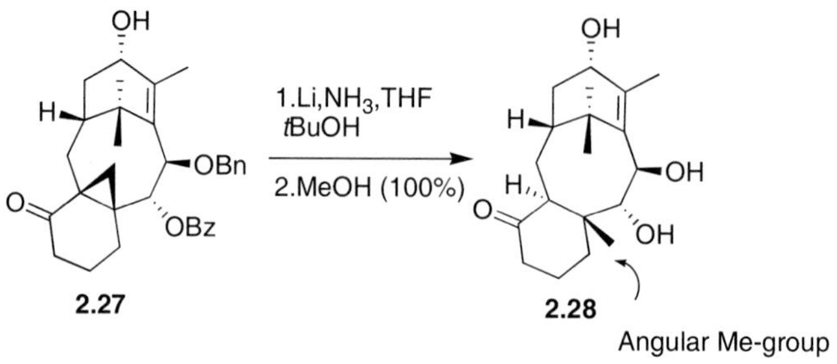

2.27

1.Li,NH$_3$,THF
*t*BuOH

2.MeOH (100%)

2.28

Angular Me-group

This example is on the borderline between the fine tuning of the conditions to effect a transformation and a tactical change. Nevertheless, the reactivity of ketone **2.20** as a consequence of its topology, requires more knowledge of the transformation before going ahead in the synthesis. The following example involves a more important tactical change.

2.2.2   (–)-Strychnine [11]

2.32 (–) strychnine

Target relevance
Considering its molecular weight, strychnine is one of the most complex natural products [12]. In fact, the molecule of strychnine has only 24 skeletal atoms assembled in seven rings with a total of six stereocenters, five of them in the core cyclohexane ring. This complexity, its pharmacological and extremely toxic properties, make strychnine a fascinating target for organic synthesis.

Synthetic plan
The easy conversion of Wieland–Gumlich aldehyde **2.33** to strychnine **2.32** [13] targets this aldehyde as a key point for the synthesis of alkaloid **2.32**. Wieland–Gumlich aldehyde **2.33** would be derived from pentacyclic alcohol **2.34**. The reduction of the ester moiety of **2.34** to an aldehyde should promote the intramolecular acetal formation. The building of the indol ring on tricyclic ketone **2.35** would, in turn, form alcohol **2.34**. The key point of this approach is the formation of the piperidine D ring of **2.35** by the intramolecular conjugate addition of propargylic silane **2.36**, derived from the alkylation of hexahydroindolone **2.37** (Scheme 2.14).

Predictable problems
The closure of the piperidine D ring of *Strychnos* alkaloids by the intramolecular conjugate addition of a propargyl silane to an enone has already been demonstrated [14]. Therefore, the use of a propargylic silane like **2.36** bearing a α-alkoxy substituent would generate **2.35** having the desired D-ring. Compound **2.35** has an alkoxyvinylidene side chain to be further elaborated into the required 20(*E*)-hydroxyethylene substituent. The remaining synthetic steps to strychnine were based in well-established methodology. Therefore, no problems, *a priori*, were envisaged.

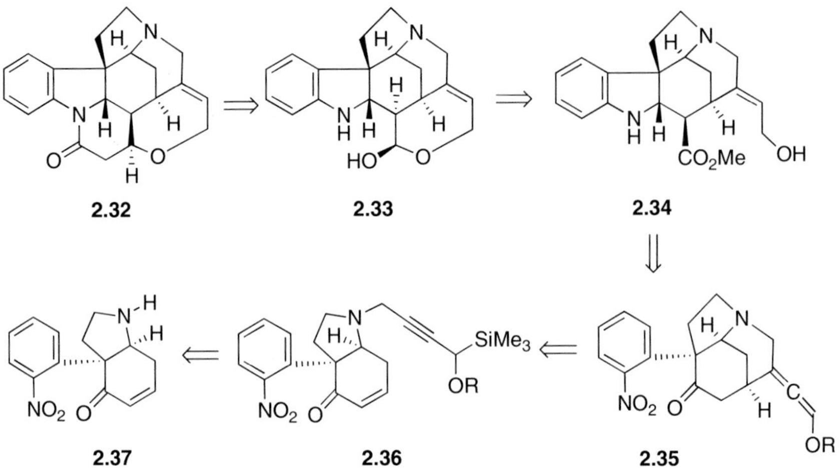

2.32 2.33 2.34

2.37 2.36 2.35

Scheme 2.14

Synthesis

Step 1. Piperidine D ring closure (Scheme 2.15):

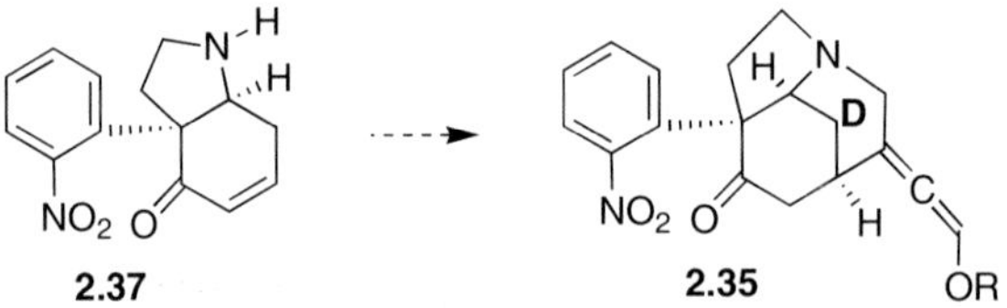

2.37 2.35

Scheme 2.15

The alkylation of racemic hexahydroindolone **2.37** with propargylic iodides **2.38** afforded the propargylic silanes **2.39**. With the exception of the desired ketone **2.35**, different tri- and tetracyclic products were obtained on treatment of silanes **2.39** with different Lewis acids. Thus, variable mixtures of alcohol **2.40**, ketone **2.41** and cyclopropane **2.42** were obtained, depending on the Lewis acid used in the reaction of **2.39** (Scheme 2.16).

The unexpected failure of silanes **2.39** to yield ketone **2.35** was a consequence of the evolution of carbocation **2.43**, formed by the Michael-type cyclization of the triple bond on the conjugated double bond of **2.39**. In the designed reaction, carbocation **2.43** should desylilate, yielding the desired **2.35**. However, intermediate **2.43** experiences instead a 1,2-H shift forming the more stable α-alkoxy cation **2.44**.

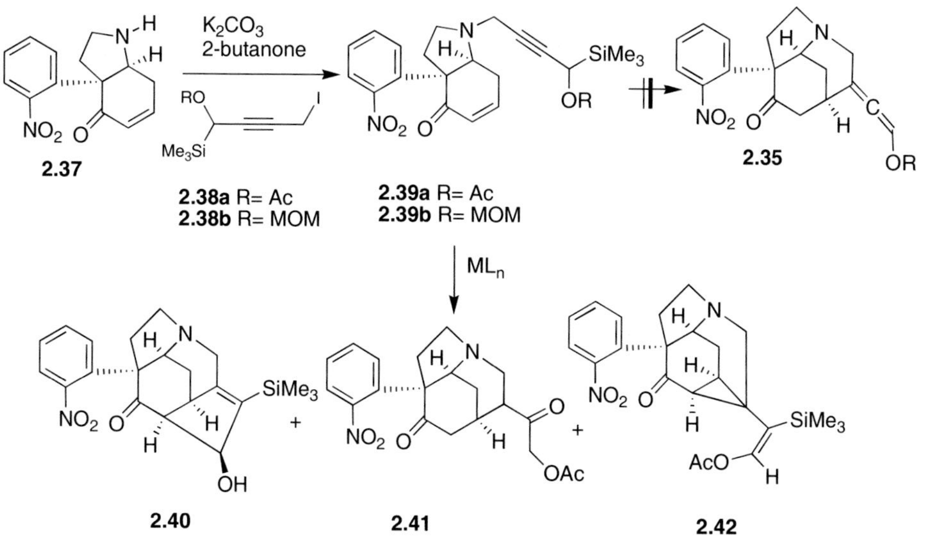

Scheme 2.16

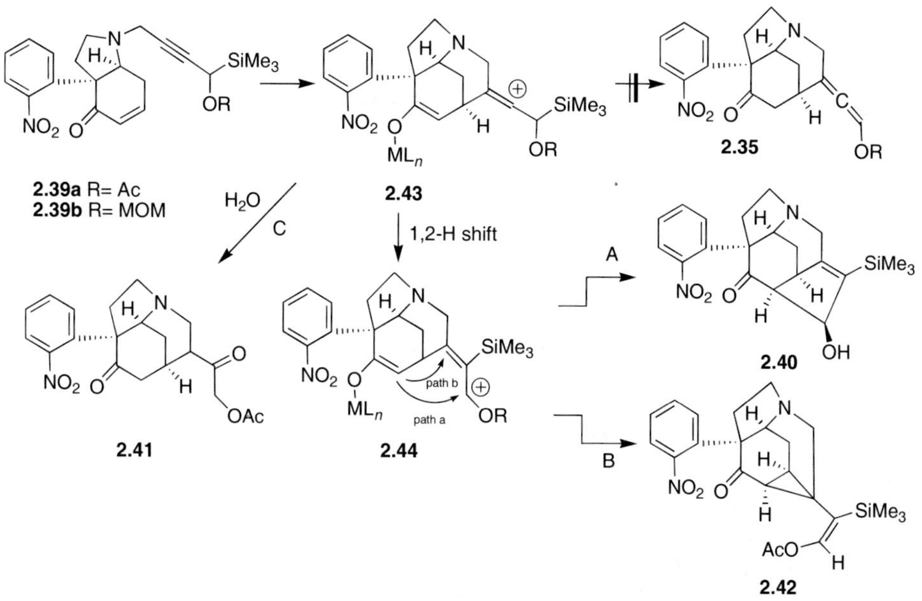

Scheme 2.17

The evolution of cation **2.44** through two competing reaction pathways, accounts for the formation of **2.40** and **2.42**. Alcohol **2.40** would be formed by attack of the enolate carbon on the less substituted carbon of the allylic cation, followed by hydrolysis of the ester moiety (path A in Scheme 2.17), while cyclopropane **2.42** would derive from the attack of the enolate carbon on the more substituted end of the allylic cation (path B in Scheme 2.17). Quenching of cation **2.43** by water would form ketone **2.41** (path C in Scheme 2.17).

Step 2. Second route to achieve the closure of the piperidine D ring (Scheme 2.18): The failure of silanes **2.39** to experience the cyclization to form the allene-ketone **2.35** was followed by a major tactical change. Now, an intramolecular Heck reaction [15] on iodide **2.45** was designed to achieve the piperidine ring closure. Iodide **2.45** would derive from the alkylation of hexahydroindolone **2.37** (this is the starting compound shared by both approaches) with bromoiodide **2.46**. The carboxymethyl group that will become the C17 of strychnine could be placed at this stage by a tandem Heck cyclization–carbonylation on **2.45**. These conditions would form **2.47** in a single synthetic step and supposedly to withdraw the allenylsilane approach to strychnine.

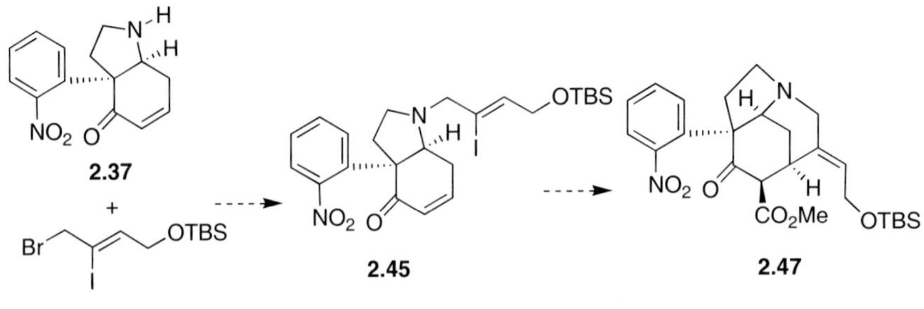

Scheme 2.18

Alkylation of the hexahydroindolone **2.37** with allylic bromide **2.46** formed the cyclization substrate **2.45** in a 74% yield. After extensive experimentation, the Heck cyclization of **2.45** to form **2.48** was achieved in 53% yield using Pd(OAc)$_2$/PPh$_3$, as the catalyst and TEA as the solvent. The tandem cyclization–carbonylation process was not feasible. Therefore, the introduction of the carboxymethyl group was effected in an additional step of the synthesis, by methoxycarbonylation of the enolate derived from **2.48** with NCCO$_2$Me to yield **2.47** (Scheme 2.19).

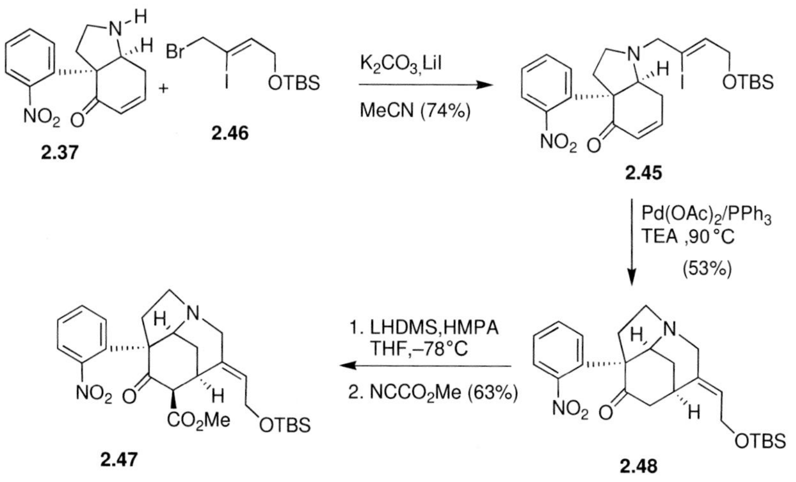

Scheme 2.19

Step 3. Completion of the synthesis of strychnine (Scheme 2.20):

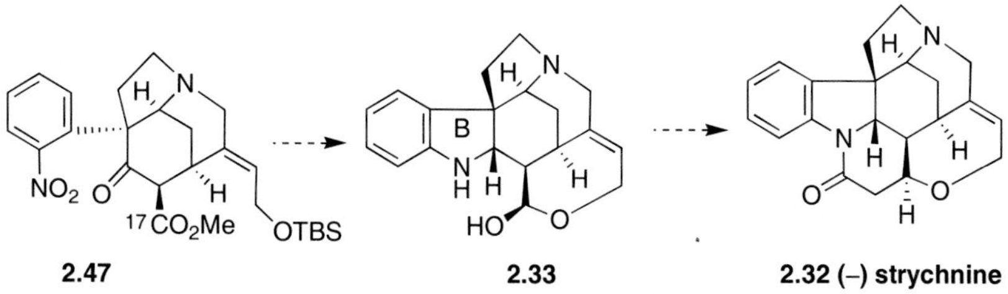

Scheme 2.20

Access to the Wieland–Gumlich aldehyde **2.33** from **2.47** requires closing the indol ring B, and the reduction of the C17 ester group to an aldehyde. This operation should promote the formation of the cyclic hemiacetal of compound **2.33**. The indol moiety was built from the *o*-nitrophenyl ketone moiety of **2.47** by reductive cyclization with Zn-dust in acid MeOH, followed by equilibration of the carboxymethyl group to the pure β-isomer **2.49**, having the natural, more stable, configuration at C16, by base treatment (NaH/MeOH). Reduction of **2.49** with DIBALH formed Wieland–Gumlich aldehyde **2.33**. Finally, strychnine was obtained using the reported conditions [13], by heating in the presence of a mixture of malonic acid/Ac$_2$O/AcOH/NaOAc (Scheme 2.21).

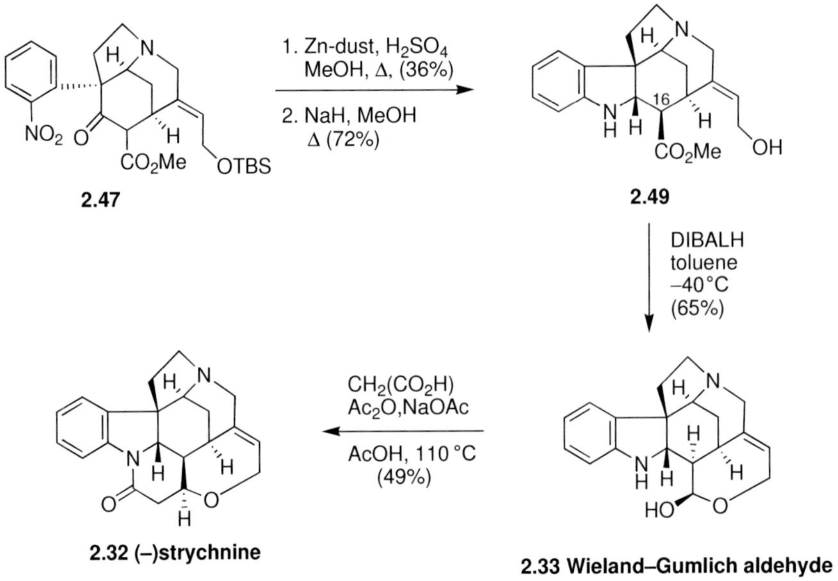

Scheme 2.21

Evaluation

The example discussed above represents a major tactical change for access to an advanced intermediate in strychnine synthesis. However, the strategic plan to synthesize strychnine remains unaltered. In fact, the synthetic scheme requires the closure of the D-ring of the strychnine nucleus, while placing an adequate side-chain to continue with the building of the molecule. The propargylic silane approach failed to produce the ring closure but the alternate Heck intramolecular ring closure succeeded.

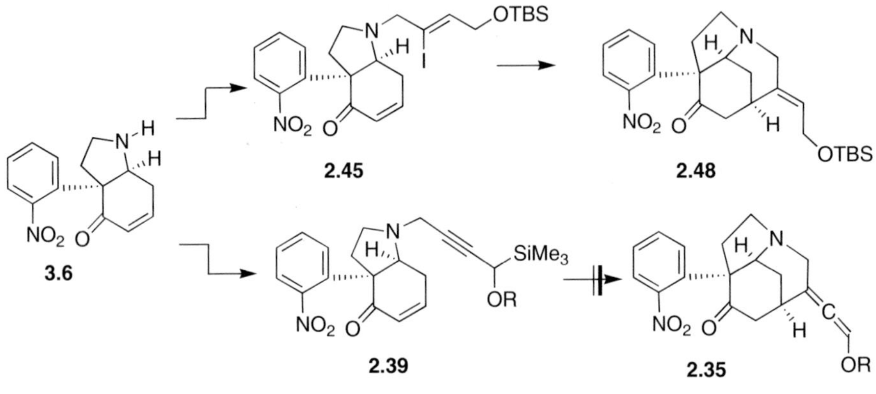

Key synthetic reactions

Heck coupling: The inter- or intramolecular coupling of a aryl- or vinyl halide and an olefin (usually an α,β-unsaturated carbonyl derivative) [15]. The double bond of the olefin is maintained during the process.

2.3   Strategic Changes

A strategic change during a synthesis occurs when the planned transformation does not occur in the desired sense or does not take place at all. However, while a tactical change irons out the problem for a deviation or a modification of the planned synthesis without affecting the overall synthetic scheme; a strategic change derives from a problem that is not resoluble within the planned scheme. That is, the synthesis reaches a dead-end. The insoluble problems emerging from either undesired transformations or the inertness of a functional group key, lead to major modifications of the synthetic planning. On occasions the problem ends with the inability to reach the synthetic target. As sketched above there are two reasons for giving up a planned synthesis:

- Undesirable result of a specific synthetic operation.
- Inert functional groups in a specific transformation.

2.3.1   The Core (2.52) of (+)-Lepicidin A Aglycon [16]

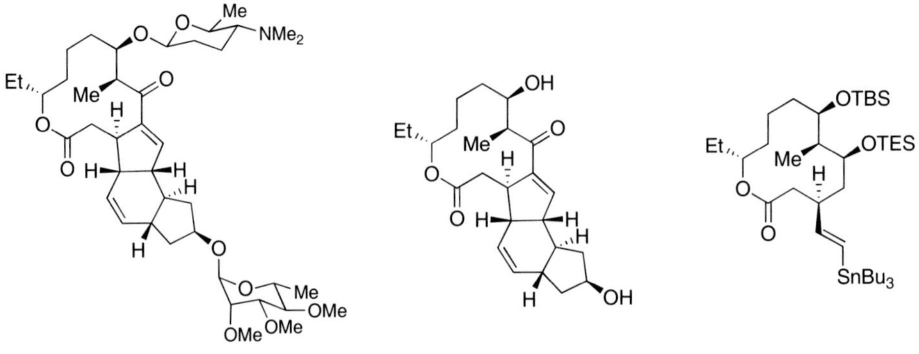

2.50 (+)-lepicidin A   2.51 (+)-lepicidin A aglycon   2.52

Target relevance

The tetracyclic macrolide lepicidin A **2.50** was isolated [17] from the fermentation broth of the soil microbe *Saccharopolyspora spinosa* and exhibits insecticidal activity, particularly against *Lepidoptera larvae*.

Synthetic plan

The synthesis of the core of (+)-lepicidin A aglycon **2.52** rests in the conjugate addition of a vinylstannane to macrolactone **2.53**. This approach is appealing because it delays the introduction of the labile stannane moiety until the last step of the synthesis of **2.52**. Access to **2.53** would be gained from fragments **2.54** and **2.55** by utilization of the γ-selective crotonate aldol reaction methodology developed by Fleming [18] (Scheme 2.22).

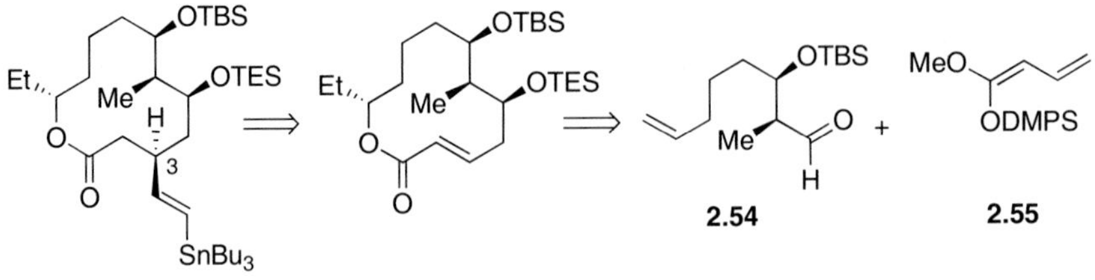

2.52   2.53   2.54   2.55

Scheme 2.22

Predictable problems
- Literature precedent for reactions of nucleophiles related to **2.55** with chiral aldehydes did not exist. Therefore, the stereochemical result of the reaction between **2.54** and **2.55** was uncertain.
- Control of the stereochemistry of the chiral center at C3 formed during the vinylstannane addition was also uncertain.

Synthesis

*Step 1. Synthesis of macrolactone **2.53** (Scheme 2.23):*

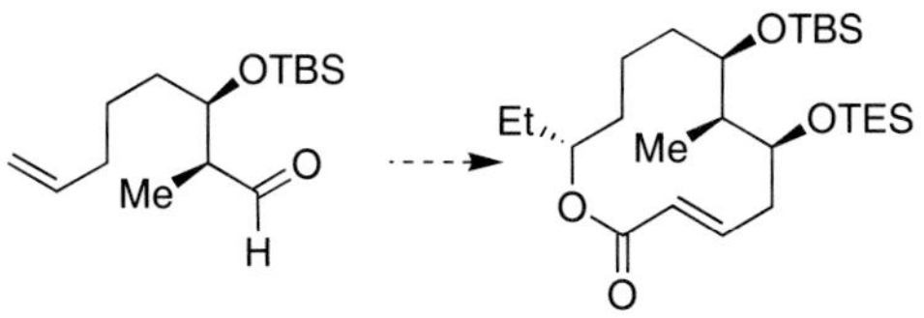

Scheme 2.23

The synthesis of macrolactone **2.53** begins with the Lewis acid-promoted reaction of aldehyde **2.54** with silyl ketene acetal **2.55**. This addition occurs in fact with good γ-selectivity, to yield the Felkin adduct **2.56**, which bears three of the four stereocentres of lactone **2.53** with the correct stereochemistry. Obtaining macrolactone **2.53** from adduct **2.56** requires sequential selective oxidation of the terminal double bond, introduction of the ethyl substituent at carbon C21 and finally macrolactonization. Selective hydroboration of **2.56** followed by oxidation of the resulting alcohol under Parikh–Doering conditions provided aldehyde **2.57**. Treatment of **2.57** with diethylzinc in the presence of (+)-*N,N*-dibutylnorephedrine yielded alcohol **2.58** as an inseparable mixture of C21 diastereoisomers. Base hydrolysis of ester **2.58** afforded the corresponding hydroxy acid that was submitted to Yamaguchi's macrolactonization [6] providing **2.53** (Scheme 2.24).

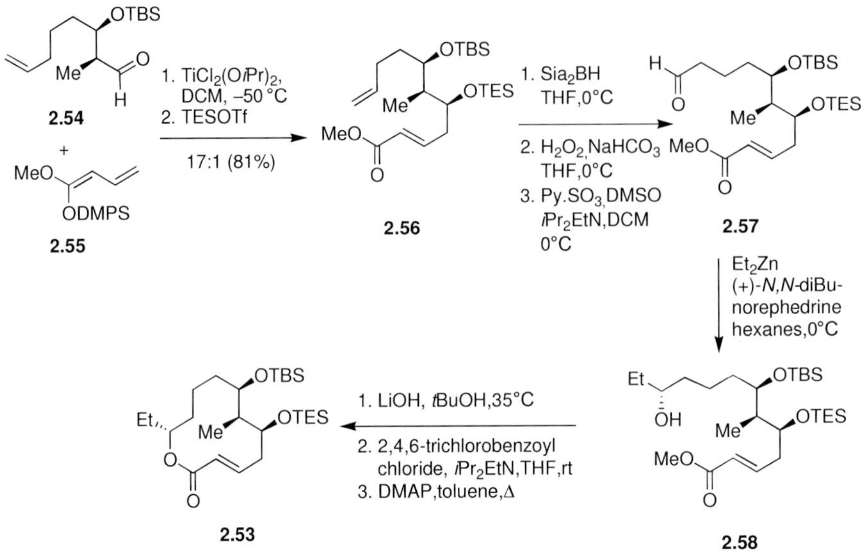

Scheme 2.24

Step 2. Introduction of the vinylstannane appendage: The conjugated addition of the vinyl cuprate **2.59** to the macrolactone **2.53** proceeded readily at –90 °C to provide **2.60** in a 64% yield and with a diastereoselection higher than 20:1. This is an exceedingly good result *except for the fact that **2.60** has the wrong stereochemistry at C3* (the expected compound was **2.52**). This is an unexpected outcome for this type of addition reaction, since the previous work [19] on the addition of Me$_2$CuLi to macrolactone **2.61** that yields exclusively **2.62**, as well as computer calculations [20], pointed to the nearly exclusive production of the desired Michael adduct **2.52** having the right β-stereochemistry (Scheme 2.25). These results forced authors to reevaluate the synthetic plan and to design a new approach to the macrocyclic stannane **2.52**.

*Step 3. A new strategy for the synthesis of macrocyclic stannane **2.66**:* In this new approach the stereocenter at C3 was generated, with the correct stereochemistry, by conjugated addition of cuprate **2.59** to an α,β-unsaturatedδ-lactone **2.63**, a reaction for which the stereochemical outcome is well documented in the literature [21]. This approach places the highly sensitive stannane moiety at the beginning of the synthesis.

Cuprate **2.59** adds to lactone **2.63** delivering **2.64** with exceptional selectivity. The lactone **2.64** having the labile stannane appendage was hydrolized (LiOH) and *in situ* esterificated with CH$_2$N$_2$ to prevent the protodestannylation. The free alcohol at C16 was further protected with TESOTf on the crude reaction mixture (the instability of the tin-moiety precludes the purification of the reaction intermediates) producing **2.65**. The ester **2.65** was transformed into vinylstannane macrolactone **2.66** following an analogous route to that used for the synthesis of lactone **2.53** (Scheme 2.26).

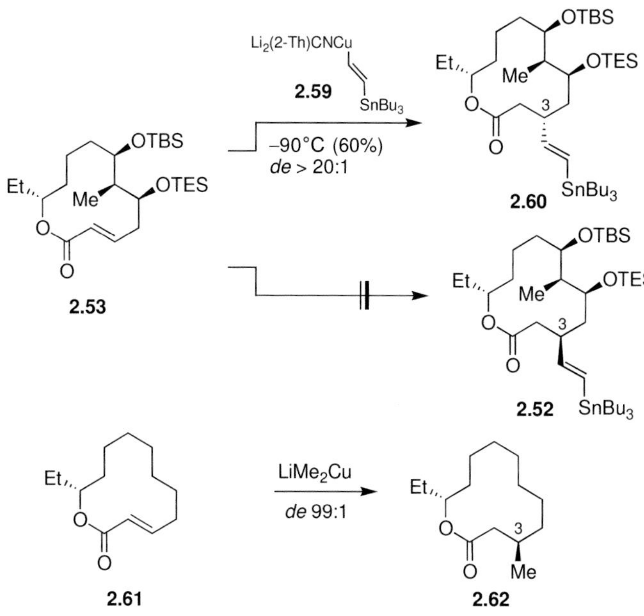

Scheme 2.25

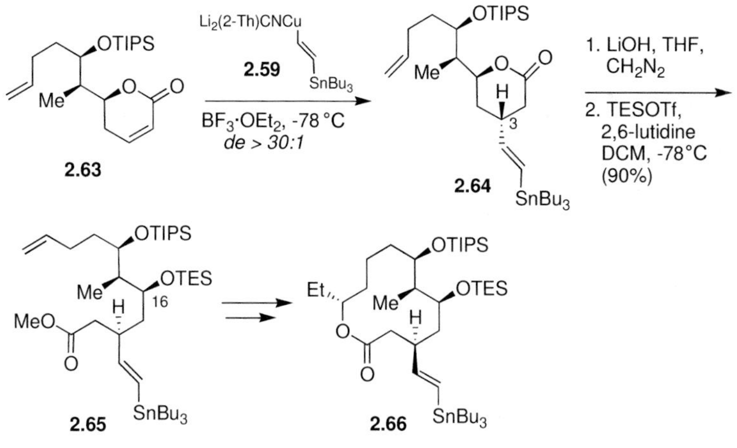

Scheme 2.26

Evaluation
Both routes to the macrolactone core **2.66** of (+)-lepicidin A are clearly very similar in efficiency, with the exception of the essential advantage of the first approach that incorporates the stannane at the last stages of the synthesis. This strategy avoids the necessity of carrying this labile group through the whole synthetic scheme. Nevertheless, *the fact that the problematic vinyl tin moiety could be incorporated at the end of the synthesis with both good yields and very high stereoselectivity, but with the wrong stereochemistry, forces a change in the strategy of the synthesis.* This is an example of how a perfectly designed strategy can be thwarted because a reaction produces an unexpected and completely unpredictable result.

The synthesis of (±)-scopadulin exemplifies a concatenation of two events leading to a substantial change in strategy. First, the wrong stereochemistry was obtained in the incorporation of a carboxy-group equivalent to the tetracyclic core of scopadulin. Second, the attachment of a carboxyl-group equivalent, through an oxy-Cope rearrangement, was attempted. The molecule in which the oxy-Cope rearrangement would be effected was inert towards this transformation. This inertia forced a major strategic change to the synthesis.

2.3.2   (±)-Scopadulin [22]

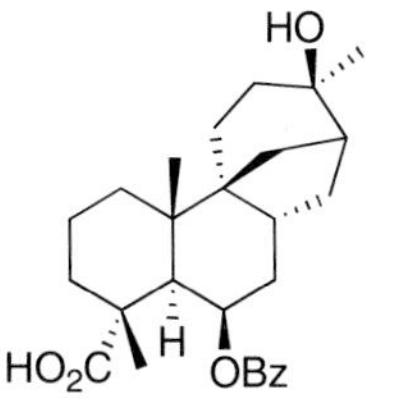

2.67 (±)-scopadulin

Target relevance

Scopadulin **2.67** is a tetracyclic aphidicolane diterpenoide isolated [23] from the widely distributed plant *Scoparia dulcis* (fam. Scrophulariaceae). The crude drug, known as Typchá-Kuratû in Paraguay, made from the whole plants of *S. dulcis* has been used as a traditional medicine for hypertension, toothaches, blennorhagia and stomach disorders, throughout South America, India and Taiwan. Isolated scopadulin **2.67** shows notable antiviral and cytotoxic activities.

Synthetic plan for the final stages of the synthesis

The key intermediate for the last stages of the synthesis of (±)-scopadulin **2.67** is enone **2.68**. The key transformation would be the stereoselective formation of the quaternary carbon center at C4. This step would be preformed by the α-face addition of a cyano nucleophile to **2.68** to form the nitrile **2.69**, in principle, easily transformable to (±)-scopadulin by standard functional group manipulation (Scheme 2.27).

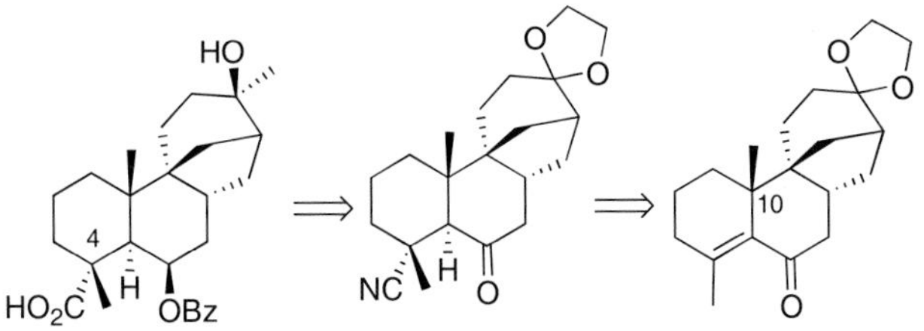

2.67   **2.69**   **2.68**

Scheme 2.27

Predictable problems

It was assumed that the angular C10-Me group of enone **2.68** would favor the addition of the cyano nucleophile by the α-face of the molecule. This is *a priori* a risky assumption since the ciano group is sterically very small and usually adds in an axial mode [24].

Synthesis

Step 1. Addition of the cyano-group to enone 2.68: Conjugate addition of Et₂AlCN to the enone **2.68** afforded the undesired Michael adduct **2.70**. Moreover, the reaction conditions used in the addition of cyanide epimerizes the *trans*-junction of the AB rings of **2.68** (the adequate stereochemistry of (±)-scopadulin **2.67**) to the *cis*-junction (the inadequate stereochemistry to access to (±)-scopadulin). Clearly, the strong bias for the axial attack of the cyanide group and its low steric bulkiness (small enough not to interact with the C10 Me-group) was responsible for the β-face attack. Further, the C10 Me-group still encourages the protonation of the enolate **2.71**, resulting from the 1,4-addition of the cyanide anion, by the α-face. Thence, the result of the reaction is the *cis*-fused AB system. The addition of bulkier reagents that can be further transformed in a carboxy-group (for example allylmagnesium bromide or methoxyallylcooper) were unrewarding. It is clear that the steric hindrance of the C4 and C10 Me-groups thwarted the synthetic approach, which uses a conjugated addition to place any carboxy-group equivalent at C4 with an α-stereochemistry (Scheme 2.28).

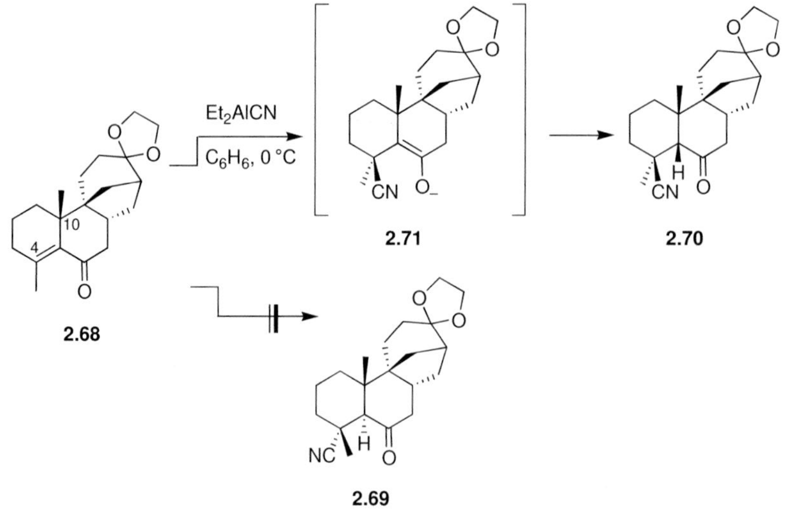

Scheme 2.28

*Step 2. Installation of an α-methoxycarbonyl equivalent in enone **2.68** by nucleophilic addition followed by oxy-Cope rearrangement:* The failure of the conjugated addition of cyanide to produce the required α-isomer **2.69** forced a revision of the synthetic planning. The overall plan remains the same but now the carboxy-equivalent would be introduced through an anionic oxy-Cope rearrangement [25], carried out on intermediate **2.72**. Intermediate **2.72** would be derived from ketone **2.68** through 1,2-addition of an allyl metal to the carbonyl group. Provided that the 1,2-addition of the allyl group in **2.68** would occur by the less hindered α-face of the ketone (this is warranted by the presence of the C10 β-Me-group), the concerted rearrangement would place the C4 allyl-group in the right α-face giving **2.73**. Compound **2.73** has now an allyl-group as a carboxy group equivalent (Scheme 2.29). *The change effected on the original synthetic planning is on the borderline between a tactical change and a strategic change. However, it does not modify the original idea and thence it can be better considered a tactical change, since the goal is still to place an α-carboxy–group equivalent onto the C4 position.*

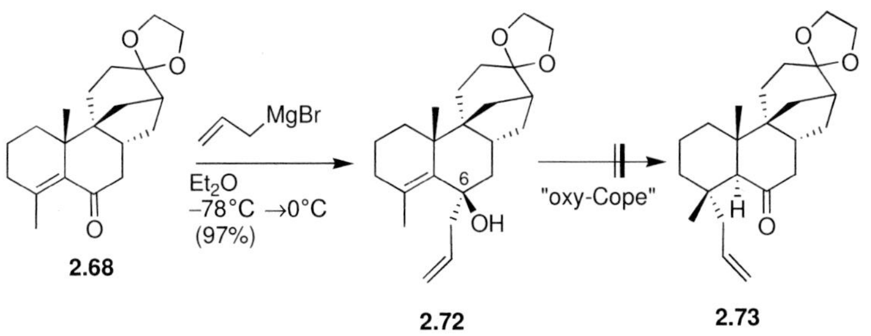

Scheme 2.29

The substrate for the anionic oxy-Cope rearrangement, alcohol **2.72**, was prepared by 1,2-addition of allylmagnesium bromide to the enone **2.68** in excellent yield. However, the different reaction conditions used to promote the oxy-Cope rearrangement in compound **2.72** met with no success. The desired product **2.73** could not be obtained (Scheme 2.29). Again, the steric hindrance due to the C4-Me substituent, coupled to the conformational bias of the quasi-equatorial pendant allyl group at C6 lead to a situation that does not favor the six-membered transition state needed for the rearrangement. These concatenations of facts are responsible for the very disappointing results. The failure of this new approach, to place a carboxy-group equivalent with the right stereochemistry at C4, lead to a complete strategic change. *Therefore, this is a clear case where a change in a synthetic route results from the inertia of an intermediate to experience a specific transformation.*

Step 3. Revised synthetic plan, access to the advanced intermediate **2.74** *having a carboxy group equivalent and the desired stereochemistry at C4:* It seems clear from the results above that the presence of the Me group at C4 is responsible for the failure of the original synthetic route to achieve the desired results. Therefore, the strategy was changed and enone **2.75**, lacking the troublesome C4-Me group, was used in the new synthetic plan. Enone **2.75** had been prepared previously [26]. Now a β-cyanide group would be placed at C4 by conjugated addition and used as the C4-Me precursor, *while the benzyloxymethyl group would be used as synthetic equivalent for the C4 carboxy-group*. (±)-Scopadulin could be gained from this new key intermediate **2.74**. Clearly, while in the two earlier attempts discussed above the topology of the molecule counteracted the synthetic effort, now the presence of the groups at C4, C6, and C10 would favor the desired stereochemistry during the alkylation of **2.76** (Scheme 2.30).

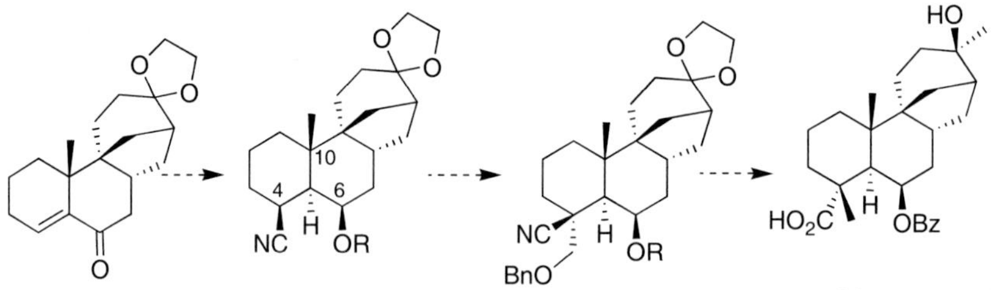

2.75   **2.76**   **2.74**   **2.67 (±)-scopadulin**

Scheme 2.30

In the intering, conjugate addition of Et₂AlCN to the enone **2.75** proceeds smoothly in the presence of TMSCl to yield the cyanide **2.77** in 81% yield. The ketone group of cyanide **2.77** was reduced to the C6-β-alcohol that was obtained as a single stereoisomer [the C10-Me efficiently directed the entry of the hydride for the less hindered α-face, as had happened before during the formation of the allyl adduct **2.72** (Scheme 2.29)]. The C6-β-alcohol was silylated (TMSCl/py/DMAP) to form **2.78**. The nitrile **2.78** was alkylated by reaction with LDA/BOMCl to yield, as expected, a single reaction product **2.79**. The cooperative efforts of C6 and C10 groups promoted the exclusive alkylation of the anion derived from **2.79** by its α-face. Reduction of the nitrile group of **2.79** and heating of the crude amine with KOH in ethylene glycol furnished diol **2.80** in 63% yield. Oxidation of the primary alcohol of **2.80** (RuCl₂(PPh₃)₃/air) was followed by deoxygenation to the 4,10-dimethylated alcohol **2.81** using the Huang Minlon reduction [27] (Scheme 2.31).

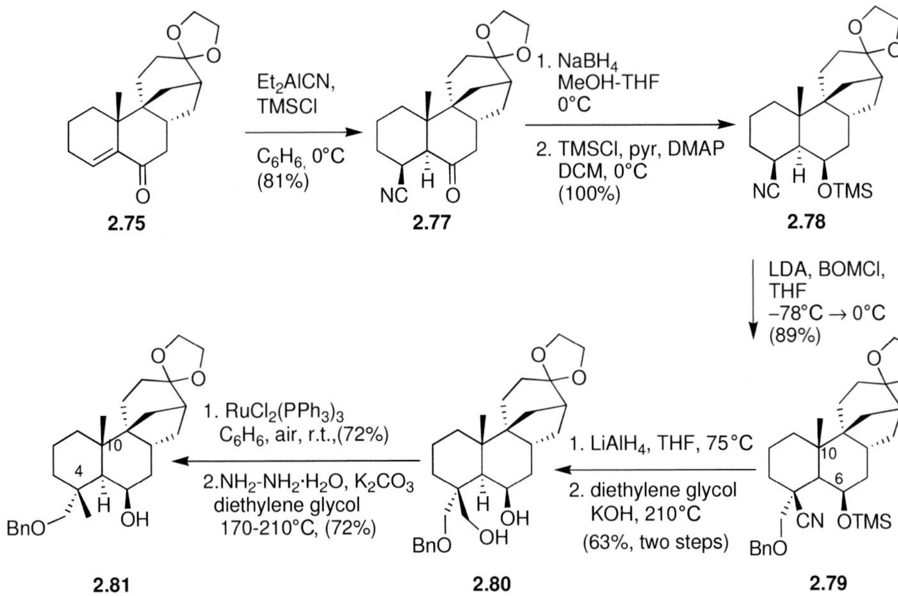

Scheme 2.31

Step 4. Synthesis of (±)-scopadulin: The final access to (±)-scopadulin **2.67** from alcohol **2.81** still requires two compromising synthetic manipulations. First, the benzoylation of the highly congested secondary alcohol at C6 and second the chemo- and stereoselective methylation at C16. Previous model studies suggested that BzOTf-2,6-lutidine system was the reagent of choice to effect the delicate benzoylation step (Figure 2.1).

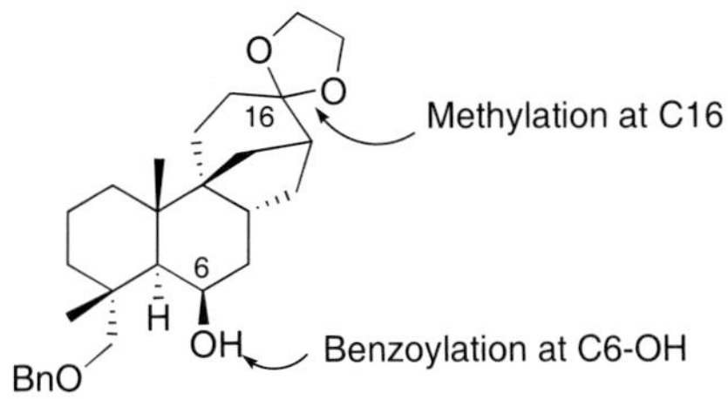

Figure 2.1

However, attempts to use this reagent in the presence of the ketal functionality at C16 in **2.81** resulted in the formation of small amounts of the desired benzoate **2.82**, together with high amounts of unreacted starting material, and mixtures of several products of unknown structure. Only under very controlled reaction conditions and after recycling of the starting material, was compound **2.82** obtained in a poor 35% yield. This is an unfortunate, hardly soluble situation, which appears occasionally at the late stages of a synthesis. This troublesome scenario creates serious logistic problems, due to the small amounts of material usually available in the final events of a multi-step synthesis. In any case, the completion of the synthesis (±)-scopadulin **2.67** requires the chemo- and regioselective incorporation of the C16 Me-group, a previous step to the oxidation of the protected primary alcohol to the carboxylic acid at C4. The hydroxyl group of (±)-scopadulin **2.67** is at an axial position. Therefore, a methylating agent with a bias for an equatorial attack is needed, with an additional requisite: it should not be reactive with the benzoate moiety. This was achieved with selectivity higher than 99:1 by reaction of MeTi(O*i*-Pr)₃ with **2.82** to yield alcohol **2.83**. Finally, **2.83** was debenzylated and the primary alcohol oxidized to the acid by RuO₄ to yield (±)-scopadulin (Scheme 2.32).

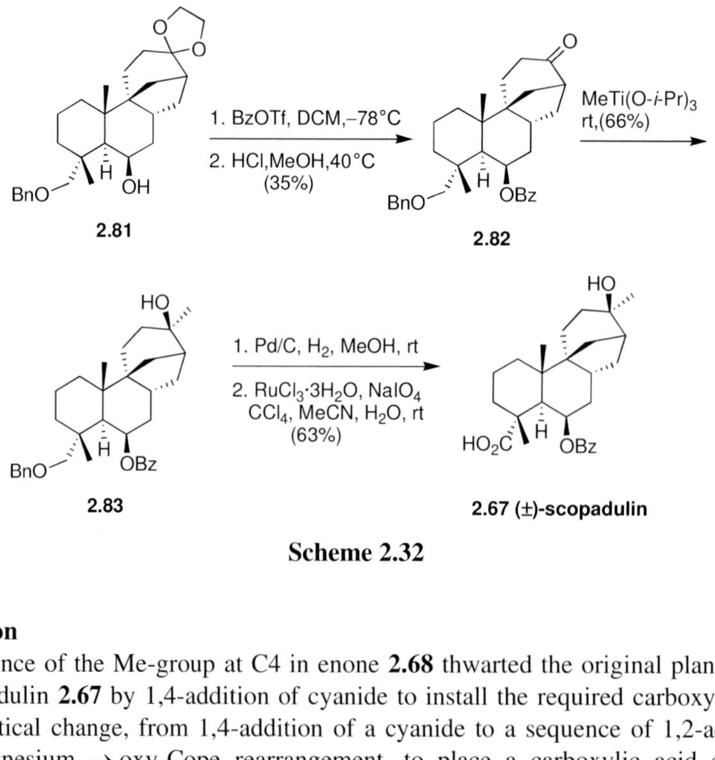

Scheme 2.32

Evaluation

The presence of the Me-group at C4 in enone **2.68** thwarted the original plan to access (±)-scopadulin **2.67** by 1,4-addition of cyanide to install the required carboxy-group. A major tactical change, from 1,4-addition of a cyanide to a sequence of 1,2-addition of allyl magnesium → oxy-Cope rearrangement, to place a carboxylic acid equivalent having the right stereochemistry in the troublesome C4 carbon, was also unsuccessful. Two consecutive failures lead to a fully redesigned route that, even if successful, has a final undesirable low-yielding step, which is always problematic in a multi-step synthesis.

2.68   2.70

2.72   2.73

Comment

Good axial versus equatorial selectivity during the nucleophilic methylation of fused cyclohexanones is a key synthetic transformation in many syntheses. The added value is that in some cases it is effected at the very end of the synthetic pathway. This compromising situation is exemplified by the reaction of ketone **2.82** with diverse methylation reagents. Thus, **2.82** reacts with MeLi leading to a nearly equimolar mixture (54:46) of alcohols **2.83** and its epimer **2.84**. The mixture MeLi–LiClO$_4$ has been reported [28] as good reaction to incorporate a Me-group through an equatorial attack. The reaction of this mixture with **2.82** only produced low yields of the mixture of epimeric alcohols **2.83** and **2.84**. Other reagents having good equatorial selectivity such as MeLi–Me$_2$CuLi [29] were not used in this case because of competitive reaction with the benzoate group. The selected reagent to transform **2.82** into **2.83** was MeTi(OiPr)$_3$ a methylating reactive which is exceptionally efficient in effecting an equatorial transfer of the Me group.

The equatorial transfer of the Me group is effective if coordination with the ketone group can occur prior to the transfer. Thus, the reagent MeLi–LiClO$_4$ is adequate to transfer the Me group in equatorial fashion to unhindered ketones. Thence, the failure of this reagent to achieve good selectivity with the highly hindered ketone **2.82** is due to the absence of coordination. In contrast, the complete selectivity of MeTi(OiPr)$_3$ is due to the steric repulsion of the bulky reagent with the axial hydrogens at C11 and C14 (Figure 2.2).

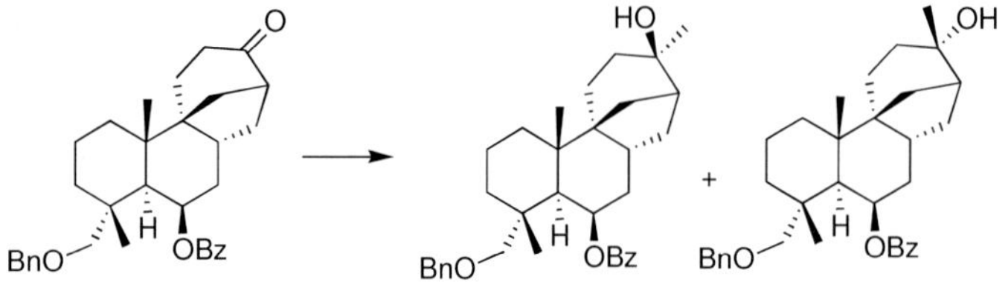

2.82 **2.83** **2.84**

MeLi, −78 °C, (80%) de = 54:46
MeLi-LiClO$_4$, −65 °C, (24%) de = 50:50
MeTi(O*i*Pr)$_3$, rt, (66%) de > 99:1

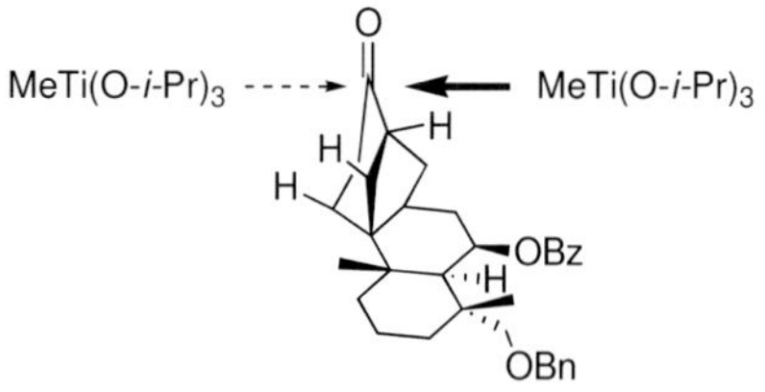

MeTi(O-*i*-Pr)$_3$ ----▶   ◀── MeTi(O-*i*-Pr)$_3$

Figure 2.2

Key synthetic transformation

Huang Minlon reduction: There are few procedures to efficiently transform a carbonyl group into CH$_2$ moiety in one step. The Huang Minlon [27] modification of the Wolff–Kisner reduction is one such. The procedure consists in heating the aldehyde or ketone with hydrazine in KOH or K$_2$CO$_3$ /ethylene glycol at temperatures above 120°C.

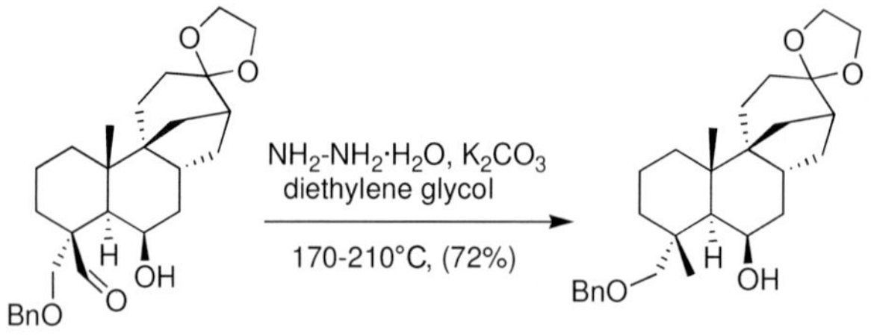

NH$_2$-NH$_2$·H$_2$O, K$_2$CO$_3$
diethylene glycol

170-210°C, (72%)

2.85 **2.81**

The synthesis discussed above has exemplified tactical or strategic changes, which are more or less important depending on the original synthetic plan. All of them have one point in common: they have attempted to surpass a pitfall encountered during the application of the planned synthesis. There are situations in which a route to a determinate molecule is modified after the initial planned route has already succeeded. This is usually made for the sake of increasing the efficiency of the already completed synthesis. For example, in the words of Danishefsky, during his synthesis of Eleutherobin [30], the reasons to pursue a new synthetic scheme were given as *...though a great deal of progress had been achieved, the cumbersome and mediocre yielding nature of our route came to be of increasing concern as we sought to bring through substantial quantities of material in anticipation of in vivo biological evaluation...* These synthetic changes are not going to be discussed in this book but are evident that tactical and strategic changes are introduced in the new route to the target in order to increase the efficiency level.

References

1.   Smith, III, A. B.; Adams, C. S.; Lodise-Barbosa, S. A.; Degnan, A. P. *J. Am. Chem. Soc.* **2003**, *125*, 350.
2.   Schmitz, F. J.; Gunasekera, S. P.; Yalamanchili, G.; Hossain, M. B.; van der Helm, D. *J. Am. Chem. Soc.* **1984**, *106*, 7251.
3.   Fusetani, N.; Sugawasara, T.; Matsunaga, S. *J. Org. Chem.* **1991**, *56*, 4971.
4.   Trost, B. M., Fleming, I., Eds. *Comprehensive Organic Synthesis;* Pergamon Press: Oxford, 1991; vol. 7.
5.   Djerassi, C. *Org. React.* **1951**, *6*, 207.
6.   Inanaga, J.; Hirata, K.; Saeki, H.; Katsuki, T.; Yamaguchi, M. *Bull. Chem. Soc. Jpn.* **1979**, *52*, 1989.
7.   Hara, R.; Furukawa, T.; Kashima, H.; Kusama, H.; Horiguchi, Y.; Kuwajima, I. *J. Am. Chem. Soc.* **1999**, *121*, 3072.
8.   Kingston, D. G. I., Molinero, A. A.,; Rimoldo, J. M.; *Prog. Chem. Org. Nat. Prod.* **1993**, *61*, 1.
9.   Holton, R. A.; Jou, R. R.; Kim, H. B.; Williams, A. D.; Harusawa, S.; Lowenthal, R. S.; Yagai, S. *J. Am. Chem. Soc.* **1988**, *110*, 6558.
10. Dauben, W.G.; Deniny, E. *J. Org. Chem.* **1969**, *25*, 2647.
11. Solé, D.; Bonjoch, J.; García-Rubio, S.; Peidró, E.; Bosch, J. *Chem. Eur. J.* **2000**, *6*, 655.
12. a) Sapi, J.; Massiot, G. in *The chemistry of Heterocyclic compounds,* Supplement to Vol. 25, Part 4, Taylor, E. C. Eds, Wiley: New York, 1994, pp 279–355; b) Bosch, J.; Bonjoch, J.; Amat, M. in *The Alkaloids*, Vol. 48 Cordell A. Ed., Academic Press: New York, 1996 pp 75–189.
13. Anet, F. A. L.; Robinson, R. *Chem. Ind.* **1953**, 245.
14. a) Solé, D.; Bonjoch, J.; García-Rubio, S.; Suriol, R.; Bosch, J. *Tetrahedron Lett.* **1996**, *37*, 5213; b) Bonjoch, J.; Solé, D.; García-Rubio, S.; Bosch, J. *J. Am. Chem. Soc.* **1997**, *119*, 7230.

15. a) Brase, S.; deMeijere, A. *in Metal-Catalyzed Cross-Coupling Reactions;* Diederich, F.; Stang, P. J. Eds.; Wiley-VCH, 1998; b) Tsuji, J. *Palladium Reagents and Catalysts;* John Wiley & Sons: Chichester, 1995; c) Heck, R. F. *Palladium Reagents in Organic Synthesis*; Academic Press: New York, 1985.
16. Evans, D. A.; Black, W. C. *J. Am. Chem. Soc.***1993**, *115*, 4497.
17. Boeck, L. D.; Chio, H.; Eaton, T. E.; Godfrey, O. W.; Michel, K. H.; Nakatsukasa, W. M.; Yao, R. C. (Eli Lilly) Eur. Pat. Appl. EP 375 316, **1990**; *Chem. Abstr.* **1991**, *114*, 80066.
18. a) Fleming, I. *Bull. Soc. Chem. Fr.* **1981**, **II**. 7-**II**.13; b) Fleming, I.; Lee, T. V. *Tetrahedron Lett.* **1981**, *22*, 705; c) Fleming, I.; Goldhill, J.; Paterson, I. *Tetrahedron Lett.* **1979**, *20*, 3205; d) Fleming, I.; Iqbal, J. *Tetrahedron Lett.* **1983**, *24*, 2913.
19. Still, W. C.; Galynker, I. *Tetrahedron* **1981**, *37*, 3981.
20. All calculation were performed with an MM2 forcefield on structures generated by a Multiconformer search using MacroModel (Version 3.5) provided by Professor W. Clark Still, Columbia University.
21. a) Takano, S.; Shimazaki, I.; Iwabuchi, I.; Ogasawara, K. *Tetrahedron Lett.* **1990**, *31*, 3619; b) Willson, T. M.; Kociensky, P.; Jarowicki, K.; Isaac, K.; Hitchcock, P. M.; Faller, A.; Campbell, S. F. *Tetrahedron* **1990**, *46*, 1767; c) Takano, S.; Shimazaki, Y.; Moriya, M.; Ogasawara, K. *Chem. Lett.* **1990**, 1177; d) Deslongchamps, P. *Stereoelectronic Effects in Organic Chemistry*; Pergamon: Oxford, 1983; pp 221–231.
22. Rahman, S. M. A.; Ohno, H.; Murata, T.; Yoshino, H.; Satoh, N.; Murakami, K.; Patra, D.; Iwata, Ch.; Maezaki, N.; Tanaka, T. *J. Org. Chem.* **2001**, *66*, 4831.
23. Hayashi, T.; Kawasaki, M.; Miwa, Y.; Taga, T.; Morita, N. *Chem. Pharm. Bull.* **1990**, *38*, 945.
24. Nagata, W. In *Org. React. (N.Y.)* **1977**, *25*, 255.
25. a) Cope, A. C.; Field, L.; MacDowell, D. W. H.; Wright, M. E. *J: Am. Chem. Soc.* **1956**, *78*, 2547; b) Hill, R. K. *Comprehensive Organic Synthesis;* Trost, B. M., Fleming, I., Eds.; Pergamon Press: Oxford, 1991,; Vol. 5, Chapter 7.1; For reviews on oxy-Cope rearrangement see: c) Paquette, L. A. *Angew. Chem., Int. Ed. Engl.* **1990**, *29*, 67; d) Wilson, S. R. *Org. React. (N. Y.)* **1993**. *43*, 93; e) Paquette, L. A. *Tetrahedron* **1997**, *53*, 13971.
26. Tanaka, T.; Okuda, O.; Murakami, K.; Yoshino, H.; Mikamiyama, H.; Kanda, A.; Kim, S.-W.; Iwata, C. *Chem. Pharm. Bull.* **1995**, *43*, 1407.
27. a) Ashby, E. C.; Nodling, S. A. *J. Org. Chem.* **1979**, *44*, 4371; b) Todd, D. *Org. React.* **1948**, *4*, 378.
28. Macdonald, T. L.; Still, W. C. *J. Am. Chem. Soc.* **1975**, *97*, 5280.
29. a) Reetz, M. T.; Westermann, J.; Steinbach, R.; Wenderoth, B.; Peter, R.; Ostarek, R.; Maus, S. *Chem. Ber.* **1985**, *118*, 1421; b) Reetz, M. T.; Steinbach, R.; Westermann, J.; Peter, R.; Wenderoth, B. *Chem. Ber.* **1985**, *118*, 1441.
30. Chen, X. T.; Bhattacharya, S. K.; Zhou, B.; Gutteridge, C. E.; Pettus, T. R. R.; Danishefsky, S. J. *J. Am. Chem. Soc.* **1999**, *121*, 6563.

Chapter 3
Working with Models

New synthetic procedures are always developed using simple models and, after the efficiency of the method has been proved, the transformations are applied to a step on a total synthesis. Often, before effecting the desired transformation in the *real synthetic intermediate*, the suitability of the reaction is determined using a model molecule.

But the question is how applicable to the real targets are the results obtained when working with models? Their usefulness in synthetic chemistry is obvious as demonstrated by the huge amount of work published, detailing the construction of model compounds of specific molecules. Thankfully for us, most of the time they are applicable, but sometimes they are not at all. Then, the synthetic plan based on reactions that have been developed for the model compound, needs to be changed, or at least thoroughly modified, to achieve the desired result. The following examples detail some cases in which the model transformation is not applicable at all to the real synthesis.

3.1 (–)-Stenine [1]

3.1 (–)-stenine

Target relevance

Extracts of *Stemona* species have been used in Chinese and Japanese folk medicine as insecticides, as drugs for treatment of respiratory diseases such as bronchitis, pertussis, and tuberculosis, and as antihelmintics [2].

Dead Ends and Detours: Direct Ways to Successful Total Synthesis
Miguel A. Sierra and María C. de la Torre
Copyright © 2004 WILEY-VCH Verlag GmbH & Co. KGaA, Weinheim
ISBN: 3-527-30644-7

Synthetic plan

The key intermediate in the total synthesis of (−)-stenine **3.1** is bicycle **3.2** derived from *N*-Cbz-*l*-tyrosine **3.3** by oxidative cyclization. After appendage of the four-carbon chain on C4 of bicycle **3.2**, the seven-membered ring of *trans*-fused perhydroindole present in **3.1** will be closed using a Mitsunobu reaction [3]. The lactone ring of **3.1** will be constructed prior to the ring closure (Scheme 3.1).

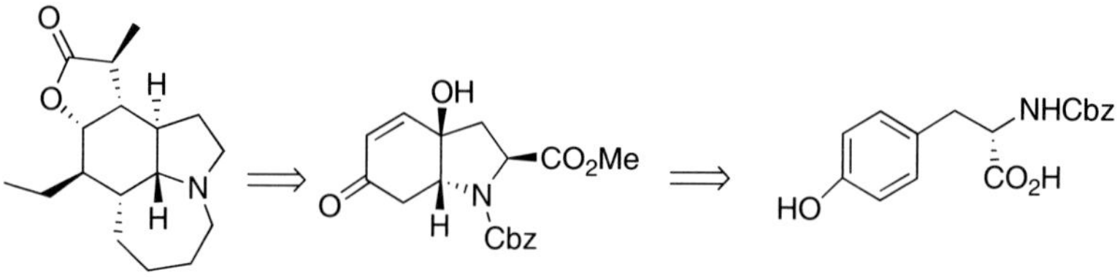

Scheme 3.1

Predictable problems

- The oxidative cyclization of *l*-tyrosine derivative **3.3** should be effected to build the *cis*-perhidroindole system of **3.2** with the right stereochemistry present in (−)-stenine **3.1**.
- The conversion of **3.2** to the *trans*-tricyclic perhydroindole moiety of (−)-stenine **3.1**.

Synthesis

Step 1. Synthesis of cis-perhydroindolone 3.4 (Scheme 3.2):

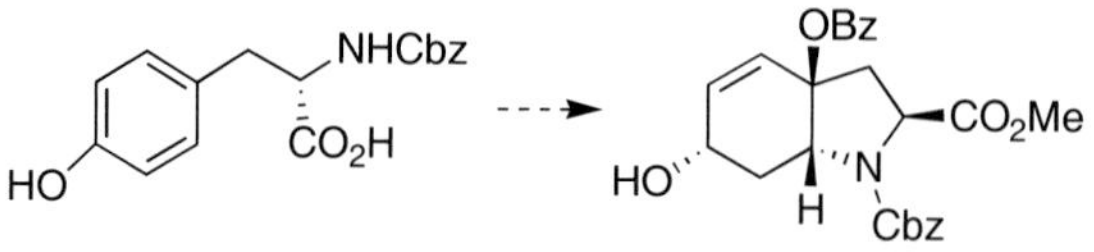

Scheme 3.2

The oxidative cyclization of the *l*-tyrosine derivative **3.3** to form **3.4** was achieved in 60% yield and with diastereoselectivities higher than 98:2, favoring the desired esteroisomer **3.4** by using $PhI(OAc)_2$ as the oxidant. The exceptional selectivity of this cyclization is due to destabilizing steric interactions in conformer **3.6**, specially due to allylic 1,3-strain ($A^{1,3}$-strain) [4] that would lead to the undesired diastereomer **3.7**.

These interactions are minimized in conformer **3.5** leading to **3.2**. Benzoylation of the tertiary alcohol of **3.2** was followed by the regioselective reduction of the ketone group using Luche conditions (NaBH$_4$/CeCl$_3$.7H$_2$O) [5] to form alcohol **3.4** as a single diastereomer (Scheme 3.3).

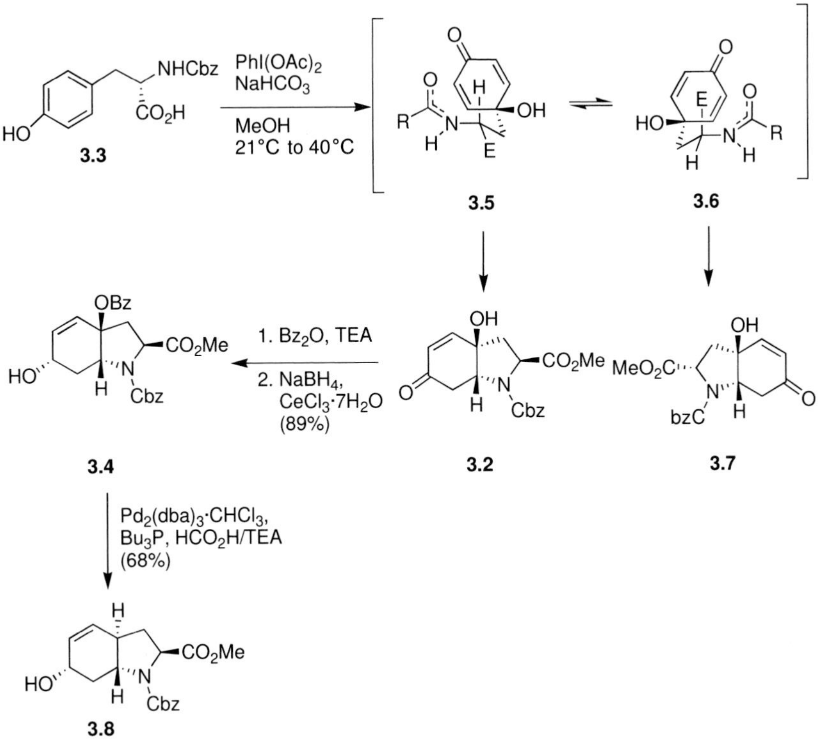

Scheme 3.3

Desoxygenation of the benzoate moiety at the bridge was achieved by reduction of the π-allyl-paladium complex obtained by treatment of **3.4** with Pd$_2$(dba)$_3$ · CHCl$_3$ at its more hindered tertiary carbon. The best conditions were found after extensive experimentation by using the mixture Pd$_2$(dba)$_3$ · CHCl$_3$/Bu$_3$P/HCO$_2$H/TEA. Under these conditions the desired allylic alcohol **3.8** was obtained in a 68% yield. The stereochemistry of the ring fusion is now *trans*. The complete selectivity obtained at this point should derive from the assistance of the hydroxyl group both to the complexation of the Pd-catalyst and to the hydride transfer from the formiate reagent. Otherwise, it would be difficult to understand the transference of the hydride group by the highly encumbered concave face of **3.4**.

Concept:
Allylic 1,3-Strain (A^{1,3}-strain) [4]

Allylic 1,3-strain refers to the control of conformation in an alkene by a *cis*-substituent. It is represented by the symbol A^{1,3}-strain and it is a determinant of the result of many asymmetric processes. The conformational preference in a simple alkene can be represented as depicted in Scheme 3.4.

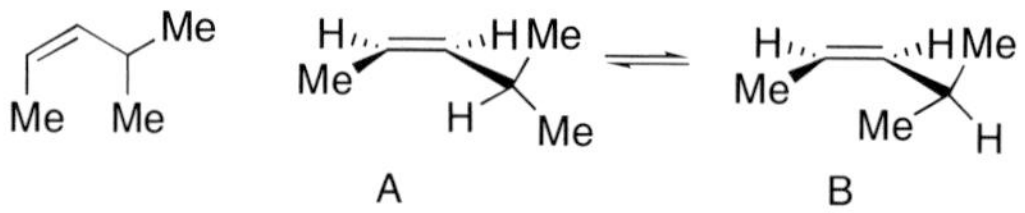

Conformer B is considerably higher in energy than conformer A because there is simply no room to place the methyl attached to the double bond and the aliphatic methyl group is eclipsed. Therefore, the conformation of these open chain compounds is usually fixed, with the hydrogen of the allylic center eclipsing the double bond, and they can be used to achieve good stereochemical results. In the case represented in Scheme 3.3 it is obvious that conformer **3.6** has a strong interaction between the carboxy-group and the quinone ring that is absent in conformer **3.5**. Therefore, the sterochemical result of the cyclization is exceptional.

Step 2. Installation of the C4 four-carbon ring chain and lactone ring closing (Scheme 3.4):

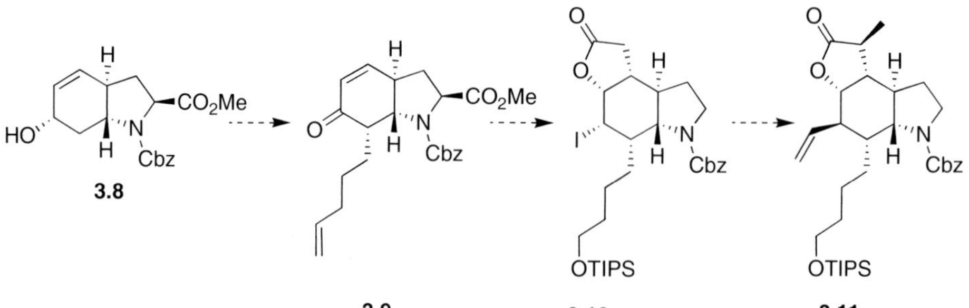

Scheme 3.4

Oxidation of allylic alcohol **3.8** with tetrapropylammonium perruthenate (TPAP) [6] generated the enone **3.12**. This enone was deprotonated with KHMDS and the enolate reacted with triflate **3.13** to afford, albeit in low yield (36%), exclusively compound **3.9**.

Compound **3.9** has the required α-oriented carbon side chain. Luche's reduction of the carbonyl group of **3.9** was followed by the Eschenmoser–Claisen [7] rearrangement to yield **3.14**. Functionalization of the double bond of the pendant carbon chain was effected before closing the lactone-ring by treatment of **3.14** with Ad-mixβ [8] and *in situ* breakage of the formed diol with NaIO₄. The resulting aldehyde was reduced to the corresponding alcohol that was protected as its TIPS derivative **3.15**. Since the carboxymethyl group of **3.15** is not present in (–)-stenine **3.1**, it was removed by reductive decarboxylation under Ireland conditions (hydrolysis, followed by treatment with PhOPOCl₂/TEA, PhSeH and reaction with TBTH/AIBN) [9] yielding **3.16** (Scheme 3.5).

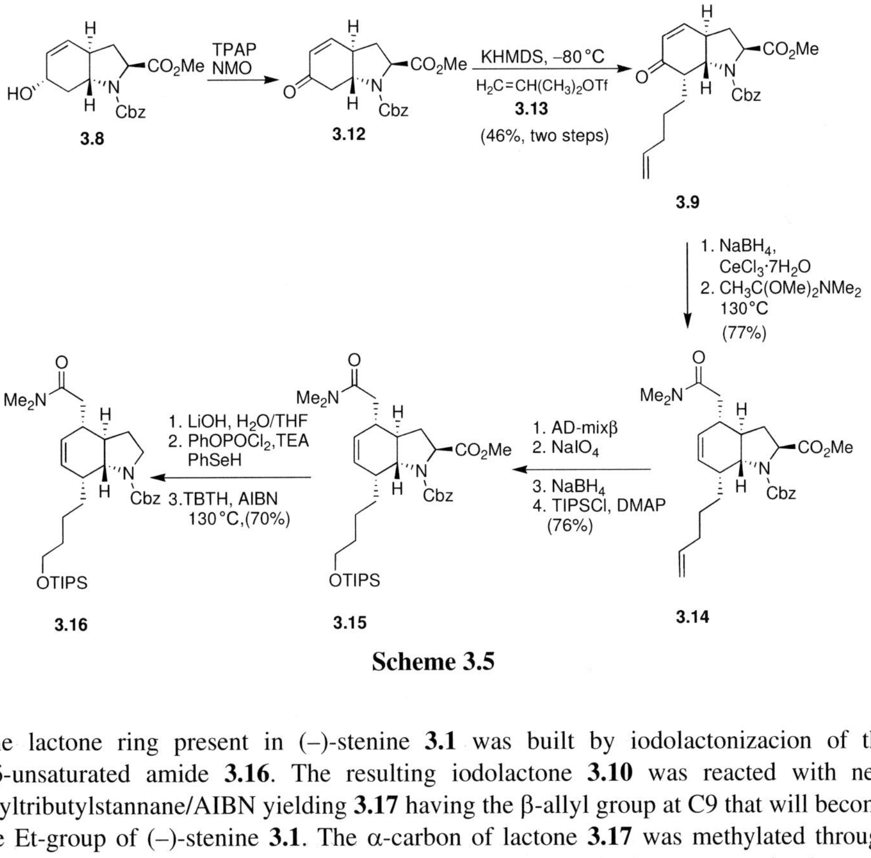

Scheme 3.5

The lactone ring present in (–)-stenine **3.1** was built by iodolactonizacion of the γ,δ-unsaturated amide **3.16**. The resulting iodolactone **3.10** was reacted with neat allyltributylstannane/AIBN yielding **3.17** having the β-allyl group at C9 that will become the Et-group of (–)-stenine **3.1**. The α-carbon of lactone **3.17** was methylated through enolization and quenching with MeI. The Me group of **3.18** was place in the most accessible convex β-face. Vinyl lactone **3.11** was formed by removing one carbon of the allyl moiety (osmilation (OsO₄), diol breakage (HIO₄), aldehyde reduction (NaBH₄)), and Grieco-elimination [10] (Scheme 3.6).

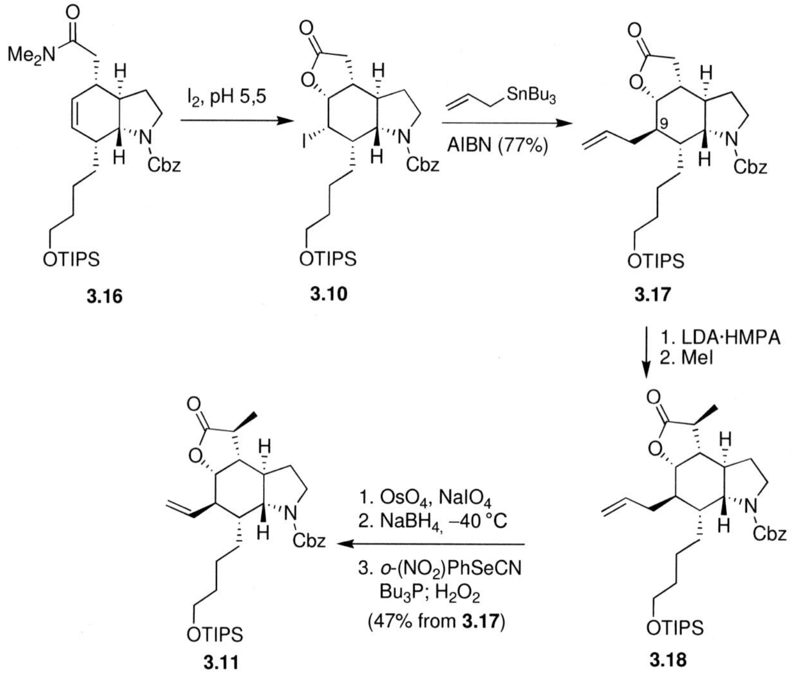

Scheme 3.6

Step 3. Closure of the seven-membered ring and completion of the synthesis (Scheme 3.7):

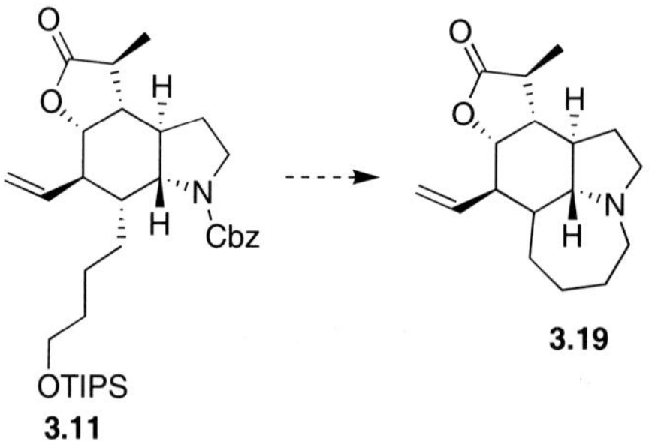

Scheme 3.7

The key step to complete the synthesis of (–)-stenine **3.1** was the Mitsunobu cyclization of amino alcohol obtained from **3.11**, followed by hydrogenation of the vinyl group (which should simultaneously eliminate the Cbz-nitrogen protecting group). The cyclization step was predictably troublesome. However, it had worked nicely on the related model **3.20** that had yielded the tricycle **3.21**, resembling the stenine core, after submission to Mitsunobu reaction conditions. Surprisingly, cyclization of alcohol **3.22**, obtained by removal of the protecting groups of **3.11** produced no (–)-stenine **3.1** (Scheme 3.8). It was argued that, since the Mitsunobu reaction strongly depends on the rate of cyclization *vs* side reactions of the activated alcohol, **3.20** should be more preorganized towards the formation of the seven-membered ring than intermediate **3.22**. This was probably the reason because the model, on which the key step of the synthetic plan was based, efficiently cyclized, but not the real intermediate. Therefore access to (–)-stenine **1.1** through a Mitsunobu cyclization was abandoned.

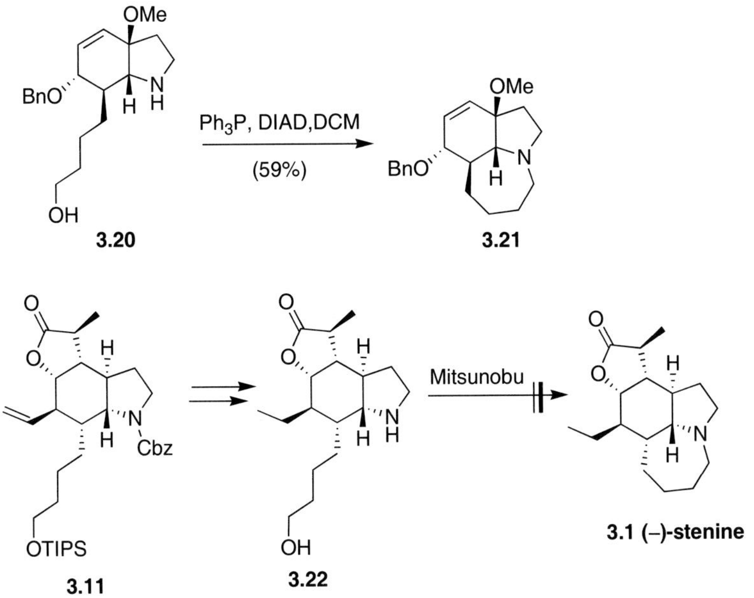

Scheme 3.8

Step 4. Re-evaluation of the methodology to effect the seven-membered ring-closure. A tactical deviation: Access to (–)-stenine **3.1** was obtained from the key intermediate **3.11**. Removal of the silyl protecting group was followed by sequential oxidation of the free alcohol to carboxylic acid and hydrogenation of the double bond to yield the acid **3.23**. Now the seven-membered ring was closed by lactone formation using pentafluorophenyl diphenylphosphinate (FDPP) as the condensing agent, to yield **3.24**. Stenine was finally obtained by the reduction of the amide group of **3.24** in a two-step sequence: reaction with Lawesson's reagent [11] and Raney-Ni desulfurization (Scheme 3.9).

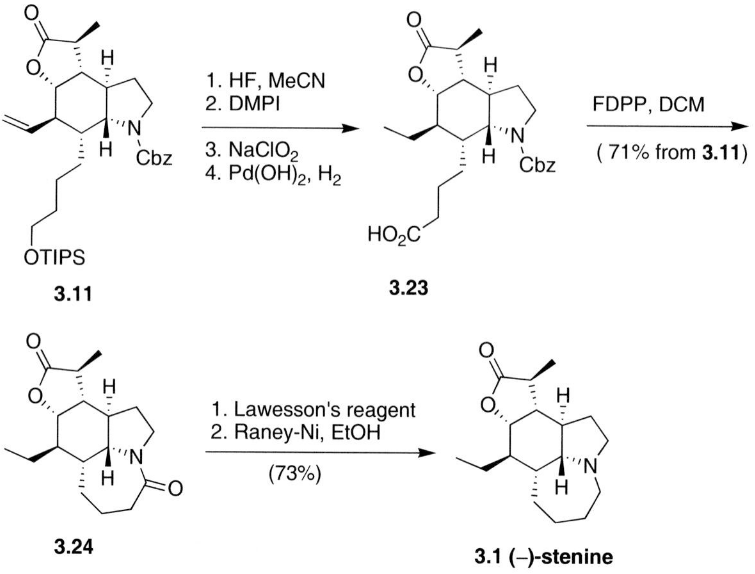

Scheme 3.9

Evaluation

- No strategic changes of the original planning have been made. The oxidative cyclization of *l*-tyrosine derivative **3.4** gave, as designed, the *cis*-perhidroindole system of **3.2** with the right stereochemistry to be transformed in the *trans*-fused system of (–)-stenine **3.1**.

- The main tactical synthetic change refers to the mode of cyclization to form the seven-membered ring of (–)-stenine **3.1**. The Mitsunobu approach was based on the nice results obtained in a model system. However, how applicable to compound **3.22** are the results obtained with **3.20**, since they are significantly different? The failure of intermediate **3.22** to cyclize may be attributable to reasons other than the absence of a hypothetical preorganization present in **3.20**. In fact, molecular mechanic analysis showed no appreciable differences between the two compounds regarding a preorganization towards cyclization [12]. The cis-anti-tricyclic system **3.21** is 7.8 Kcal mol^{-1} more stable than **3.25** (this compound would derive from the cyclization of **3.26**), which harbors the ring stereochemistry present in the target compound **3.1**. It is probable that the energetically more demanding cyclization to yield the trans-syn-fused system present in (–)-stenine **3.1** is responsible of the failure of compound **3.22** to cyclize under Mitsunobu conditions.

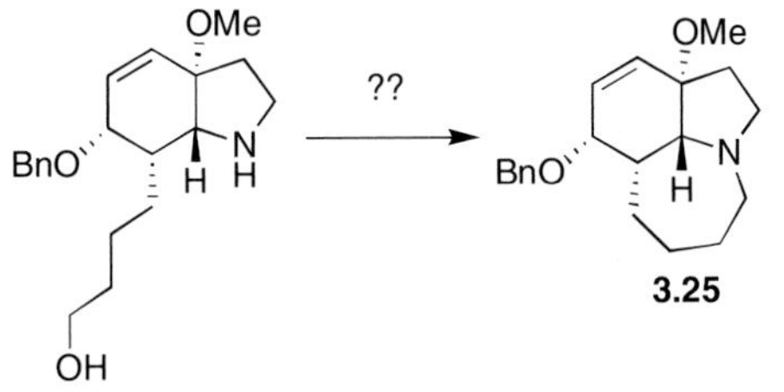

3.26   **3.25**

Nevertheless, the seven-membered ring of (−)-stenine **3.1** was closed with a minor tactical change that added just one additional step to the original plan.

Key synthetic reactions

Eschenmoser–Claisen rearrangement: In situ formation of an *O*-allyl-aminoketene acetal by reaction of an allylic alcohol with MeC(OMe)$_2$NMe$_2$ followed by a Claisen rearrangement. The concerted mechanism accounts for the 1,3-chirality transfer with the simultaneous formation of a C–C bond and the incorporation of an amide functional group [7].

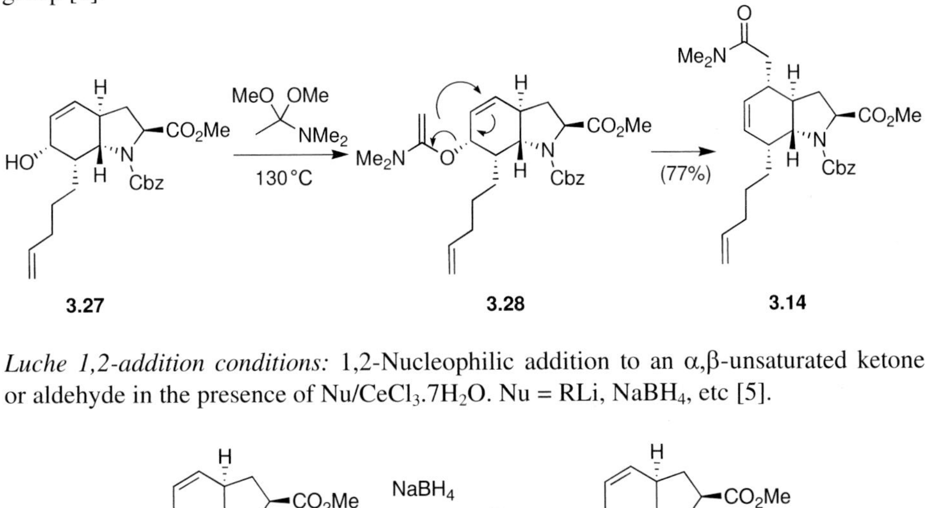

3.27   **3.28**   **3.14**

Luche 1,2-addition conditions: 1,2-Nucleophilic addition to an α,β-unsaturated ketone or aldehyde in the presence of Nu/CeCl$_3$.7H$_2$O. Nu = RLi, NaBH$_4$, etc [5].

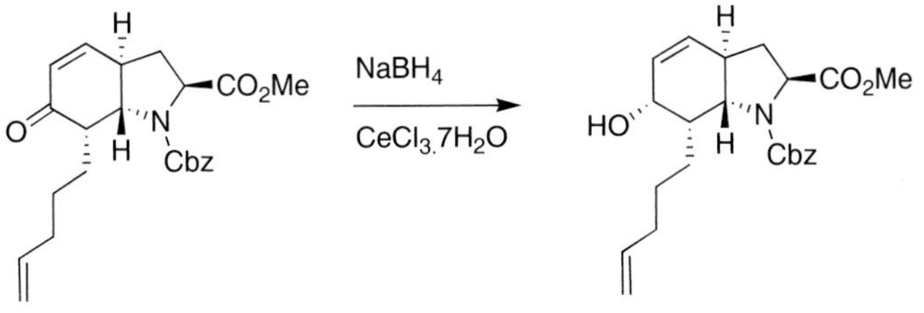

3.9   **3.27**

Ireland reductive decarboxylation: Removal of a carboxylic group by formation of an acylseleno ester followed by treatment with TBTH/AIBN,Δ [9].

3.15   **3.29**

3.30   **3.16**

Especially interesting reagents

Ad-mixβ: (Sharpless asymmetric dihydroxylation) [8]. Enantioselective *syn*-dihydroxylation reagent for olefins. It consists of a mixture of $K_2OsO_2(OH)_4$ (the catalytic hydroxylating agent), $K_3Fe(CN)_6$ (the stoichiometric oxidating agent) and $(DHQD)_2$-PHAL **3.31** (the chiral ligand). Usually a base such as K_2CO_3 is also added.

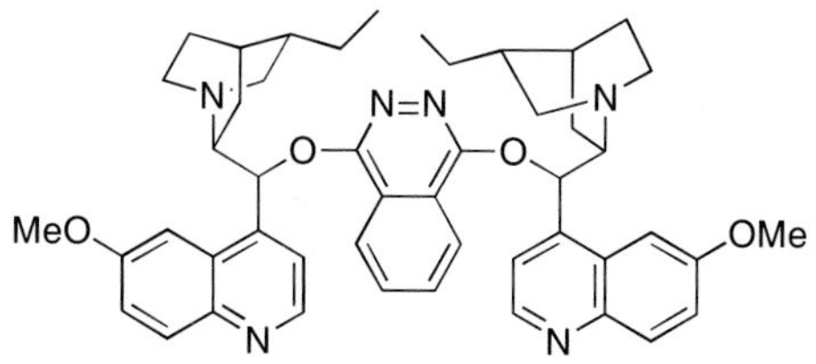

(DHQD)₂-PHAL

3.31

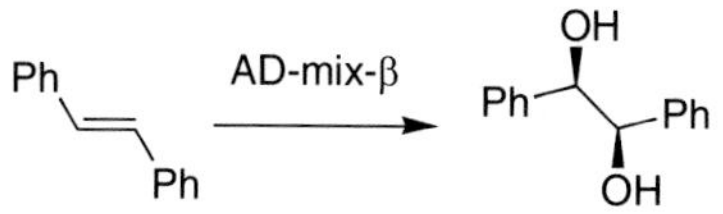

Lawesson's reagent: 2,4-bis(4-methoxyphenyl)-1,2,3,4-dithiadiphosphetan-2,4-dithione
3.32 [11]. Transform a carbonyl group from an ester, amide, etc. into the corresponding thiocarbonyl group.

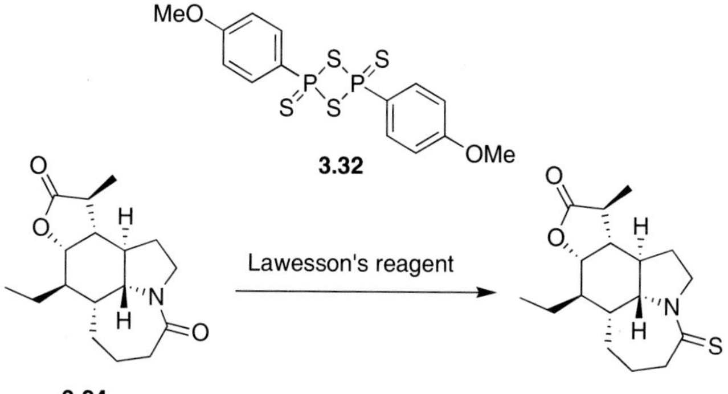

3.32

Lawesson's reagent

3.24   **3.33**

The example of the failure to reproduce the Mitsunobu ring closure, effected in the model compound **3.20**, in the actual intermediate **3.22** during the synthesis of (–)-stenine **3.1** may be attributed to the noticeable differences between the model compound and the synthetic intermediate. Sometimes the structurally closest compound to the planned intermediate does not produce satisfactory results. However, the desired results are obtained with a structurally distant, yet synthetically equivalent, product. This situation arises quite often and may be exemplified by Boger's synthesis of (+)-piperazinomycin.

3.2   (+)-Piperazinomycin [13]

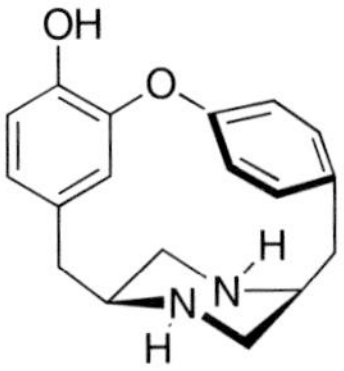

3.34 (+)-piperazinomycin

Target relevance
(+)-Piperazinomycin **3.34** is a macrocyclic piperazine isolated as a minor metabolite of *Streptoverticillium olivoreticuli* subsp. *neoenacticus* [14], which constitutes, to date, the simplest naturally occurring agent having the parent 14-membered *p*- and metacyclophane diaryl ether subunit. This structural feature is the pharmacophore of cytotoxic and antitumoral agents like bouvardin, deoxybouvardin, and RA-I-X.

Synthetic plan
The key step of the synthetic approach to (+)-piperazinomycin **3.34** is the Ulmann reaction [15] of diketopiperazine **3.35** that would be accessed from the dipeptide **3.36**. The advantage of this approach rests in the use of readily available amino acids to construct the appropriately functionalized diaryl ethers, which effect the diketo-piperazine formation and the Ulmann cyclization (Scheme 3.10).

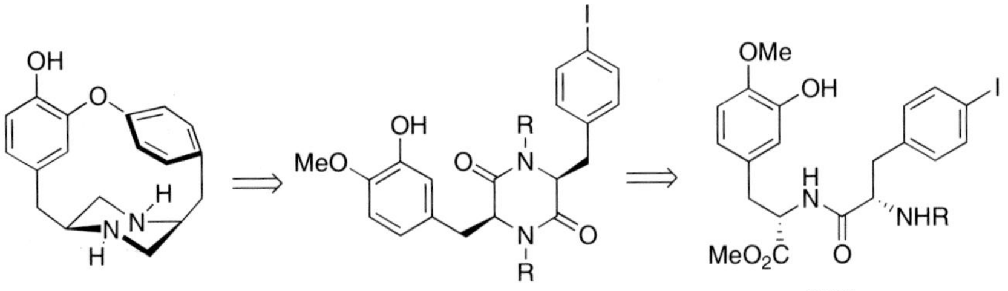

3.34   **3.35**   **3.36**

Scheme 3.10

The synthetic plan was based on the exhaustive study effected on the cyclization of a number of compounds **3.37**. The common feature of these models was the piperazine nucleus present in the natural product. Compounds **3.37** systematically failed to produce the desired macrocycles **3.38**. However, compounds **3.37** were potentially better substrates for the cyclization to piperazinomycin than diketopiperazines **3.35**, as no possibility exists for their racemization under the conditions employed to close the ring (Scheme 3.11).

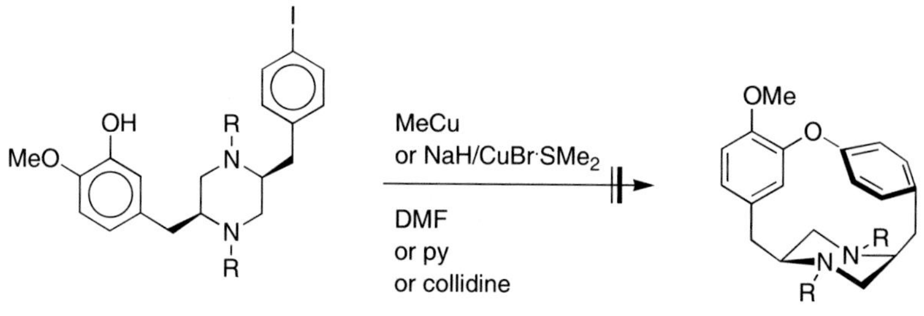

3.37a R = H
3.37b R = COOMe
3.37c R = Cbz

3.38

Scheme 3.11

Predictable problems

Racemization of the two chiral centers of diketopiperazine **3.35** during the cyclization process is the main predictable problem.

Synthesis

*Step 1. Cyclization of diketopiperazine **3.39** (Scheme 3.12):* The synthesis of diketo-piperazine **3.39** was achieved by using conventional amino acid chemistry from *l*-tyrosine derivative **3.40**. Ullmann macrocyclization of diketopiperazine **3.39** was carried out in very controlled conditions by treatment with NaH (4 equiv.) and CuBr · SMe$_2$ in dry boiling DMF under moderately diluted reaction conditions. In these strict reaction conditions 53% of compound **3.41** was obtained. The Ullmann reaction of diketopiperazine **3.39** is conducted under conditions where the secondary amides are deliberately deprotonated prior to exposure to the thermal reaction conditions. Subsequent racemization of the tris-anion **3.42** requires anion generation α- to and cross-conjugated with the amide anion. It is clear that the tris-anion **3.42** is not further deprotonated under the Ulmann cyclization conditions and the level of racemization is maintained below 3%. Reduction of macrocyclic diketopiperazine **3.41** with BH$_3$. THF and removal of the Me-group (HBr-AcOH) proportioned (+)-piperazinomycin **3.34** in a 60% yield (Scheme 3.12).

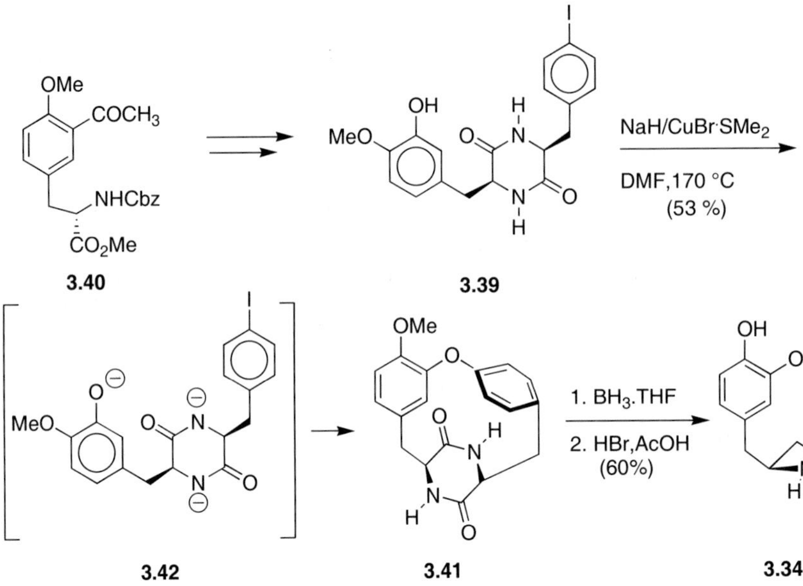

Scheme 3.12

Evaluation

The synthesis of (+)-piperazinomycin **3.34** was achieved as planned after having examined and extensively studied the cyclization of piperazines **3.37**. However, *the question in this synthesis is why compounds **3.37** that were potentially better substrates for the cyclization than diketopiperazine **3.39** (since there is no possibility of their racemization during the conditions employed to close the ring) failed to cyclize under the Ullmann conditions?*

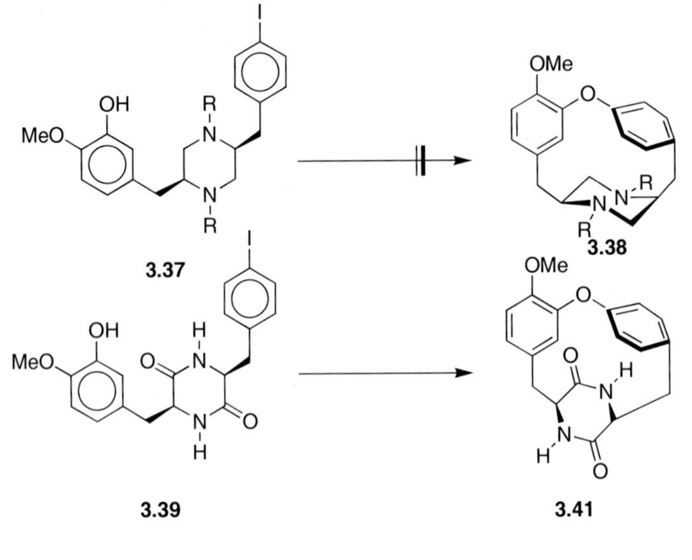

Modeling compounds **3.39** and **3.37a** clearly explain the observed differences found during the cyclization step. Both compounds must adopt a boat conformation in order to cyclize. This conformational switch is energetically more demanding for piperazines **3.37** (E_{boat}–E_{chair}= 6.6 Kcal mol^{-1}) than for the flatter diketopiperazine **3.39**, for which the preferred conformation is already boat-like (E_{boat}–E_{chair} = –0.4 Kcal mol^{-1}).

Boger synthesis of (+)-piperazinomycin illustrates that it is not always the closer model to the real intermediate that is the best. Roush's synthesis of (+)-damavaricin D **3.43** illustrates the risk of using the stereoisomer of the synthetic intermediate as a model to develop specific reaction conditions which are to be applied during the synthesis.

3.3   (+)-Damavaricin D [16]

3.43 (+)-damaravicin D

Target relevance
(+)-Damavaricin D **3.43** is a biosynthetic precursor of the streptovaricin antibiotic family [17]. It acts as an inhibitor of RNA-directed DNA polymerase, and some of its derivatives have potential usefulness as antiviral agents and for treatment of adult T-cell leukemia.

Synthetic plan
The synthesis of (+)-damavaricin D **3.43** was designed to use the coupling of the highly functionalized naphthalene **3.44** and the ansa-chain aldehyde **3.45**. The elaboration of the pendant chain on intermediate **3.45**, followed by macrolactonization and functional group manipulation of **3.46**, should form the target molecule **3.43** (Scheme 3.13).

Predictable problems
The densely functionalized (+)-damaravicine requires a fine-tuning of the different protecting groups employed in its synthesis; the compatibility of all of them along the synthetic sequence may be difficult to ensure.

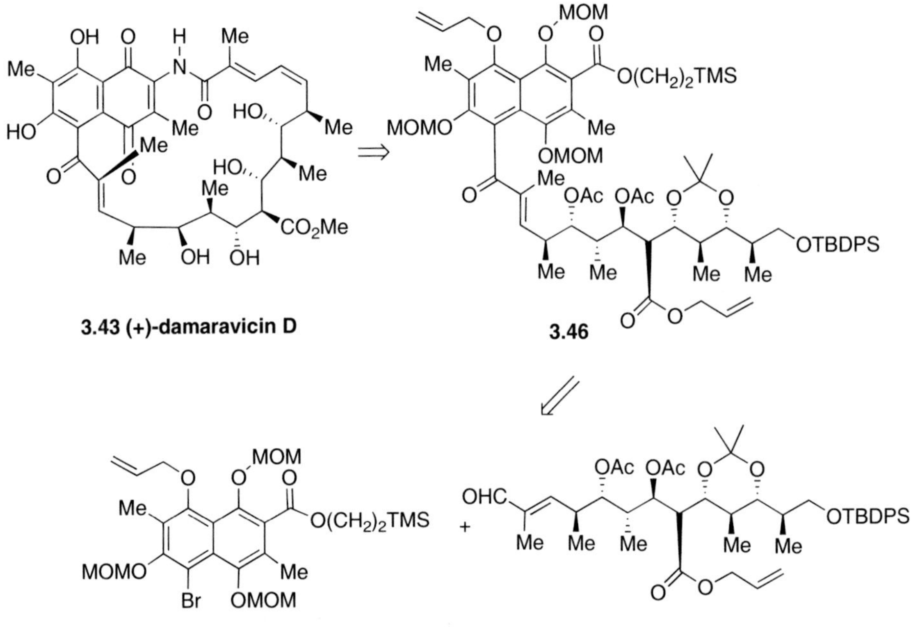

3.43 (+)-damaravicin D **3.46**

3.44 **3.45**

Scheme 3.13

*Step 1. Coupling of the ansa-chain **3.45** and naphthalene **3.44** and initial elaboration of the side-chain* (Scheme 3.14). The coupling of the aryllithium derived from naphthalene bromide **3.44** with aldehyde **3.45**, provided a mixture of epimeric alcohols, the oxidation of which, with the Dess–Martin periodinane, yielded enone **3.46** as a mixture (1:1) of atropisomers. Elongation of the ansa-chain requires the terminal aldehyde **3.47**, which was obtained by sequential removal of the TBS-protecting group followed by Swern oxidation [18]. Two additional carbons were introduced at this stage *via* Still's (Z)-selective olefination procedure, providing **3.48** in 72% yield, along with small amounts of the *E*-isomer (Scheme 3.14).

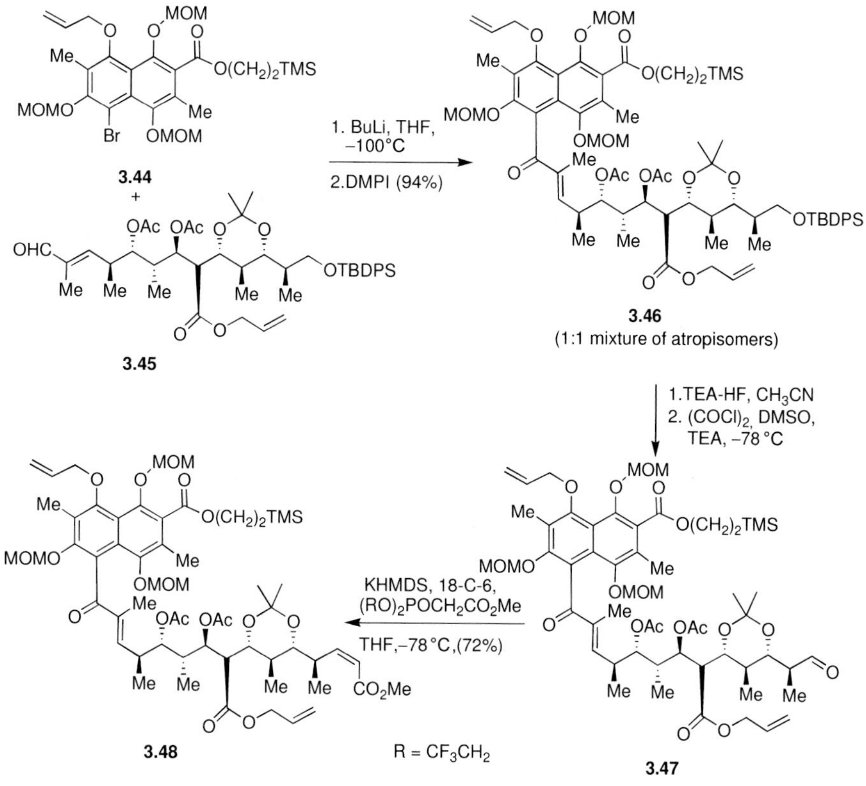

Scheme 3.14

Step 2. Instalation of the carbamate functional group in the naphthalene ring, two carbon extension of the appendage and macrolactonization (Scheme 3.15): Prior to effecting the macrocyclization step, intermediate **3.48** has to be elaborated to incorporate two more carbon atoms in the ansa-chain, the carbamate functional group that will become the amino group, as well as to close the macrocycle. The aromatic ester of **3.48** was hydrolyzed by TBAF treatment and the corresponding acid submitted to Curtius rearrangement in the presence of TMSCH$_2$CH$_2$OH. Carbamate **3.49** was obtained under these conditions.

Scheme 3.15

Next, a very compromising selective reduction of the Z-enoate to the corresponding aldehyde has to be achieved to build the full-carbon skeleton of the ansa-chain. It is evident that a very careful control of the reaction conditions is required to prevent competitive reduction of the C11 and C13 acetate moieties. A good solution to this problematic situation could not be found. Thus, treatment of **3.49** with DIBALH produced the mixture of the desired aldehyde **3.50** and the alcohol **3.51**, together with high amounts of unreacted **3.49**. After a tedious and time consuming separation and recycling process, the required aldehyde **3.50** was obtained in 82% yield. Aldehyde **3.50** was elaborated to the *E,Z*-dienoate **3.52** by Horner–Wadsworth–Emmons olefination. To complete the synthesis, only unmasking of the amine and carboxylic acid groups, for macrocyclic amide ring closure, and the final removal of the protecting groups, remained. The crude amino acid, obtained from **3.52**, yielded the macrolactam **3.53** in 76%, under the Mukaiyama salt macrolactamization protocol [19], as a single atrop-isomer (Scheme 3.15).

Step 3. Removal of protecting groups on 3.53 and failed access to (+)-damavaricin D (Scheme 3.16): The deprotection steps to be applied at the final stages of (+)-damavaricin D **3.43** were developed in the model compound **3.54**. Removal of the MOM and acetonide groups was successfully achieved by acid treatment to yield **3.55** [20]. However, when the analogous conditions were applied to compound **3.53**, the precursor of (+)-damavaricin D, this resulted in its complete decomposition. Therefore, which compound **3.53** is a very advanced intermediate in the synthesis of (+)-damaravicine **3.43** could not be converted in these conditions in the target molecule. Attempted oxidative demethylation of **3.53** produced compound **3.56**, which could not be elaborated either, to (+)-damaravicine.

Step 4. Tunning-up the hydrolysis protocol and the achievement of the synthesis of damavaricin 3.43 (Scheme 3.17): Clearly, compound **3.54** was a poor model for the synthesis of damavaricin **3.43**. Macrocycles **3.53** and **3.54** belong to opposite atropisomer series and their reactivity is substantially different. The ultimate reasons why these two structurally similar compounds have such different conformational characteristics and chemical reactivity still remains a mystery.

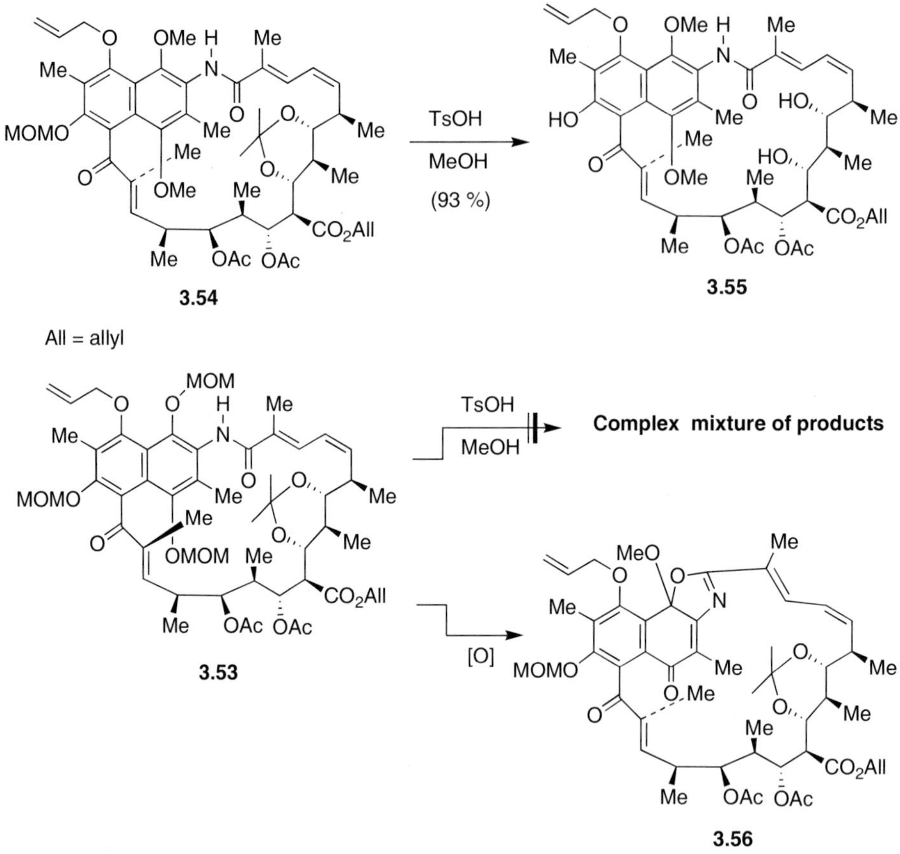

Scheme 3.16

The authors did not significantly modify the synthetic route in order to access the target molecule. However, they were forced to use very strict reaction conditions to eliminate the protecting groups of **3.53**. Thus, acid hydrolysis of **3.53** yielded **3.57** together with a mixture of partially deprotected derivatives and considerable amounts of recovered starting material **3.53**. Compounds other than the desired **3.57** were resubmitted to several cycles and the final yield of **3.57** was 54%. Removal of the allyl protecting groups was achieved by treatment of **3.57** with TBTH in the presence of catalytic Pd(PPh$_3$)$_4$. Base hydrolysis followed by esterification successfully yielding **3.58**, which by hydrolysis with TFA and air oxidation finally provided (+)-damaravicine **3.1** (Scheme 3.17).

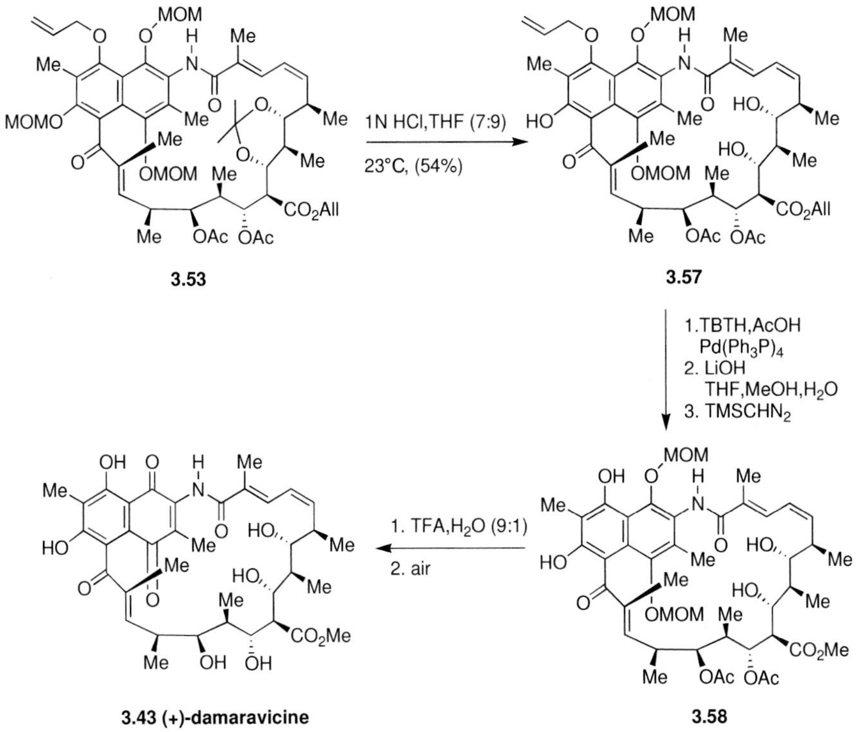

Scheme 3.17

Evaluation

As noted before, compound **3.54** was a poor model for the synthesis of damavaricin **3.43**. Macrocycles **3.53** and **3.54** belonging to opposite atropisomer series presented substantially different reactivity. The ultimate reasons why these two structurally similar compounds have such different conformational characteristics and chemical reactivity still remains a mystery. Therefore, although the consequences on the overall planning were disturbing, the ultimate reasons for this failure, based on the assumption of similar reactivity of the atropisomeric series, are not known.

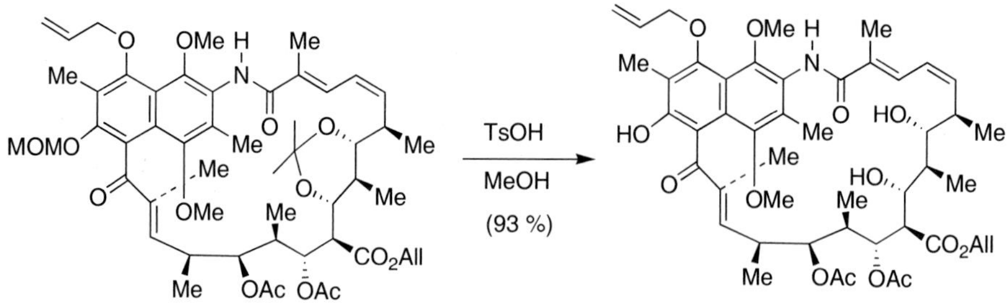

3.54

All = allyl

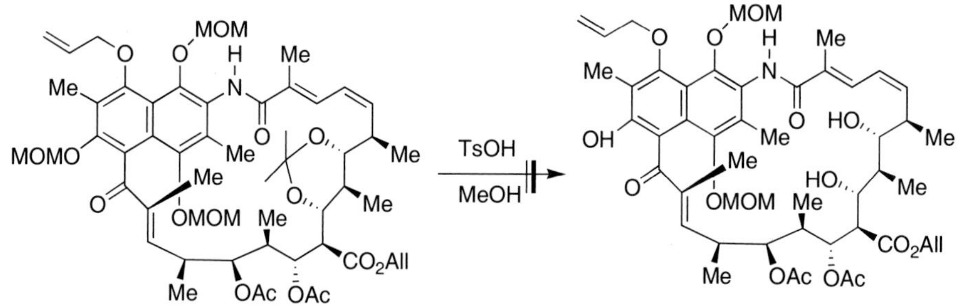

3.53

3.55

3.57

Key synthetic reactions

Swern oxidation: The oxidation of an alcohol to an aldehyde or ketone using an oxalyl chloride/DMSO/TEA system in DCM as the solvent. The method is very mild and compatible with different functional groups [18].

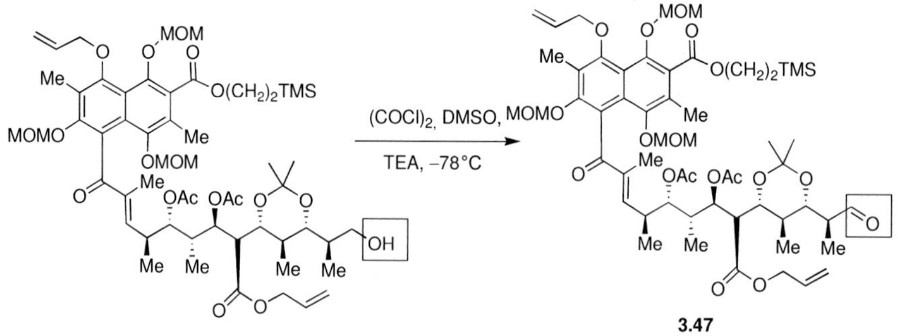

3.47

To end this chapter, we describe an interesting situation, which arises when a reaction tested in a model system (or in an earlier precursor in a synthetic route) does not work at all, but is thereafter successfully applied to the final synthetic route. Unfortunately, it is not easy to ascertain how often this situation arises due to the lack of reported instances. Nevertheless, even against the odds and coupled with negative results from models in hand, to try out a nice idea on a few milligrams of an advanced intermediate may be appealing. One of these examples is found in Danishefsky's synthesis of dynemicin A.

3.4 Dynemicin A [21]

3.59 dynemicin A

Target relevance
Dynemicin A is a member of the endiyne family of antibiotics isolated from *Micromonospora chersina* [22]. It effects single and double stranded DNA cleavage, and shows *in vitro* antitumor activity.

Synthetic plan
The synthetic plan to prepare dynemicine A **3.59** foresees the synthesis of intermediate **3.60**, which contains the cyclic endiyne function of the final product. The coupling of a suitable *syn*-dyine precursor **3.61** with an adequately functionalized ethylene fragment, could access this key intermediate. The tricyclic core of **3.61** would derive from the quinoline **3.62** (Scheme 3.18).

Predictable problems
The synthetic approach depicted in Scheme 3.18 is highly demanding in terms of stereochemistry. Intermediate **3.61** requires a *cis*-relationship of the C2, C4 and C7 substituents. While the *cis*-relationship of C4 and C7 will be ensured by the estereochemical bias of the Diels–Alder reaction used during its preparation, the introduction of a substituent at C2 by the same β-face was expected to be troublesome.

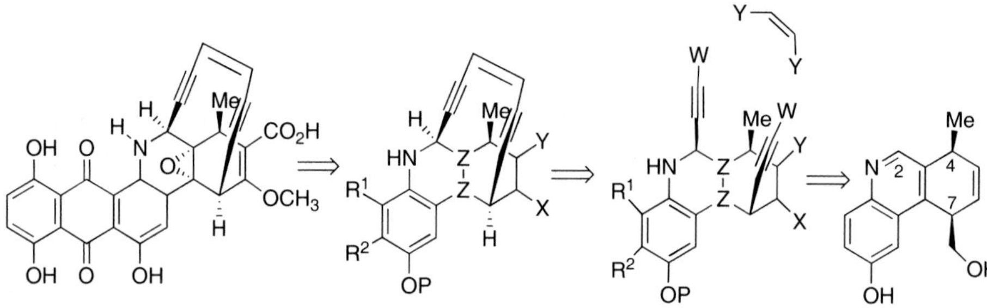

3.59 dynemicin A 3.60 3.61 3.62

Scheme 3.18

Synthesis

*Step 1. Synthesis of the key intermediate **3.64** (Scheme 3.19):*

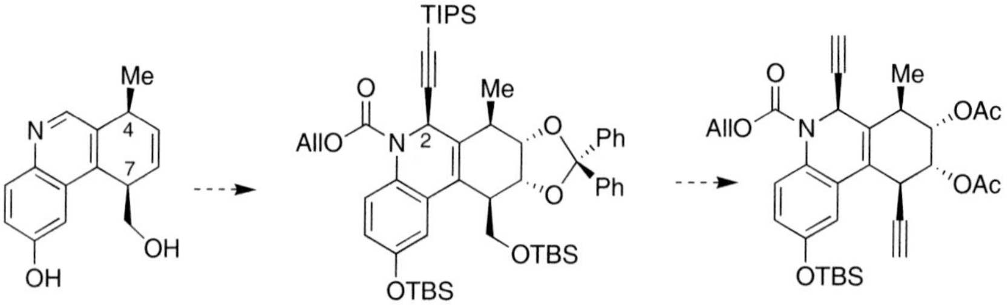

3.62 3.63 3.64

Scheme 3.19

Access to **3.64** required the introduction of the ethynyl substituent at carbon C2 of quinoline **3.62** by means of an intermolecular Reissert reaction [23]. The new substituent must be placed on the β-face of C2 having *cis*-relationship with respect to the already existing substituents at C4 and C7. As indicated above, this may be troublesome. It was anticipated that the intermolecular Reissert reaction on **3.62** would occur from the α-face of the molecule, avoiding the β-substituents at C4 and C7. This troublesome situation arises from the topology of the substrate that dictates the stereochemical outcome of the reaction. In the end, this situation will result in the production of the undesired stereoisomer. One obvious solution is to use a new substrate whose topology works in the required sense, albeit that *this would mean a detour in the synthesis.*

A thoughtful and elegant solution is to block the α-face of **3.62**. In this situation the new substrate for the intermolecular Reissert reaction having the α-face more hindered than the β-face, would render the stereochemistry needed for desired key intermediate **3.64**. For this purpose diol **3.66** was obtained by dihydroxylation of **3.65**.

Compound **3.65** had been obtained from **3.62** by using very bulky silylating agents (TBS-groups). In the author's words, *the hope was to engage these alcohols in a protective devise which would render the α-face ....more hindred than the β-face.* The diol linkage was engaged as ketal by treatment of **3.66** with dimethoxybenzophenone to form **3.67**. Submission of **3.67** to treatment with allylchloroformate, in the presence of the Grignard reagent derived from triisopropilsilylacetilene, yielded a 9:1 mixture!! favoring the desired β-addition product **3.63**. Thus, the major problem of the *syn* required stereochemistry relating C2, C4 and C7 has been masterly solved. The price paid was the lost of the vinilogous carbonate function of the natural target that must be restored at a later stage of the synthesis. (Scheme 3.20).

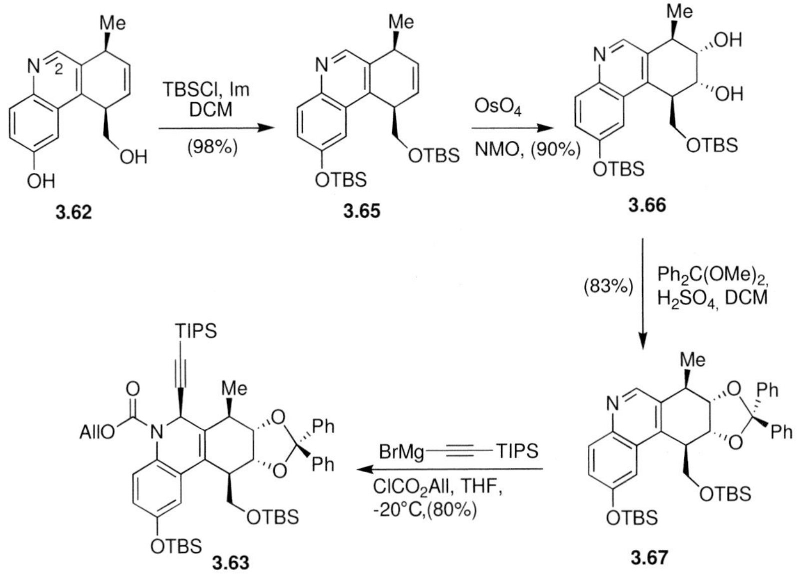

Scheme 3.20

Transformation of **3.63** into the cyclization precursor **3.64** requires installment of the terminal acteylene at carbon C7. Removal of the TBS-protecting group of the primary alcohol, followed by Swern oxidation [18] yielded the labile aldehyde **3.68**. This intermediate was submitted to Corey–Fuchs protocol [24] delivering diyne **3.69**. Removal of the TIPS-group from the alkynyl fragment with TBAF caused as well deprotection of the phenol hydroxyl. The benzophenone ketal linkage was removed by acidic treatment. Protection of the phenolic hydroxyl again as TBS-derivative followed by acetylation of the secondary alcohols yielded diyne **3.64** (Scheme 3.21).

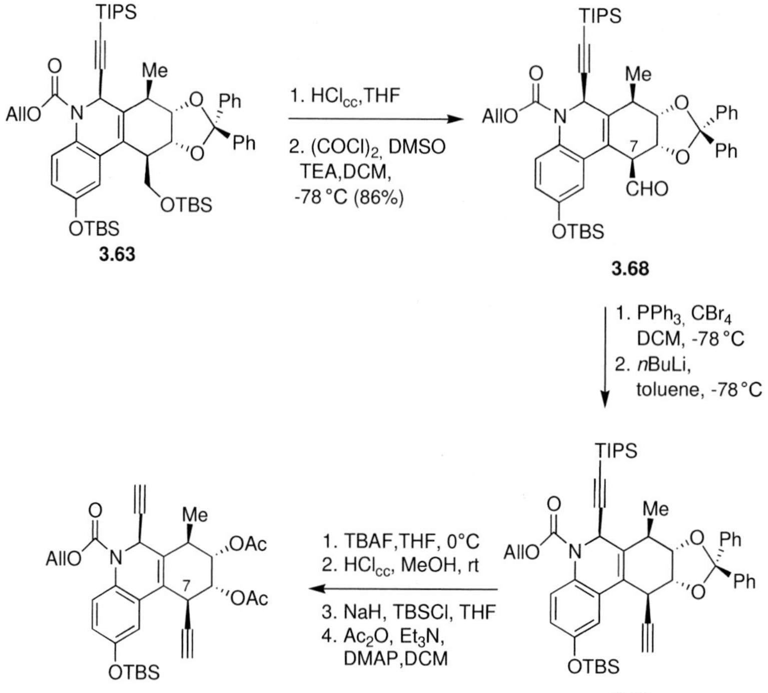

3.63 3.68 3.69 3.64

Scheme 3.21

*Step 2. Synthesis of the endyine key intermediate **3.71** (Scheme 3.22):*

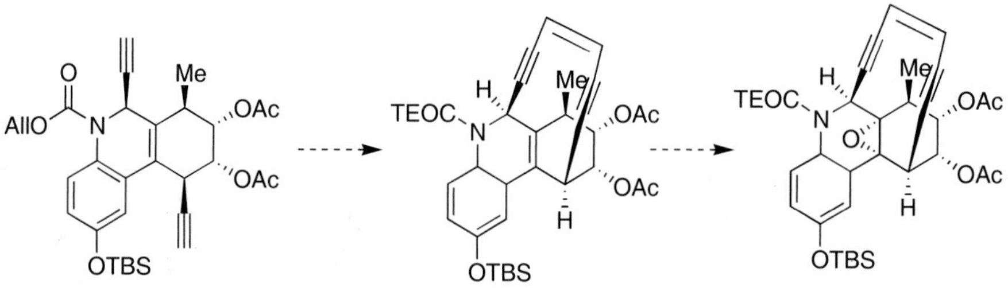

3.64 3.70 3.71

Scheme 3.22

The installation of the endiyne system is one of the most sensitive steps in the synthesis of dynemicin A. The insertion of the ethylene fragment, using a Sonogashira coupling [25], was first tested in the model compound **3.72**, having the tetrasubstituted epoxide of dynemicin A in place. Submission of **3.72** to the action of Pd(PPh₃)₄/CuI in the presence of *cis*-1,2-dichloroethylene did not yield the expected endiyne **3.73**. The Stille's cross-coupling [26] between the bis-iodoalquine **3.74** and *cis*-diestannyl ethylene **3.75** did not produce the desired endiyne **3.70**. Products arising from two intermolecular couplings were isolated instead (Scheme 3.23).

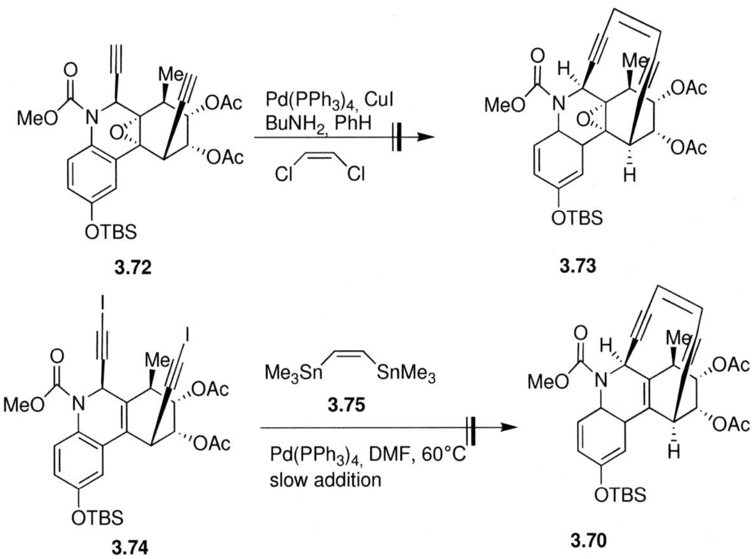

Scheme 3.23

The refusal of the model compounds **3.72** and **3.74** to yield the coupled endiyne derivatives **3.73** and **3.70**, respectively, led to a rethinking of the way of installing the endiyne moiety. In the words of the authors: *though the interpolation of the ethylene unit into **3.72** and **3.74** was unsuccessful, it was hoped that cyclization could be accomplished if the epoxide were in place. The C8-C9 epoxide might serve to shorten the approach of the two ethynyl units while providing some relief from the projected strain in the cyclization product.* Therefore, based on this hypothesis and against the results obtained with the models, the transformation was undertaken.

Step 3. Installation of the ethylene moiety using the Stille coupling (Scheme 3.31):

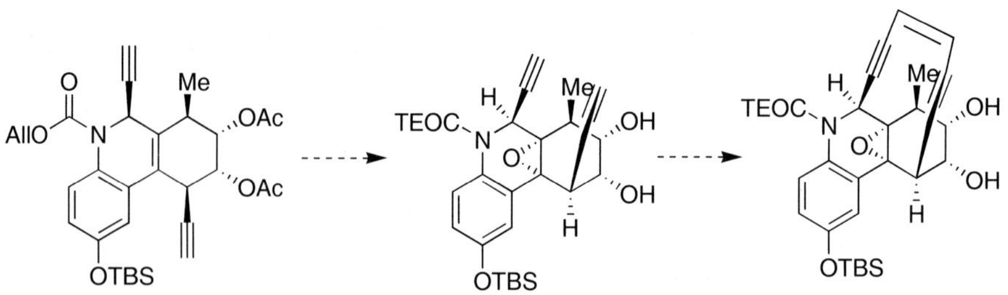

Scheme 3.24

Epoxide **3.77** was prepared from diyne **3.64** by removal of the alloc-group and *in situ* acylation of the free aniline nitrogen with TEOCCl. The acetates were hydrolyzed on TEOC-protected **3.78** with ammonia and the free diol stereospecifically epoxidized with MCPBA yielding epoxide **3.76**. Iodination of the alkyne was achieved by the action of NIS catalyzed by silver, and the diodide **3.79** submitted to the Stille's cross coupling with tin reagent **3.75**. The endiyne **3.77** was obtained in 81% yield (Scheme 3.32). Effectively, the presence of the C8-C9 epoxide shortens the approach of the two ethynyl groups, allowing for the success of the ethylene bridging. As a matter of fact, the introduction of this strain may be what forces the Bergman cyclization upon epoxide opening [27]. The success of the bis-Stille cyclization step opens doors to achieve the total synthesis of dynemicin A. The completion of the synthesis involved the development of the C5-C6 vinilogous carbonate framework, for which differentiation of the two hydroxyl groups is needed. Treatment of **3.77** with Tf_2O resulted in selective triflation of the equatorial C5 alcohol. Dess–Martin periodinane oxidation followed by reduction with $CrCl_2$, afforded ketone **3.80**. α-Carboxylation of **3.80** proved to be problematic and was achieved using the Rathke protocol [28]. Thus, treatment of ketone **3.80** with $MgBr_2$/TEA/CO_2 produced a highly unstable β-ketoacid that was isolated as the MOM ester **3.81**, the key intermediate from where dynemicin A will be finally accessed (Scheme 3.25).

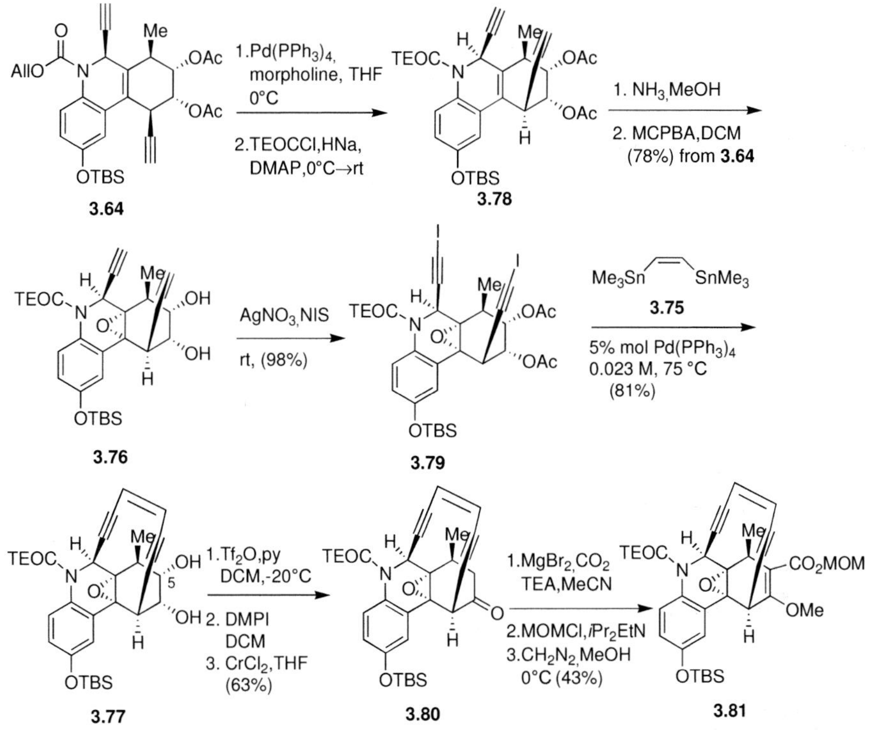

Scheme 3.25

Step 4. Incorporation of the trihydroxyanthraquinone moiety and completion of dynemicin A synthesis (Scheme 3.26):

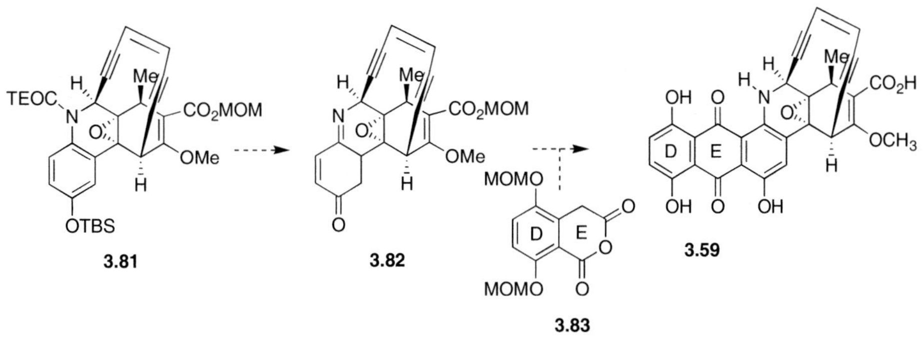

Scheme 3.26

Incorporation of the DE region of dynemicine A was designed to use the protocol developed by Tamura [29]. Homophthalic anhydride **3.83** would be attached to the quinone imine **3.82** to form the rings DE of dynemicin A, following a well-established methodology to construct a naphtoquinone ring under very mild and efficient conditions. In the event, intermediate **3.81** was oxidized to the needed quinone imine **3.82** by removal of the TEOC-group followed by periodinane oxidation. Deprotonation of **3.83** with LHMDS generates a species that adds to **3.82** to yield adduct **3.84**, presumably through an unstable no-aromatic intermediate that is immediately oxidized with PIFA. Exposure of **3.84** to daylight under air gives the corresponding antraquinone, which on treatment with MgBr$_2$ on ether, yielded (±)-dynemicine **3.59** in 15% yield (Scheme 3.27).

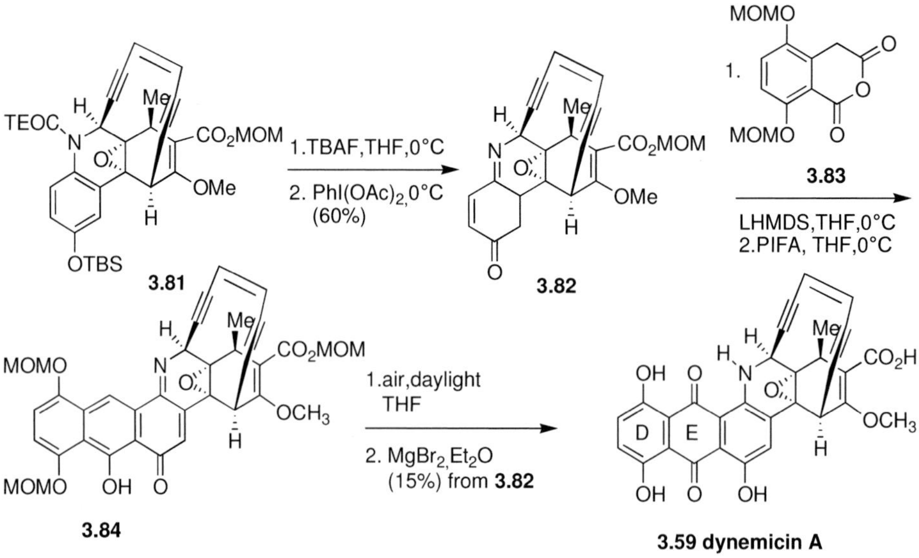

Scheme 3.27

This is a low-yielding sequence attributable, according to the authors, to *the instability of the intermediates, and the difficulty associated with the isolation and purification of **3.59**, rather than to failings in the chemical reactions per se.*

Evaluation

The installation of the ethylene subunit of the endyine bridge of dynemicin A was unsuccessful in both model reactions attempted, using the Sonogashira and Stille couplings. However, and in spite of these hardly encouraging results *it was hoped that cyclization could be accomplished if the C8-C9 epoxide were in place. The C8-C9 epoxide might serve to shorten the approach of the two ethynyl units while providing some relief from the projected strain in the cyclization product.* Therefore, based on this hypothesis and against the results obtained with the models, the Stille coupling was undertaken and successfully achieved. Now, even against the odds, new approaches may succeed based on the experience of the negative results obtained on previous models.

Key synthetic reactions

The Stille coupling: the coupling of an organotin reagent (usually a vinyl stannane) with an aryl-, vinyl halide or triflate in the presence of a Pd catalyst [26].

3.79   **3.77**

References

1.   Wipf, P.; Kim, Y.; Goldstein, D. M. *J. Am. Chem. Soc.* 1995, *117*, 11106.

2.   a) Götz, M.; Strunz, G. M. *Alkaloids*; Wiesner, K., Ed.; London, 1973 Vol 9, pp 143-160; b) Lee, H. M.; Chen, K. K.; *J. Am. Pharm. Assoc.* 1940, *29*, 391.

3.   a) Hughes, D. L. *Org. React.* 1992, *42*, 335; b) Bernotas, R. C.; Cube, R. V. *Tetrahedron Lett.* 1991, *32*, 161. c) Mitsunobu, O. *Synthesis* 1981, 1.

4.   Hoffman, R. W. *Chem. Rev.* 1989, *89*, 1841.

5.   Luche, J.-L.; Gemal, A. L. *J. Am. Chem. Soc.* 1979, *101*, 5848.

6.   Griffith, W. P.; Ley, S. V. *Aldrichimica Acta* 1990, *23*, 13.

7.   a) Wick, A. E.; Felix, D.; Steen, K.; Eschenmoser, A. *Helv. Chim. Acta* 1964, *47*, 2425; b) Chen, C.-Y.; Hart, D. J. *J. Org. Chem.* 1993, *58*, 3840; c) Chen, C.-Y.; Hart, D. J. *J. Org. Chem.* 1990, *55*, 6236.

8.   Sharpless; K. B.; Amberg, W.; Bennani, Y. L.; Crispino, G. A.; Hartung, J.; Jeong, K.-S.; Kwong, H.-L.; Morikawa, K.; Wang, Z.-M.; Xu, D.; Zhang, X.-L. *J. Org. Chem.* 1992, *57*, 2768.

9.   Ireland, R. E.; Norbeck, D. W.; Mandel, G. S.; Mandel, N. S. *J. Am. Chem. Soc.* 1985, *107*, 3285.

10.   Grieco, P. A.; Gilman, S.; Nishizawa, M. *J. Org. Chem.* 1976, *41*, 1485.

11.   a) Scheibye, S.; Pedersen, B. S.; Lawesson, S.-O. *Bull. Soc. Chim. Belg.* 1978, *87*, 229. b) Cava, M.; Levinson, M. I. *Tetrahedron* 1985, *41*, 50.

12.   Molecular mechanic calculations were carried out using the MM2 force-field as implemented in Macromodel 4.5; Mahamadi, F.; Richards, N. G. I.; Guida, W. C.; Liskamp, R.; Canfield, C.; Chang, G.; Hendrickson, T.; Still, W. C. *J. Comput. Chem.* 1990, *11*, 440.

13.   Boger, D. L.; Zhou, J. *J. Am. Chem., Soc.* 1993, *115*, 11426.

14. Tamai, S.; Kaneda, M.; Nakamura, S. *J. Antibiot.* 1982, *35*, 1130.

15. Knight, D. W. *Comprehensive Organic Synthesis*, Trost, B. M.; Flemin, I. Eds., Pergamon Press: New York, 1991, Vol. 3, p. 481.

16. Roush, W. R.; Coffey, D. S.; Madar, D. J. *J. Am. Chem. Soc.* 1997, *119*, 11331.

17. a) Rinehart, K. L.; Shield, L. S. *Prog. Chem. Org. Nat. Prod.* 1976, *33*, 231; b) Rinehart, K. L. Jr.; Antosz, F. J.; Deshmukh, P. V.; Kakinuma, K.; Martin, P. K.; Milavetz, B. I.; Sasaki, K.; Witty, T. R.; Li, L. H.; Ruesser, F. *J. Antibiot.* 1976, *29*, 201.

18. a) Tidwell, T. T. *Synthesis* 1990, 857. b) Tidwell, T. T. *Org. React.* 1990, *39*, 297. c) Mancuso, A. J.; Huang, S.-L.; Swern, D. *J. Org. Chem.* 1979, *44*, 4148.

19. a) Meng, Q.; Hesse, M. *Top. Curr. Chem.* 1991, *161*, 107; b) Bald, E.; Saigo, K.; Mukaiyama, T. *Chem. Lett.* 1975, 1163.

20. Roush, W. R.; Coffey, D. S.; Madar, D. J.; Palkowitz, A. D. *J. Brazil. Chem. Soc.* 1996, *7*, 327.

21. Shair, M. D.; Yoon, T. Y.; Mosny, K. K.; Chou, T. C.; Danisefhsky, S. M. *J. Am. Chem. Soc.* 1996, *118*, 9509.

22. a) Konishi, M.; Ohkuma, H.; Matsumoto, K.; Tsuno, T.; Kamei, H.; Miyaki, T.; Oki, T.; Kawaguchi, H.; VanDuyne, G. D.; Clardy, J. *J. Antibiot.* 1989, *42*, 1449. b) Konishi, M.; Ohkuma, H.; Tsuno, T.; Oki, T.; VanDuyne, G. D.; Clardy, J. *J. Am. Chem. Soc.* 1990, *112*, 3715.

23. Popp, F. D. *Chem. Heterocycl. Comp.* 1982, *32*, 353.

24. Corey, E.; Fuchs, P. L. *Tetrahedron Lett.* 1972, *13*, 3769.

25. Sonogashira, K. *in Comprehensive Organic Synthesis,* Trost, B. M.; Flemin, I. Eds., Pergamon Press: New York, 1991, Vol. 3, p. 521.

26. Farina, V.; Krishnamurthy, V.; Scott, W. *J. Org. React.* 1997, *50*, 1.

27. a) A similar observation was reported by Myers et al., see Myers, A. G.; Fraley, M. E.; Tom, N. J.; Cohen, S. B.; Madar, D. J. *Chem. Biol.* 1995, *2*, 33; b) *Enediyne Antibiotics as Antitumor Agents,* Borders, D. B.; Doyle, T. W. Eds, Dekker, New York, 1995.

28. Tirpak, R. E.; Olsen, R. S.; Rathke, M. W. *J. Org. Chem.* 1985, *50*, 4877.

29. Tamura, Y.; Fukata, F.; Sasho, M.; Tsugoshi, T.; Kita, Y. *J. Org. Chem.* 1985, *50*, 2273 and references cited therein.

Chapter 4
The Unexpected Reactivity or Inertia
of Common Functional Groups

How predictable is the reactivity or the inertia of a certain functional group in a densely functionalized molecule? Taken at face value, this seems to be a silly question since, for instance, a ketone group is still a ketone group even in the more complex molecular environment. Nevertheless, even the simplest transformation of a functional group can be a nightmare when the reaction outcome is different from that expected, or when the functional group does not react at all. The examples discussed below demonstrate the necessity of gathering more information concerning the reactivity of densely functionalized molecules, and of learning more about the behavior of commonly used reagents in these systems.

4.1   Pseudotabersonine [1]

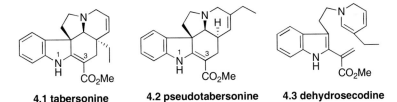

4.1 tabersonine   4.2 pseudotabersonine   4.3 dehydrosecodine

Target elevance

Tabersonine **4.1** was first isolated in 1954 from *Amsonia tabernaemontana* [2]. Soon after the initial report, the alkaloid was isolated from several other natural sources. Tabersonine is the biosynthetic predecessor of several members of the *Aspidosperma* family, and it is important not only for its biosynthetic relationship to the *Vinca* alkaloids, but also because it is the chemical progenitor of these alkaloids. Pseudotabersonine **4.2** is another *Aspidosperma* alkaloid and its biogenesis is closely related to tabersonine **4.1**. In fact, both alkaloids are postulated to proceed from

Dead Ends and Detours: Direct Ways to Successful Total Synthesis
Miguel A. Sierra and María C. de la Torre
Copyright © 2004 WILEY-VCH Verlag GmbH & Co. KGaA, Weinheim
ISBN: 3-527-30644-7

the common intermediate dehydrosecodine **4.3**. The biological properties of these compounds, their intricate pentacyclic structure and their biogenetic relationship, make these products exceptional targets for biomimetic total synthesis [3].

Synthetic plan

The biomimetic synthetic approach to pseudotabersonine **4.2** rests in the intermediacy of a dehydrosecodine derivative **4.4** accessible from oxindole **4.5**. Compound **4.5** may be in turn prepared by a tandem retro Diels–Alder/intramolecular aza Diels–Alder sequence effected on **4.6**. This product will be accessed by the reaction of spiroaziridinium triflate **4.7** and indolone **4.8** (Scheme 4.1).

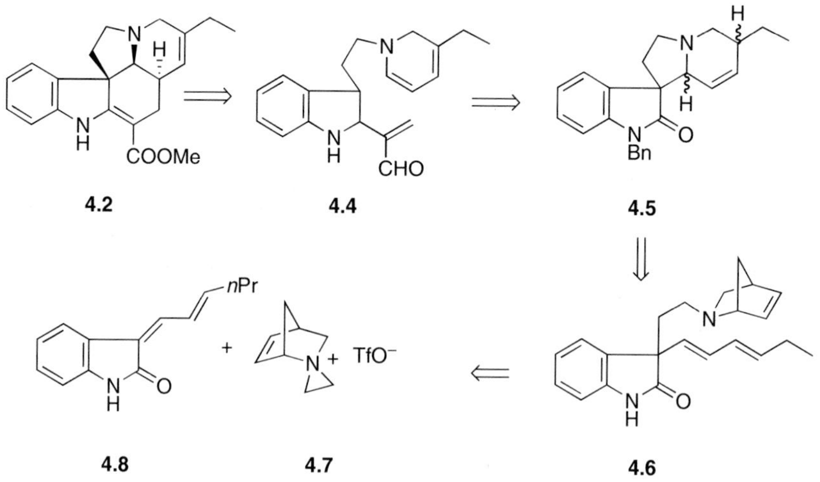

Scheme 4.1

Predictable problems

The synthesis of pseudotabersonine **4.2** features three potentially troublesome key transformations:

- The use of spiroaziridinium salt **4.7** to prepare **4.6** through Ad_E to indolone **4.8** may have regioselective problems.
- The unique tandem retro Diels–Alder/intramolecular aza Diels–Alder sequence to gain access to oxindole **4.5**.
- The unprecedented transformation of oxindole **4.5** to dehydrosecodine derivative **4.4**.

Synthesis

*Step 1. Synthesis of oxindole **4.5** (Scheme 4.2):*

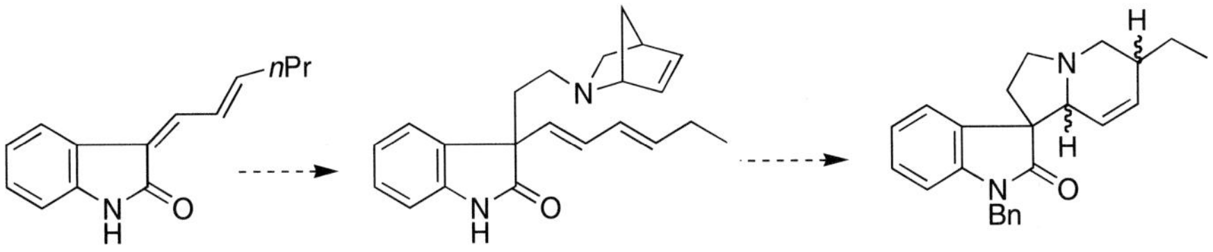
nPr
N
H
O
N
N
H
O
H
N
H
N
Bn
O
4.8
4.6
4.5

Scheme 4.2

Indolone **4.8** was reacted with KN*i*-Pr_2 in THF at – 78°C and subsequently with spiroaziridinium salt **4.7** to yield the alkylated oxindole **4.6**. The tandem retro Diels–Alder/intramolecular aza Diels–Alder sequence will be effected on intermediate **4.12** arising from the unraveling of the azanorbornene fragment of **4.6**. The oxindole nitrogen of **4.6** was benzylated prior to undertaking this transformation to yield **4.9**. Exposure of **4.9** to $BF_3.Et_2O$ in toluene at 100°C formed a diastereomeric mixture (1.5:1) of **4.10** and **4.11** (Scheme 4.3). The isolation of the mixture of adducts is of no relevance since both can be taken to dehydrosecodeine.

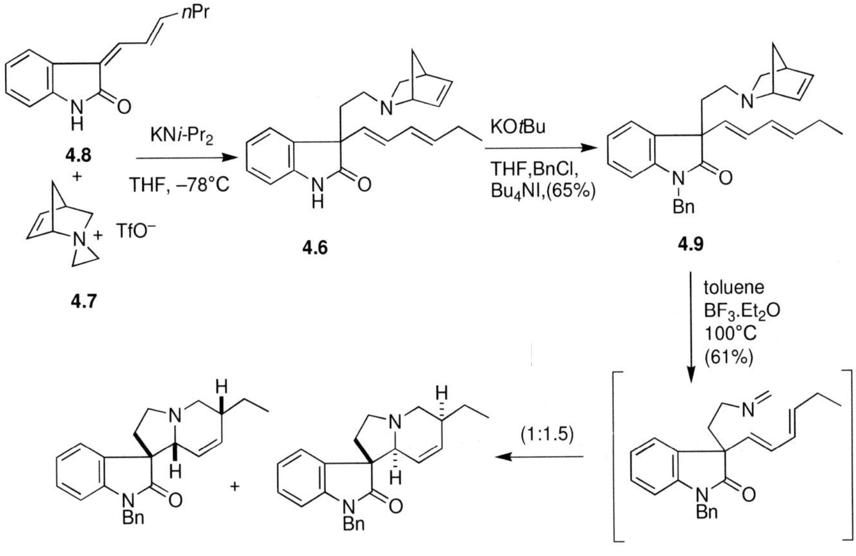
nPr
4.8
+
TfO⁻
4.7
KNi-Pr2
THF, –78°C
4.6
KOtBu
THF,BnCl,
Bu4NI,(65%)
Bn
4.9
toluene
BF3.Et2O
100°C
(61%)
(1:1.5)
4.10
+
4.11
4.12

Scheme 4.3

*Step 2. Formation of dehydrosecodine derivative **4.13** and failed entry to pseudotabersonine (Scheme 4.4):*

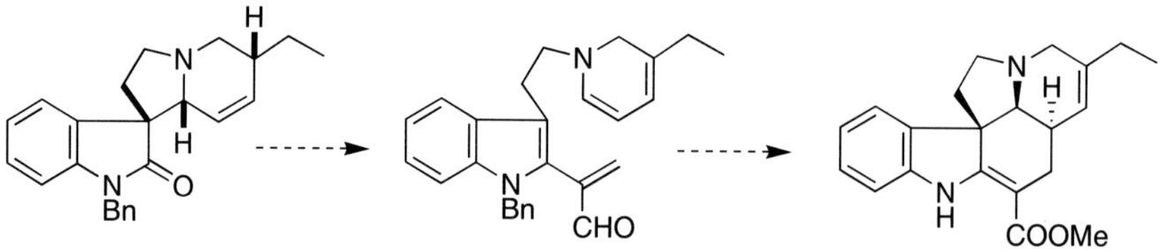

Scheme 4.4

Oxindole **4.10** was reacted with 2-lithio-1,1-diethoxy-2-propene to form carbinolamine **4.14**. The *in situ* formation of *N*-benzyl dehydrosecodine **4.13** was effected by acidic treatment (H$_2$O/TsOH) of **4.14**. Under these conditions the intermediate **4.13** was formed and cyclized to the pentacyclic alkaloid **4.15**. This compound possesses the full pentacyclic carbon skeleton of pseudotabersonine. To complete the synthesis of **4.2** the transformation of the formyl group at C3 of **4.15** into a carbomethoxy unit and the removal of the benzyl group at N1 is necessary. All attempts to oxidize the formyl group meet with no success (Scheme 4.5).

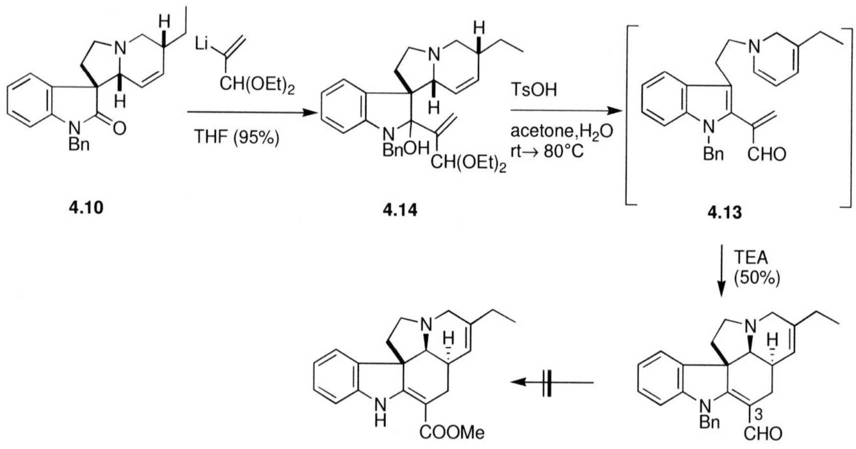

Scheme 4.5

Step 3. Sequential access to pseudotabersonine from aldehyde **4.15** *(Scheme 4.6):*

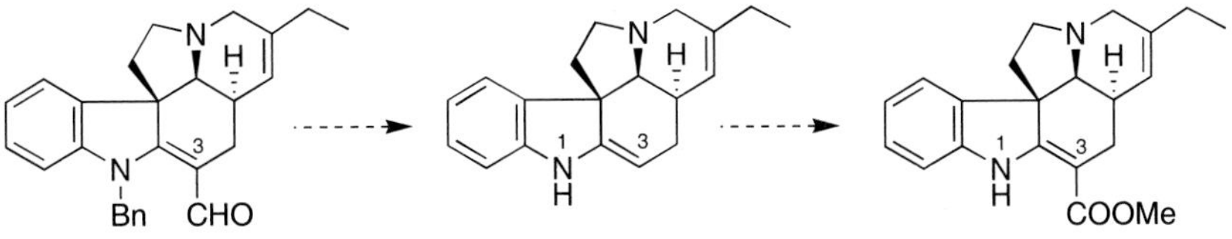

Scheme 4.6

The bottle-neck caused by the unusual inertia of the formyl group at C3 of **4.15** was surpassed by the expedite method of its elimination and subsequent reinstallation as the desired methoxycarbonyl group. Thus, hydrolytic deformylation of **4.15** formed **4.17**, which was debenzylated after many trials by treatment with a large excess of lithium 4,4′-di-*tert*butylbiphenylide to yield **4.16**. The installation of the C3-carbomethoxy group was realized in only 35% yield by lithiation of **4.16** and treatment with excess of MeOCOCl (Scheme 4.7). In this way pseudotabersonine **4.2** was prepared.

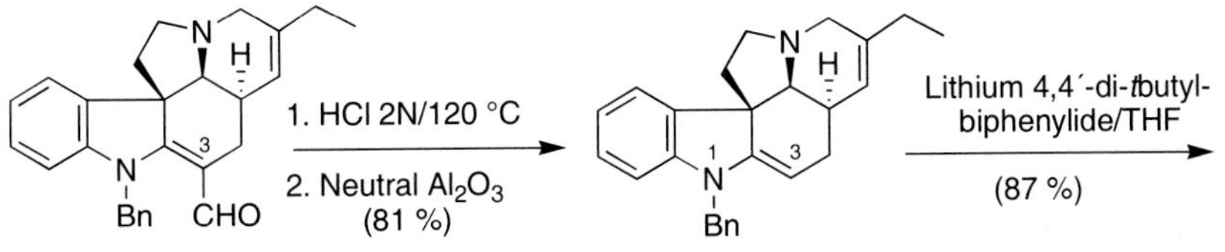

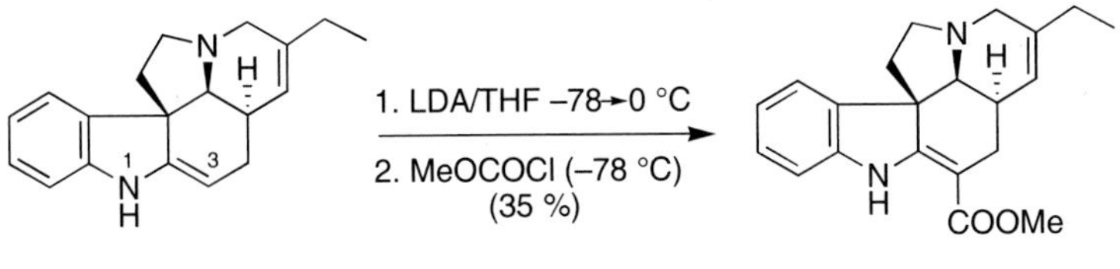

Scheme 4.7

Evaluation

- No strategic changes have been made during the synthesis of pseudotabersonine **4.2**. The synthetic plan presented in Scheme 4.1 has been implemented.
- The failure of a simple transformation like the oxidation of the aldehyde group of **4.15** to a carboxyl group led to a major tactical detour. In fact, the inert group had to be removed and reinstalled in a four-step sequence (Scheme 4.7).

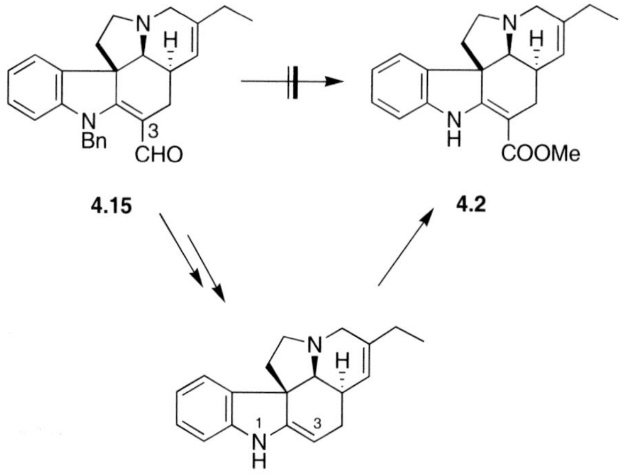

4.15   **4.2**

4.16

- It is intriguing to note the inertia of the formyl group of **4.15** towards oxidation, while the related compound **4.18** was oxidized uneventfully to **4.19** in the final stages of Magnus synthesis of (+)-16-methoxytabersonine [4]. The vinilogous amide nature of the formyl group in **4.15** may be responsible for the failure, while the carbamate group used to block the aniline nitrogen of **4.18** may attenuate this amide-like character, allowing for its oxidation. Overman has made analogous observations in his synthesis of (±)-akuammicine [5].

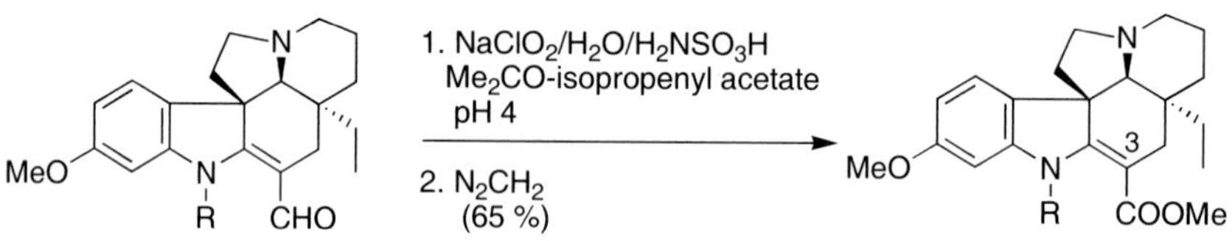

1. NaClO$_2$/H$_2$O/H$_2$NSO$_3$H
   Me$_2$CO-isopropenyl acetate
   pH 4

2. N$_2$CH$_2$
   (65 %)

4.18

R = COOMe

4.19

4.2   Octalactins A and B [6]

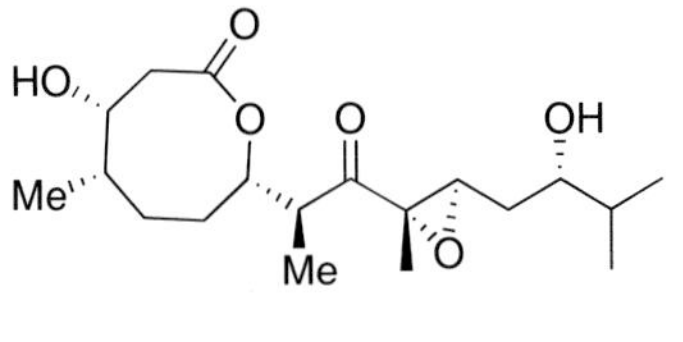

4.20 (–)-octalactin A 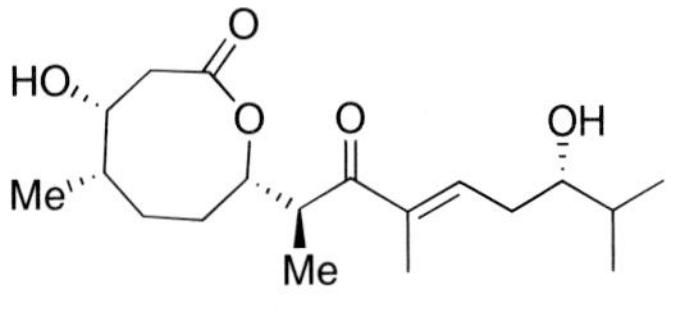   **4.21 (–)-octalactin B**

Target relevance

Octalactin A **4.20** and B **4.21** were isolated [7] from the marine bacterium *Streptomyces sp.* Octalactin A showed strong cytotoxicity toward B-16-F10 murine melanoma and HCT-116 human colon tumor cell lines while octalactin B was totally inactive in these assays.

Synthetic plan

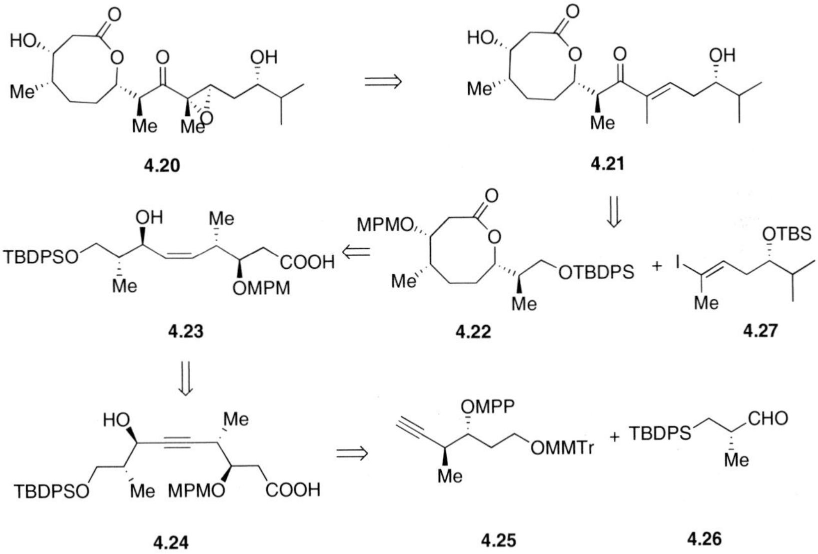

Scheme 4.8

The access to octalactins has as key point the construction of lactone **4.22**. The strategy to construct this saturated eight-membered ring lactone **4.22** envisioned the facile lactonization of the unsaturated hydroxy carboxylic acid **4.23** followed by hydrogenation of the *cis*-olefin. Acid **4.23** will be accessed from alkyne **4.24** formed by the coupling of terminal alkyne **4.25** and aldehyde **4.26** (Scheme 4.8).

After completing the building of **4.22**, the exocyclic appendage will be fully elaborated by joining first the fragment **4.27**. Further functional group manipulation and unmasking of the protected groups would lead to octalactin B **4.21**. Epoxidation of **4.21** would form octalactin A **4.20** (Scheme 4.8).

Predictable problems
No predictable problems are devised *a priori*. The geometric restriction imposed by the double bond of alkene 4.23 makes the success of the key lactonization step very plausible.

Synthesis

*Step 1. Synthesis of hydroxy carboxylic acid **4.23** (Scheme 4.9):*

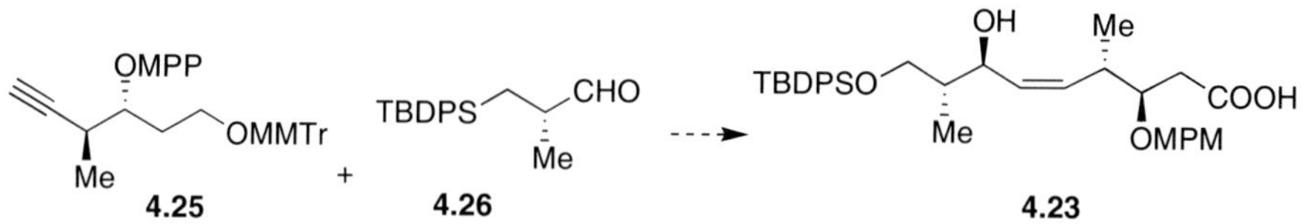

Scheme 4.9

Iodination of alkyne **4.25** using I_2/morpholine formed the terminal iodide **4.28** in 90% yield. The iodide was coupled with aldehyde **4.26** using the Ni(II)/Cr(II) protocol [8] to give an inseparable mixture of alcohols **4.29**. Hydrogenation of the triple bond using the Lindlar's catalyst [Pd(CaCO$_3$), quinoline, rt], followed by acetylation, and removal of the MMTr-ether (PPTS/DCM-MeOH) gave 85% of a separable mixture of products, of which alcohol **4.30** was the desired product.

Two-stage oxidation of the primary alcohol using the Dess–Martin periodinane [9] followed by NaClO$_2$ and finally deacetylation provided carboxylic acid **4.23** (Scheme 4.10).

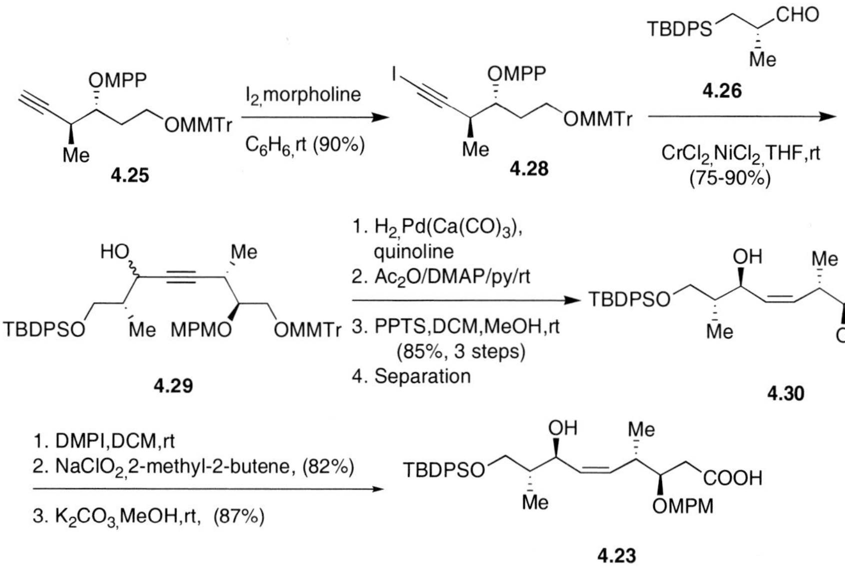

Scheme 4.10

*Step 2. Lactonization of acid **4.23** and failed hydrogenation of a double bond (Scheme 4.11):*

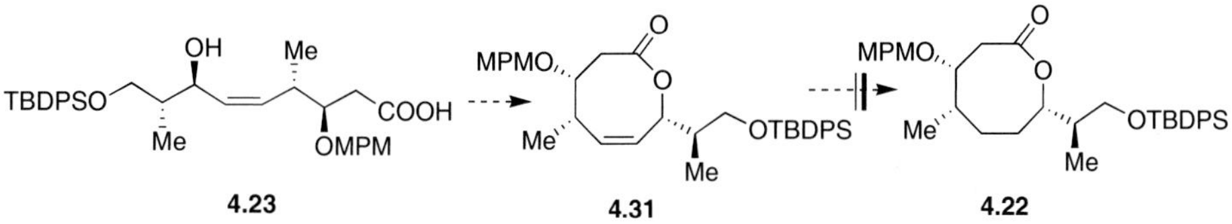

Scheme 4.11

Lactonization of **4.23** was achieved by using the Corey double activation method (2,2'-pyridine disulfide/PPh₃/DCM followed by heating in the presence of AgBF₄) in 63–75% yield [10]. Lactone **4.31** was thus obtained. Now, one of the first reactions learned at the introductory organic chemistry level, the standard olefin hydrogenation, was attempted. Unsaturated, eight-membered ring lactone **4.31** has the cyclic core of octalactin and it is properly functionalized to be transformed to the desired target. Unfortunately, in the words of the authors, *the seemingly prosaic task of reducing the double bond could not be carried out under the many conditions tried, including heterogeneous and homogeneous catalytic hydrogenation, diimide reduction ...* (Scheme 4.12). Therefore, the planned route to octalactins had to be re-evaluated.

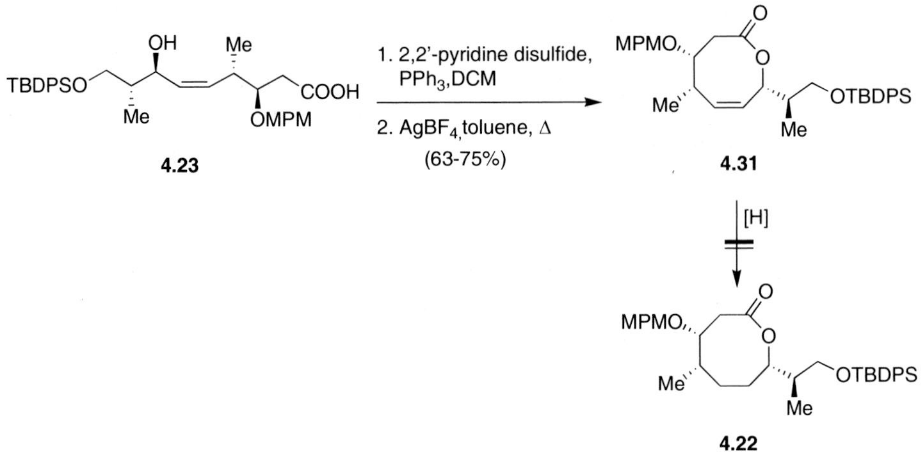

Scheme 4.12

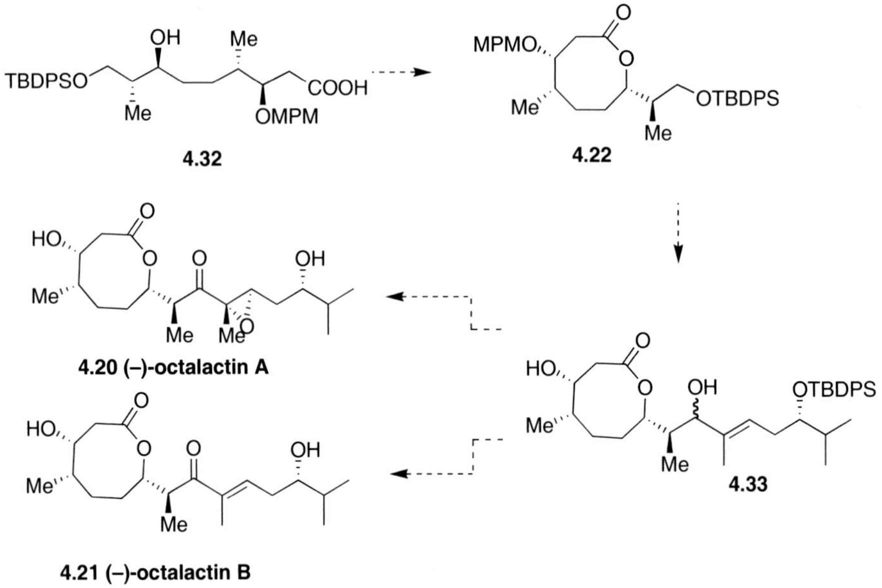

Scheme 4.13

Step 3. The successful route to octalactins: Taking a worthy risk (Scheme 4.13):
After the "simple" hydrogenation step failed in converting 4.31 to 4.22, the building of a saturated eight-membered lactone ring was devised through intramolecular esterification of the saturated hydroxy carboxylic acid 4.32. This maneuver would effect the compromising hydrogenation step on the open chain precursor 4.32. The literature offered little encouragement for this strategy. Nevertheless, based on the assumption that pre-organization of 4.32 due to stereochemical arrangements and steric factors, may approach the alcohol and acid termini, this route was pursued.

Hydrogenation (Pd(C)) of **4.23** formed **4.32** that was lactonized to **4.22** by using 2,2′-pyridine disulfide/PPh$_3$/DCM followed by heating in the presence of AgBF$_4$ in 73% yield (Scheme 4.15). Therefore, the *a priori* difficult operation was achieved in high yields. The side-chain of octalactins was elaborated after the successful assemblage of lactone **4.22**. Desylilation and oxidation, followed by coupling of the resulting aldehyde with the vinyl iodide **4.27** (NiCl$_2$/CrCl$_2$(excess)/DMSO, rt), gave a 1.5:1 separable mixture of diasteromeric alcohols **4.33** in 74% yield. Oxidation of **4.33** yielded a single ketone **4.34**, which gave synthetic octalactin B **4.21** in 88% yield by desilylation and oxidative removal of the MPM group (Scheme 4.14).

To synthesize (–)-octalactin A, the *syn* allylic alcohol *syn*-**4.33**, obtained during the addition of **4.27** to the aldehyde derived from **4.23**, was reacted with TBHP /VO(acac)$_2$ in benzene at rt to afford a single epoxide, which was oxidized (Dess–Martin periodinane) to give the protected octalactin A **4.35** in 95% yield. Removal of the protecting groups following a sequence analogous to that used for the synthesis of octalactin B, afforded octalactin A in 77% yield (overall from *syn*-**4.33**) (Scheme 4.15).

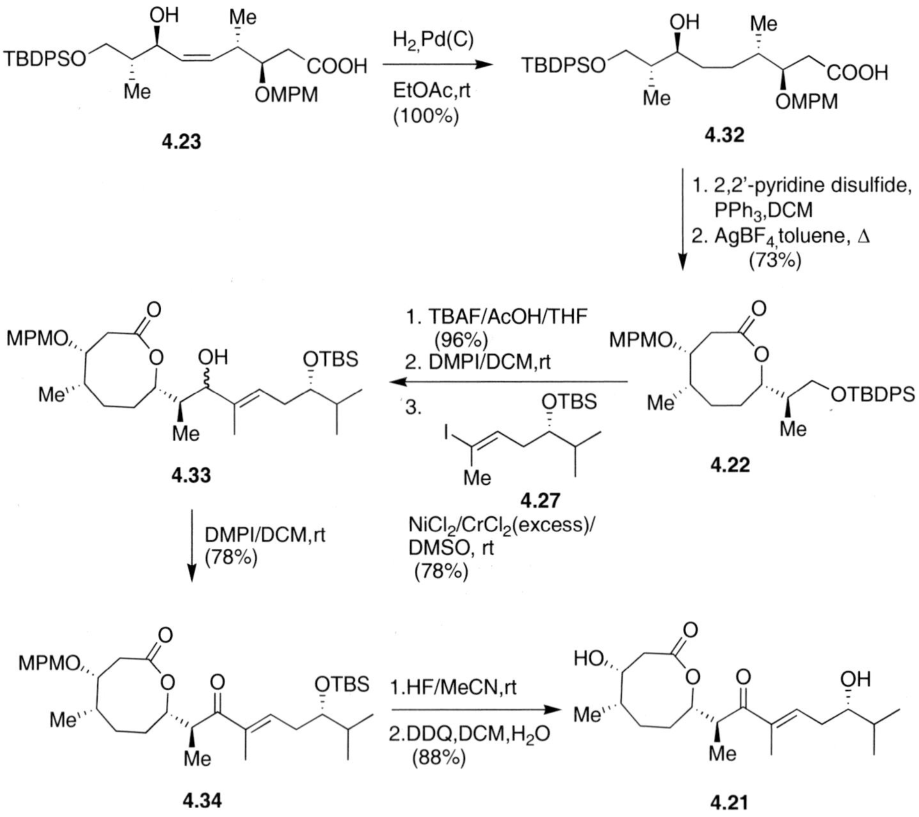

Scheme 4.14

Evaluation

The evaluation of the detours caused by failure in a *simple process* like the hydrogenation of a double bond is perfectly resumed by the words of the authors: *the seemingly prosaic task of reducing the double bond could not be carried out under the many conditions tried, including heterogeneous and homogeneous catalytic hydrogenation, diimide reduction…* The conclusion is that the behavior of simple reactions in complex systems is far away to be *simple*.

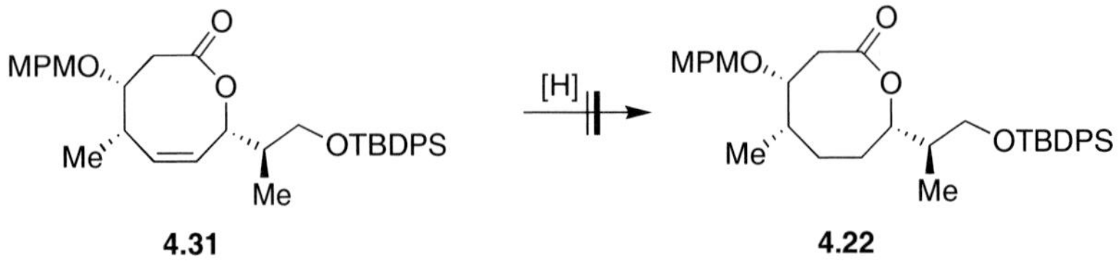

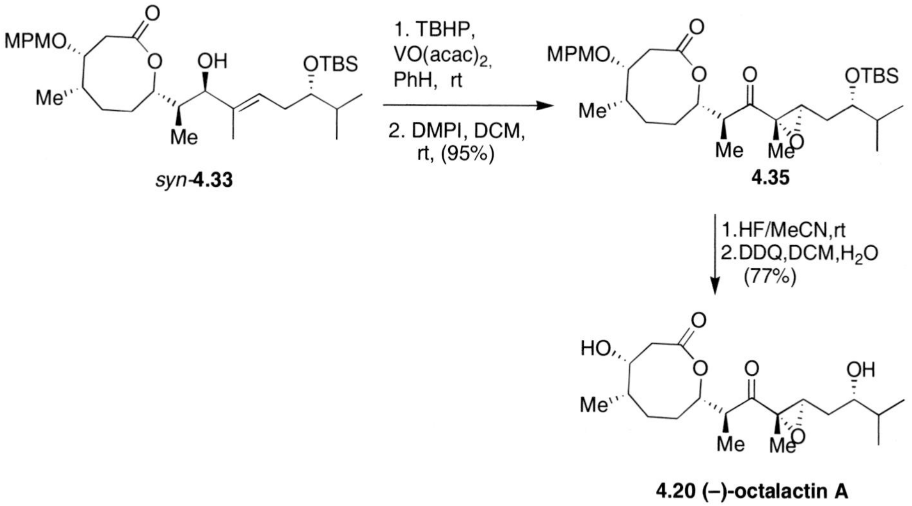

Scheme 4.15

Key synthetic reactions

The Nozaki–Takai–Hiyama–Kishi coupling: The Ni(II)/Cr(II)-mediated coupling of a vinyl iodide or alkynyl iodide and an aldehyde [8]. The process can be catalytic both in Ni and Cr salts [11].

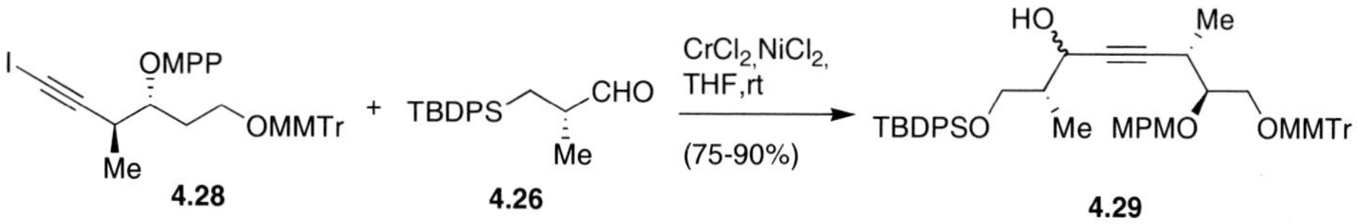

Specially interesting reagents

Dess–Martin periodinanes: The oxidation of a primary alcohol to aldehyde or a secondary alcohol to ketone by using a periodinane [9]. The reagent is compatible with a wide variety of functional groups and it effects the oxidation in smooth reaction conditions.

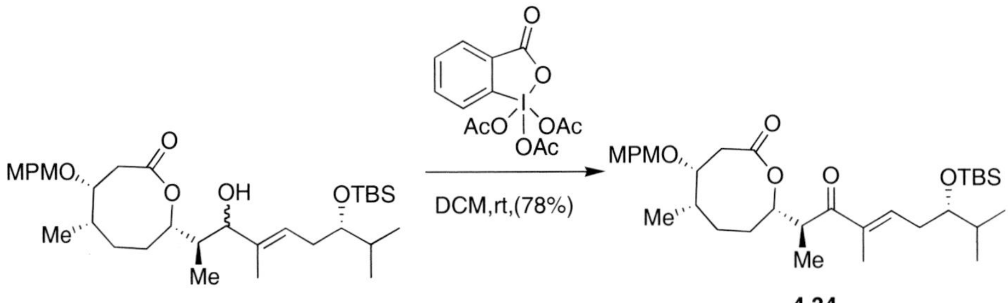

4.33 → DCM,rt,(78%) → **4.34**

4.3   (−)-Polycavernoside A [12]

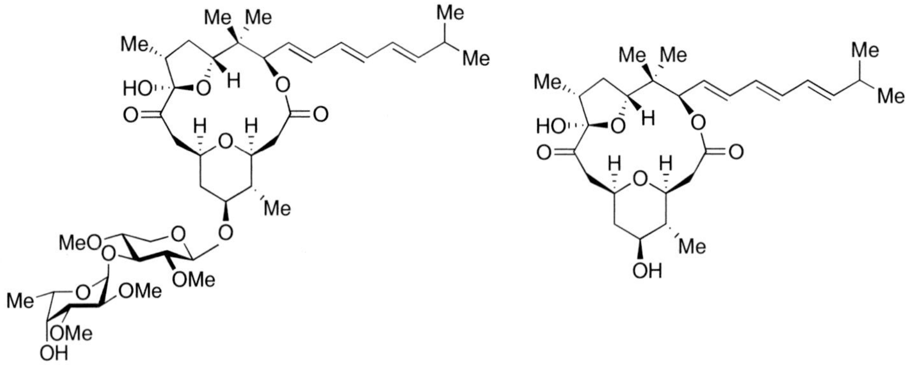

4.36 (−)-polycavernoside A   4.37 (−)-polycavernoside A aglycon

Target relevance

Polycavernoside A **4.36** is an algal metabolite isolated from the widely consumed red alga *Polycavernosa tsudai* [13]. This compound is a powerful human toxin that is produced seasonally only during the months of April and May [14].

Synthetic plan

The plan to construct (−)-polycavernoside A aglycone **4.37** rests on the building of the 14-membered lactone core **4.38** from the seco-acid precursor **4.39**. Evidently, the acid **4.39** has to be in the adequate conformation to favor the macrocycle ring closure. The sensitive polyolefinic appendage would be introduced late in the synthesis once the lactone core was constructed. The fragments **4.40** and **4.41** are going to be joined during the synthesis of **4.39**. Coupling of both fragments would be achieved by anion condensation (Scheme 4.16).

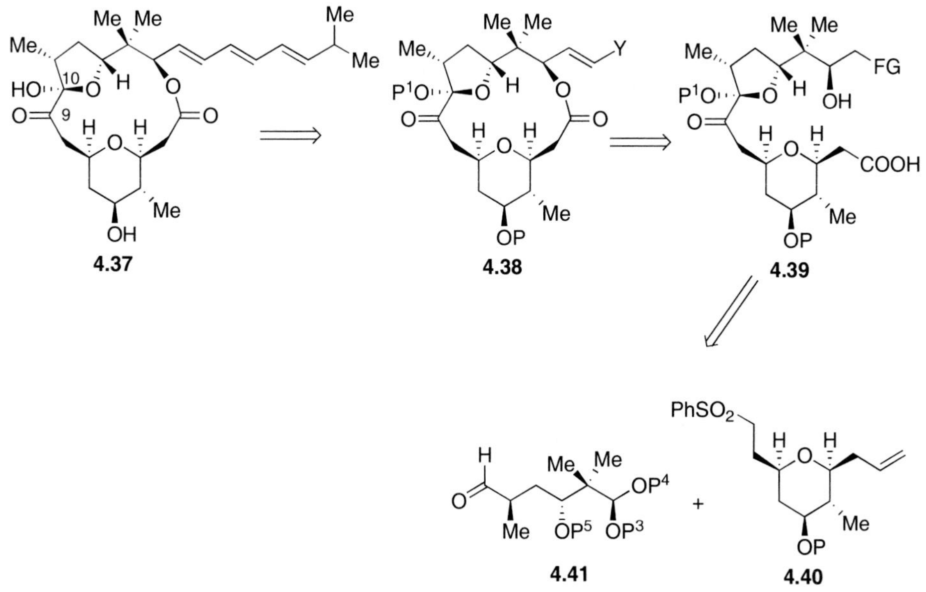

Scheme 4.16

Predictable problems

- The construction of 14-membered lactone **4.38** depends critically on the attainment by the seco-acid precursor **4.39** of a conformation properly suited to effect the ring closure.
- The establishment of the α-dicarbonyl moiety (C9-C10).
- The introduction of the sensitive trienyl functionality.

Synthesis

*Step 1. Synthesis of seco-acid **4.42** (Scheme 4.17):*

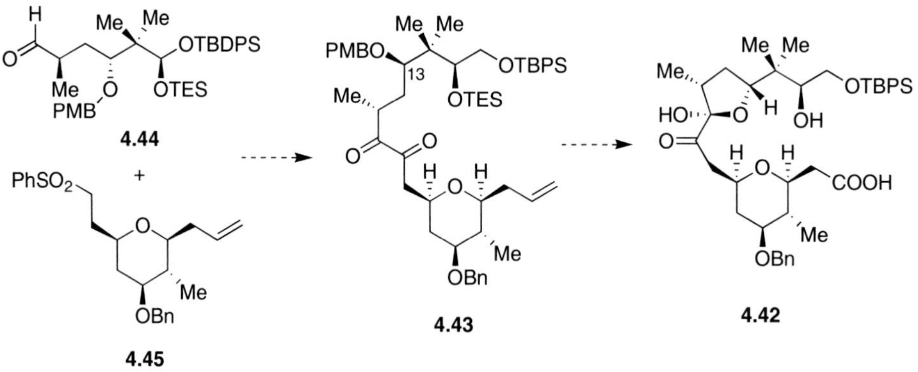

Scheme 4.17

Sulfone **4.45** was deprotonated with BuLi and reacted at –78°C with the aldehyde **4.44** to yield an alcohol whose oxidation with the Dess–Martin periodinane reagent delivered keto sulfone **4.46**. The unmasking of the α-dicarbonyl moiety and the further deprotection of the alcohol at C13 would then form the five-membered ring hemiacetal. In this regard, sulfone **4.46** was deprotonated with KO*t*Bu and the resulting anion oxidized with Davis oxaziridine **4.47** [15]. Oxidative desulfonylation took place in this way to yield **4.43** exclusively. Subsequent treatment of **4.43** with DDQ removed the PMB-protecting group to form the five-membered ring lactol **4.48** without traces of any other product. To access to the next intermediate **4.49**, the chemoselective removal of the triethylsilyl group was attempted. None of the methods used were effective. Alternatively, since the origin of the problem may be the unprotected lactol moiety present in **4.48**, the methylation of the free OH was pursued. Compound **4.50** could not be obtained under any of the conditions used. Therefore, access to seco-acid **4.42** was thwarted by the failure of the "simple" removal of a silyl protecting group in **4.48** (Scheme 4.18).

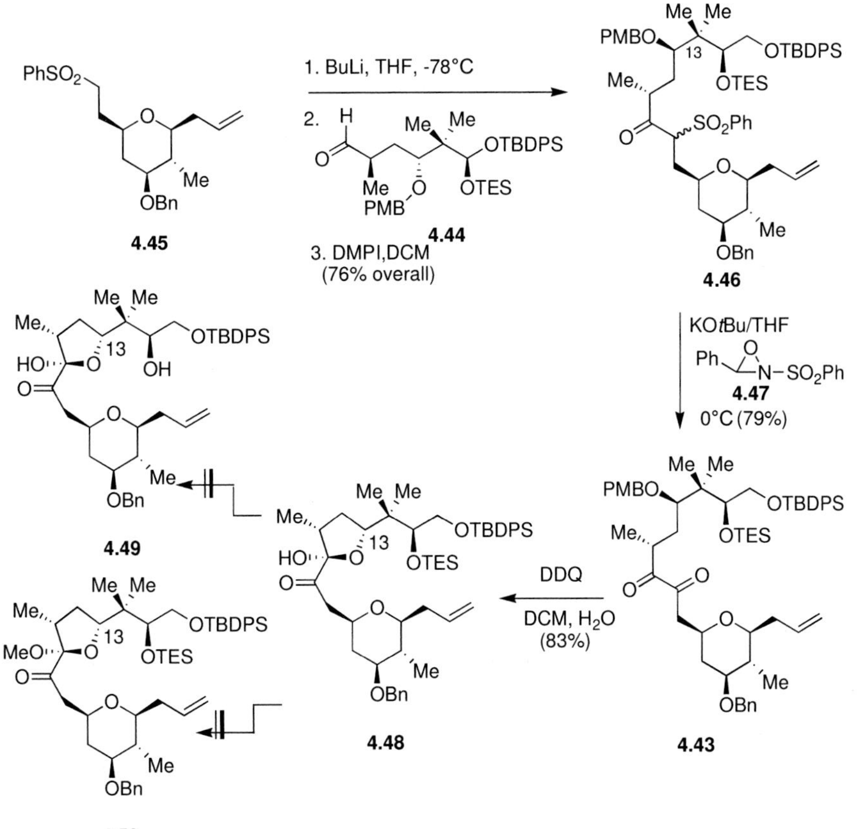

Scheme 4.18

*Step 2. A tactical change to prepare seco-acid **4.53** and its macrolactonizacion to form **4.54***: The failure to access to the seco-acid **4.42** precursor **4.43** because of the reluctant removal of an "easily" removable TES-group forced to use an alternative route. Now, the removal of this protecting group was effected on sulfone **4.46**. The standard chemoselective proteolisis (TsOH) of the TES-group was straightforward, leading to alcohol **4.51**. In this new scenario the double bond was elaborated onto the corresponding aldehyde **4.52** that was further oxidized uneventfully to the carboxylic acid **4.53** using buffered NaClO$_2$ in the presence of 2-methyl-2-butene. Macrolactonization was effected under modified Yamaguchi conditions (2,4,6-Cl$_3$C$_6$H$_2$COCl/Et$_3$N/DMAP/ boiling toluene) with excellent 82% yield, to afford lactone **4.54** (Scheme 4.19).

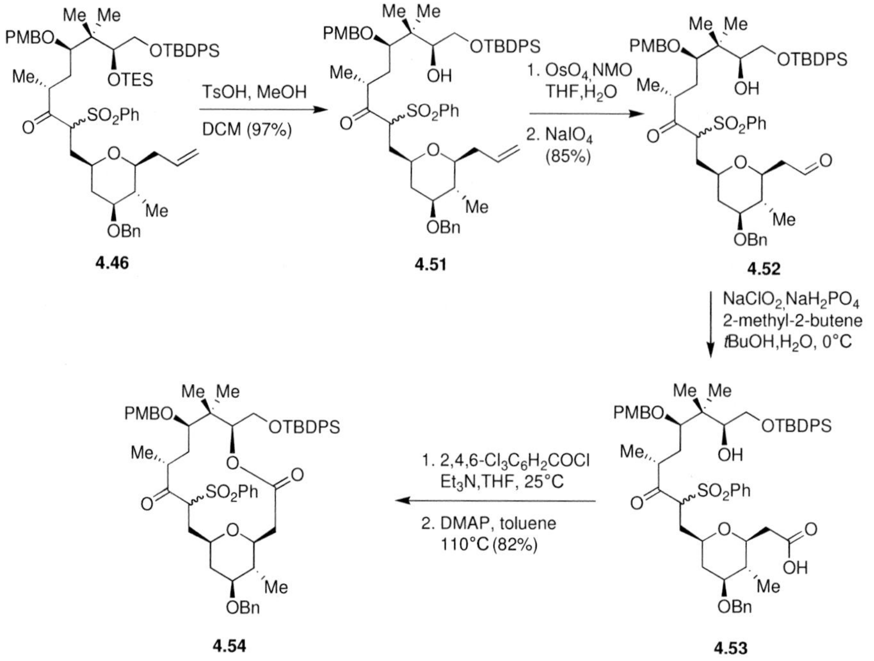

Scheme 4.19

Step 3. Desulfonation **4.54**, *lactol installation and preparation of aldehyde* **4.55**, *the key intermediate to join the polyene appendage (Scheme 4.20):*

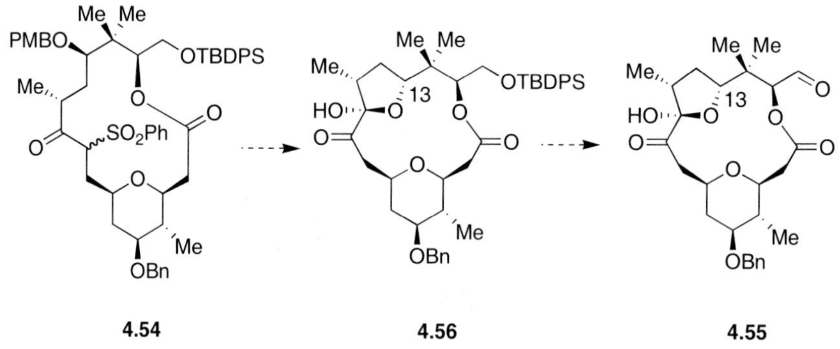

Scheme 4.20

The successful synthesis of macrolactone **4.54** opens the way to prepare the new key intermediate, aldehyde **4.55**. In fact, if the oxidative desulfuration protocol used to prepare diketone **4.48** is effective to desulfurate **4.54**, the preparation of aldehyde **4.55** would be just two "easy" steps away: silicon protecting group removal and alcohol to aldehyde oxidation. In fact, α-diketone **4.57** was smoothly prepared from **4.54** using the enol-generation–oxidation protocol as above (see Scheme 4.18), and the PMB-group protecting the C13-OH was removed by oxidation with DDQ. This resulted in the smooth formation of the five-membered lactol **4.56** as a single estereoisomer. The remaining silyl protecting group was removed using HF-py to form alcohol **4.58**. However, the oxidation of the primary alcohol of **4.57** to the desired aldehyde **4.55** totally destroyed the starting material. Apparently, the presence of the hemiacetal functionality renders the macrocycle vulnerable to their oxidation and thwarted the apparently easy access to the desired product **4.55** (Scheme 4.21).

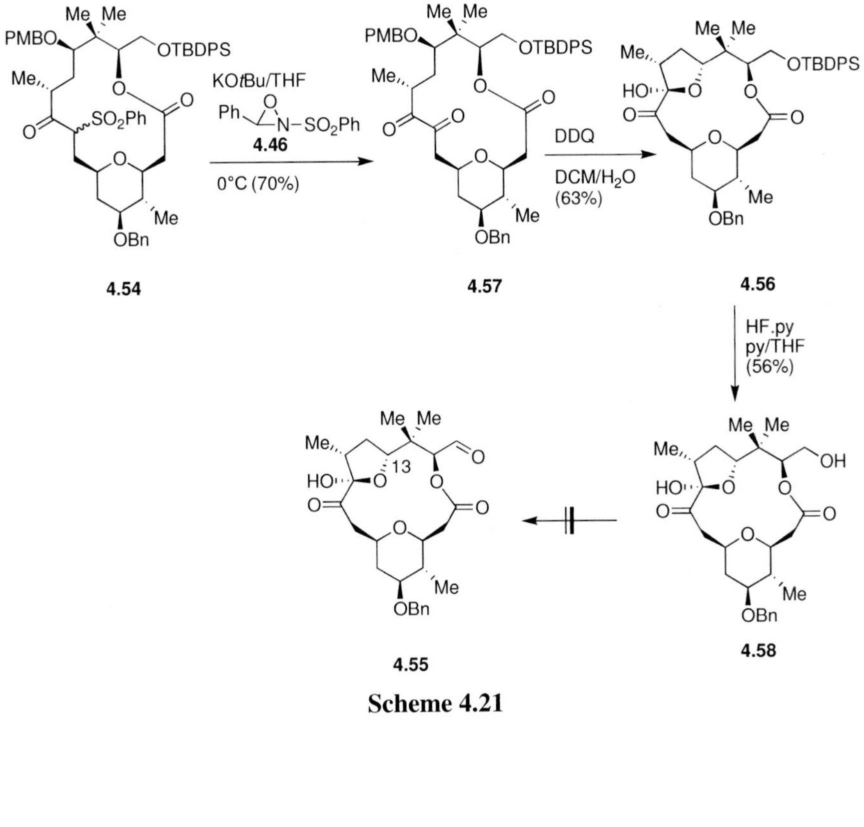

Scheme 4.21

Step 4. Re-evaluation of the strategy for the synthesis of polycavernoside A aglycon 4.37. Synthesis of vinyl-iodide 4.61 (Scheme 4.22):

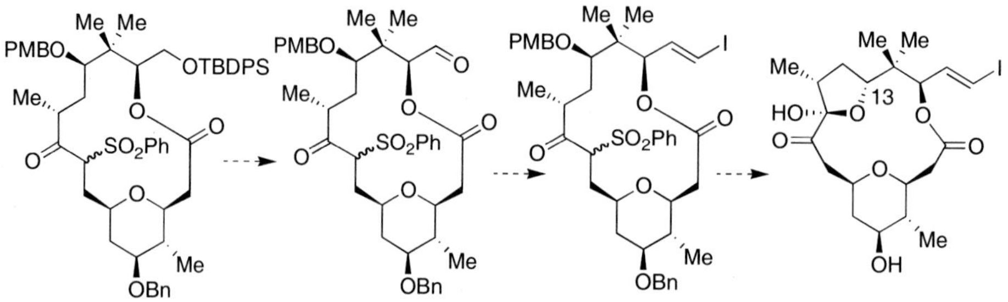

Scheme 4.22

The approach to (–)-polycavernoside A aglycone **4.36** was re-evaluated at this point. In fact, the unpredictable failure of the primary alcohol of the macrocycle **4.58** to experience oxidation, without the concomitant destruction of the molecule, was attributed to the lactol moiety (Scheme 4.21). The sensibility of the lactol moiety was also claimed to be responsible for the failure of the first undertaken approach, the drawback being in that case due to the impossibility of deprotecting a silyl group (Scheme 4.18). This extreme sensitivity of the lactol ring forced to delay its introduction to the late stages of the synthesis. In fact, the unmasking of the lactol moiety would be effected on compound **4.60** having all the anchors needed to attach the remaining moieties of (–)-polycavernoside A **4.36** in its structure.

Thus, macrocycle **4.54** was deprotected to yield the alcohol **4.62** that was now smoothly oxidized in 90% yield to the aldehyde **4.59** using Dess–Martin periodinane. Clearly, the lack of the lactol moiety allows for this oxidation to occur without decomposition of the starting material. Further iodomethylenation of the aldehyde **4.59** with the Takai reagent (CrCl$_2$/CH$_3$I) [16] afforded vinyliodide **4.60**, which was used as substrate for the oxidative elimination of the sulfone functionality to form diketone **4.63**. Now, the removal of the PMB group at C13 resulted in the formation of **4.61**. Therefore, the access to an appropriately functionalized lactol to pursue the synthesis of (–)-polycavernoside A was finally achieved (Scheme 4.23).

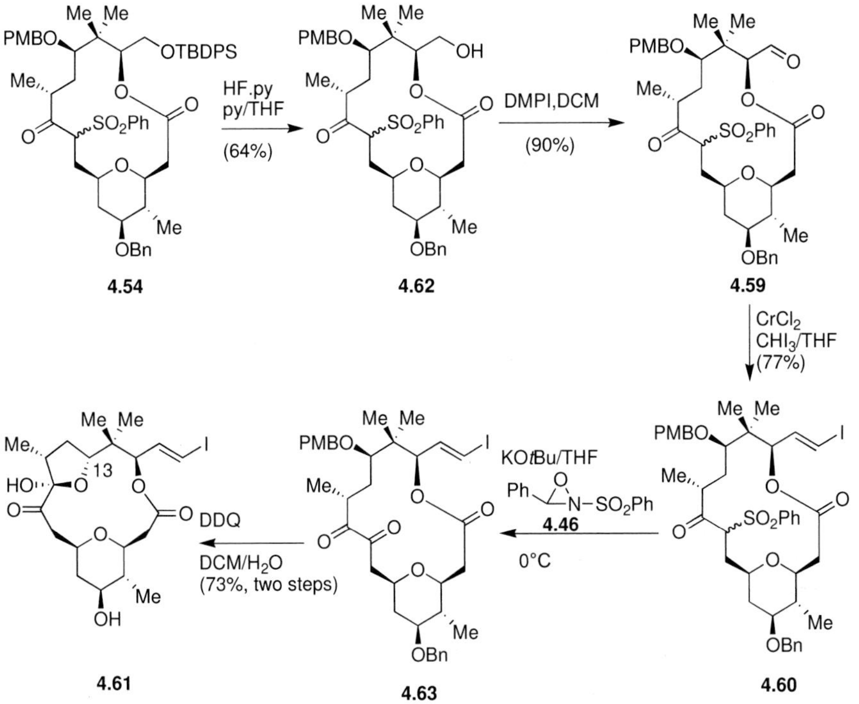

Scheme 4.23

Step 5. Completion of the synthesis: Glycosidation of **4.61** with sugar **4.64** by means of Nicolaou methodology (NBS, 4Å-Ms, MeCN) [17] formed **4.65**. The sugar PMB-protecting group was oxidatively removed to yield **4.66** to which the dienyl appendage was incorporated through the Stille coupling with (*E,E*)-1-iodo-5-methyl-1,3-hexadiene. This yielded (–)-polycavernoside A **4.36** (Scheme 4.24).

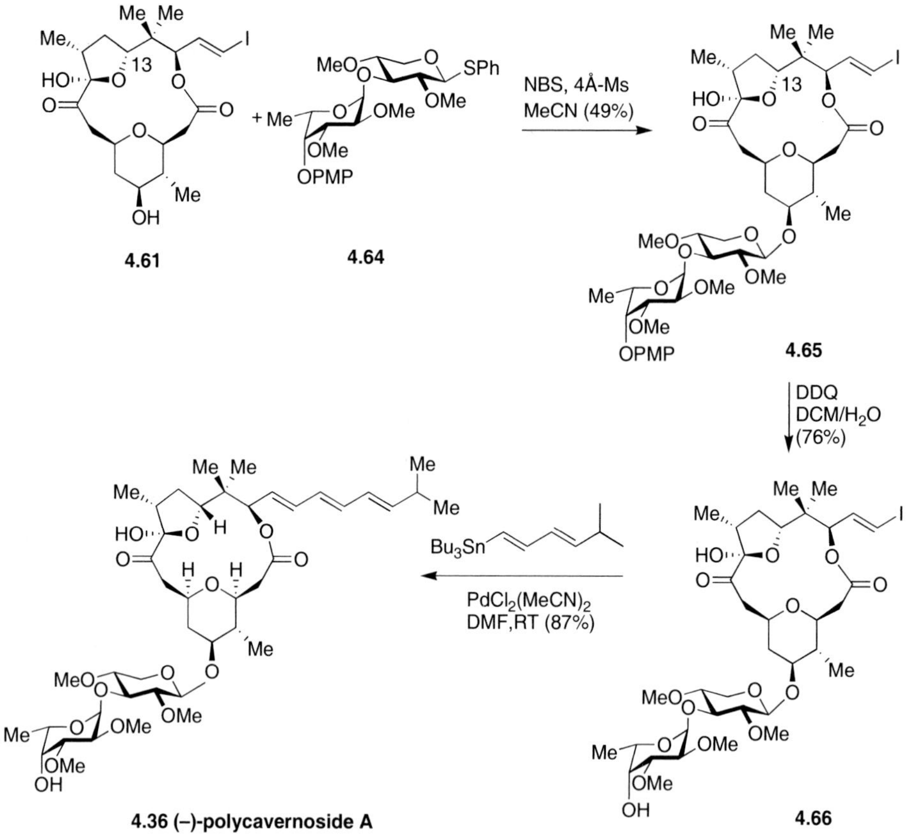

Scheme 4.24

Evaluation

- The (–)-polycavernoside A synthesis is an intriguing case of how the predictable problems are beautifully solved without further complication of the synthesis, while very simple reactions like the removal of a silyl protecting group or the oxidation of an alcohol to the corresponding aldehyde resulted in several synthetic drawbacks. The conclusion of this example is clear: the reactivity of a functional group on a densely functionalized molecule is many times unpredictable by the presence of other reactive groups.

4.48   **4.49**

- The above considerations result in a tactical change in the order of the sequence of events in the synthesis of the key seco-acid is going to be achieved. The mayor detours during this process were due simply to the impossibility of removing a TES-protecting group without affecting the rest of the molecule.
- Most dramatic is the decomposition of compound **4.58** during its oxidation to the target aldehyde **4.55**. This failure to effect the oxidation resulted in the re-evaluation of the synthetic strategy. Once identified, the lactol moiety of **4.49** and **4.58**, which are responsible for the troubles encountered during the synthesis, the unmasking of this moiety was delayed until the last stages of the synthesis, which was then successfully accomplished.

4.58   **4.55**

The examples presented in this section are just a pale reflection of how matters can become complicated when, at first glance, simple transformations are carried out in complex molecules. Evidently, this does not mean that chemical reactions always have unpredictable outcomes. The conclusion here is that much more research is needed to learn how a "simple transformation" occurs in a densely functionalized molecule. Modern organic chemistry lacks the necessary knowledge of this kind of chemistry.

References

1. Carroll, W. A.; Grieco, P. A. *J. Am. Chem. Soc.* **1993**, *115*, 1164.
2. Janot, M.-M.; Pourrat, H.; Le Men, J. *Bull. Soc. Chim. Fr.* **1954**, 707.
3. For an overview of the state of the art of biomimetic synthesis, see: de la Torre, M. C.; Sierra, M. A. *Angew. Chem. Int. Ed.* **2004**, *43*, 160.
4. Cardwell, K.; Hewitt, B.; Ladlow, M.; Magnus, P. *J. Am. Chem. Soc.* **1988**, *110*, 2242-2248.
5. a) Angle, S. R.; Fevig, J. M.; Knight, S. D.; Marquis, R. W.; Overman, L. E. *J. Am. Chem. Soc.* **1993**, *115*, 3966-3976; b) J. M. Fevig, R. W. Marquis, L. E. Overman, *J. Am. Chem. Soc.* **1991**, *113*, 5085-5086.
6. Buszek, K. R.; Sato, N.; Jeon. Y. *J. Am. Chem. Soc.* **1994**, *116*, 5511.
7. Tapiolas, D. D.; Roman, M.; Fenical, W.; Clardy, J. *J. Am. Chem. Soc.* **1991**, *113*, 4682.
8. a) Jin. H., Uenishi, J.; Christ, W. J.; Kishi, Y. *J. Am. Chem. Soc.* **1986**, *108*, 5644; b) Takai, K.; Tagashira, M.; Kuroda, T.; Oshima, K.; Utimoto, K.; Nozaki, H. *J. Am. Chem. Soc.* **1986**, *108*, 6048.
9. a) Dess, D. B.; Martin, J. C. *J. Org. Chem.* **1983**, *48*, 4155; b) Dess, D. B.; Martin, J. C. *J. Am. Chem. Soc.* **1991**, *113*, 7277; c) Wirth, T.; Hirt, U. H. *Synthesis* **1999**, 1271; d) Varvoglis, A.; Spyroudis, S. *Synlett* **1998**, 221; e) Kitamura, T.; Fujiwara, Y. *Org. Prep. Proced. Int.* **1997**, *29*, 409; f) Stang, P. J.; Zhdankin, V. V. *Chem. Rev.* **1996**, *96*, 1123; g) Moriarty, R. M.; Vaid, R. K. *Synthesis* **1990**, 431; h) Varvoglis, A., Ed. *Hypervalent Iodine In Organic Synthesis*; Academic Press: San Diego, 1996; p 256.
10. a) Corey, E. J.; Nicolaou, K. C. *J. Am. Chem. Soc.* **1974**, *96*, 5614; b) Nicolaou, K. C.; McGarry, D. G.; Somers, P. K.; Kim, B. H.; Ogilvie, W. W.; Yiannikouros, G.; Prasas, C. V. C.; Veale, C. A.; Hark, R. R. *J. Am. Chem. Soc.* **1990**, *112*, 6263; c)Gerlach, H.; Thalman, A. *Helv. Chim. Acta* **1974**, *57*, 2661.
11. a) Fürstner, A.; Shi, N. *J. Am. Chem. Soc.* **1996**, *118*, 12349; b) Fürstner, A.; Shi, N. *J. Am. Chem. Soc.* **1996**, *118*, 2533.
12. Paquette, L. A.; Barriault, L.; Pissarnitski, D.; Johnston, J. N. *J. Am. Chem. Soc.* **2000**, *122*, 619.
13. Yotsu-Yamashita, M.; Haddock, R. L.; Yasumoto, T. *J. Am. Chem. Soc.* **1993**, *115*, 1147.
14. Haddock, R. L.; Cruz, O. L. *Lancet* **1991**, *338*, 195.
15. a) Williams, D. R.; Robinson, L. A.; Amato, G. S.; Osterhout, M. H. *J. Org. Chem.* **1992**, *57*, 3740; b) Davis, F. A.; Chen, B. C. *Chem. Rev.* **1992**, 92, 919.
16. Takai, K.; Nitta, K.; Utimito, K. *J. Am. Chem. Soc.* **1986**, *108*, 7408.
17. Nicolaou, K. C.; Seitz, S. P.; Papahatjis, D. P. *J. Am. Chem. Soc.* **1983**, *105*, 2430.

Chapter 5
The Influence of Remote Substituents

In the design of a target molecule, for each single step, many different possibilities are available, which are critically and carefully evaluated. In many cases, the key transformations are tested in model compounds and, except in those syntheses in which the objective is to demonstrate the usefulness of a new methodology, well-established synthetic procedures are generally chosen. However, it may happen that a very well tested transformation fails in the real synthetic intermediate due to the presence of an offending remote substituent. This may happen more frequently than believed by two different, albeit sometimes overlapping, situations: a remote substituent acting as a blocking group or the remote functional group that interferes with the desired reaction, producing undesired structures. Let us study the first of these two situations.

5.1 "Alive" Dead Steric Volume

The influence of apparently remote inert substituents thwarting a well-designed and well-tested reaction is a serious situation, which often entails undesired drawbacks to earlier steps on the synthesis in order to repeat the synthetic scheme without the troublesome group. Evidently, subsequent steps are required after the transformation has been achieved in order to introduce the troublesome remote substituent. The following two examples clearly illustrate this point.

Dead Ends and Detours: Direct Ways to Successful Total Synthesis
Miguel A. Sierra and María C. de la Torre
Copyright © 2004 WILEY-VCH Verlag GmbH & Co. KGaA, Weinheim
ISBN: 3-527-30644-7

5.1.1 (±)-Lubiminol [1]

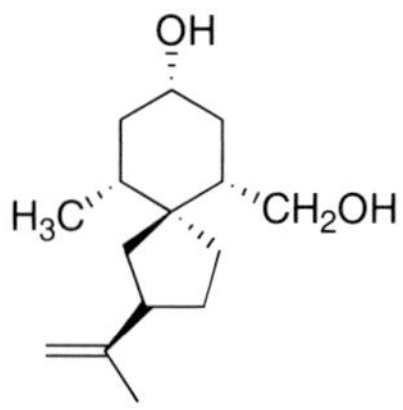

5.1 (±)-lubiminol

Target relevance
Lubiminol is a phytoalexin produced by potato tubers infected with the fungi *Phytophthora infestans* or *Glomeralla cingulata* [2]. Lubiminol **5.1** is one of the most structurally complex oxygenated spirovetivane phytoalexins, and it is implicated in the toxicity of infected plants to humans and livestock.

Synthetic plan
The key intermediate in the synthesis of lubiminol **5.1** is the tricyclic cyclobutane **5.2**, which should render the spirocyclic ketone **5.3** through a tandem radical fragmentation/Dowd–Beckwith ring expansion [3]. Intermediate **5.2** may be derived from the photochemical intramolecular [2+2] cycloaddition on enone **5.4**.

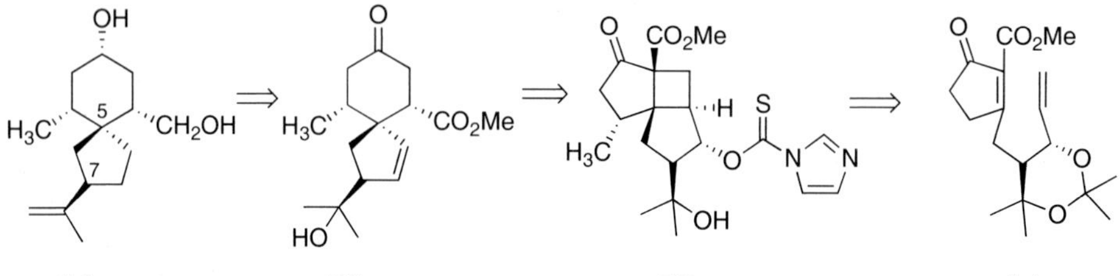

Scheme 5.1

Predictable problems
- High levels of asymmetric induction, in the intramolecular [2+2] photocycloaddition of **5.4**, are required, in order to settle the configuration at carbons C5 and C7. The relative stereochemistry of the two stereogenic centers of the tether will influence the stereochemical outcome of the reaction, but no information about this point was available at that time and examples relating to the specific substitution pattern needed for lubiminol synthesis could not be found in the literature.

- The authors had previously studied the key tandem radical fragmentation/Dowd–Beckwith ring expansion for obtaining medium rings, which should be applicable to this case.

Synthesis

Step 1. Photocycloaddition of **5.4**. *Preparation of the precursor for the fragmentation/Dowd–Beckwith ring expansion* **5.2** *(Scheme 5.2):*

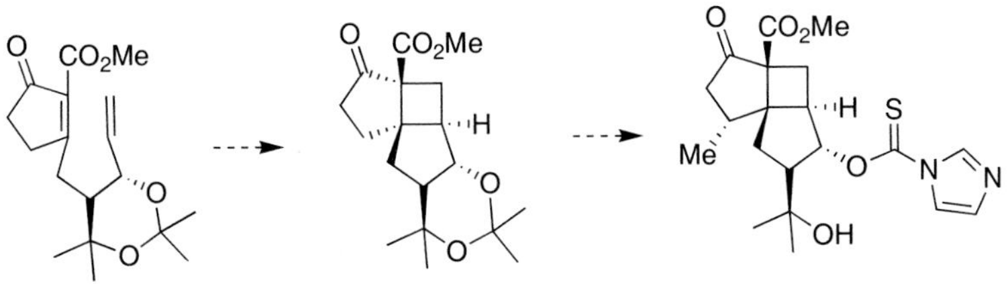

Scheme 5.2

Irradiation of **5.4** yielded the single adduct of tetracyclic keto ester **5.5**, the stereochemistry of which matches the relative stereochemistry at C5 and C7 of lubiminol (Scheme 5.3). Cycloadduct **5.5** contained all the carbons for lubiminol except for the C4 Me-group. Therefore, the next task of the synthesis is the complexion of the lubiminol backbone. Cyclobutane **5.5** was transformed to enone **5.6** by treatment with ethyl trimethylsilyl acetate (ETSA) in the presence of a catalytic amount of TBAF, followed by oxidation with Pd(OAc)$_2$. Addition of Me$_2$CuLi takes place from the β-face of enone **5.6** yielding **5.7**. It can be seen that the Me-group of **5.7** has the wrong stereochemistry at C4. Other organometallic reagents used to introduce the Me-group, as well as the use of additives, produced the same estereochemical outcome. Therefore, a detour in the synthetic planning was taken, to obtain the right stereochemistry at C4. Transformation of cyclopentanone **5.6** into enone **5.8** was achieved, connecting the oxidation of the sylil enol ether with the conjugate addition step of Me$_2$CuLi. Hydrogenation of **5.8** proceeds exclusively through the convex face of the molecule, with concomitant cleavage of the acetonide miety, yielding **5.9** that has the required α-stereochemistry at C4 (Scheme 5.3). Completion of the synthesis of the precursor for the radical rearrangement thiocarbamate **5.2**, was achieved by chemoselective esterification of the secondary alcohol of diol **5.9** with (Im)$_2$C=S in the presence of DMAP.

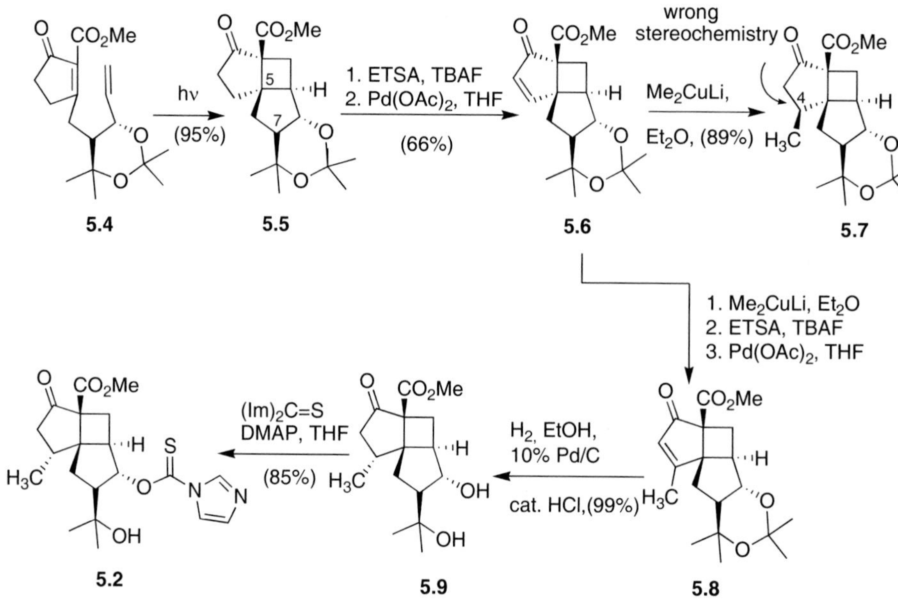

Scheme 5.3

Step 2. Tandem radical fragmentation/Dowd–Beckwith ring expansion: This step was the cornerstone of the planned lubiminol synthesis. The tandem radical fragmentation/Dowd–Beckwith ring expansion was attempted on thiocarbamate **5.2** by heating in benzene in the presence of TBTH and AIBN. Disappointingly, a mixture of three products **5.10–5.12** was obtained. None of the obtained compounds was the planned lubiminol precursor **5.3** (Scheme 5.4).

Step 3. Re-evaluation of the radical fragmentation/Dowd–Beckwith ring expansion step: It was reasoned that the failure of **5.2** to undergo the planned radical fragmentation/Dowd–Beckwith rearrangement, was probably due to the interaction of the primary radical generated on **5.2**, with the apparently distal C4 Me-group. This unpredicted interaction produces a very congested transition state for the addition of the radical to the carbonyl group in the Dowd–Beckwith rearrangement (see below Scheme 5.10). Alternatively, it can promote a hydrogen transfer from the Me-radical which also would inhibit the rearrangement. Interestingly, the C4-epimeric compound **5.13**, when it was submitted to the above conditions, yielded the expected product **5.14**, although in low yield (Scheme 5.5).

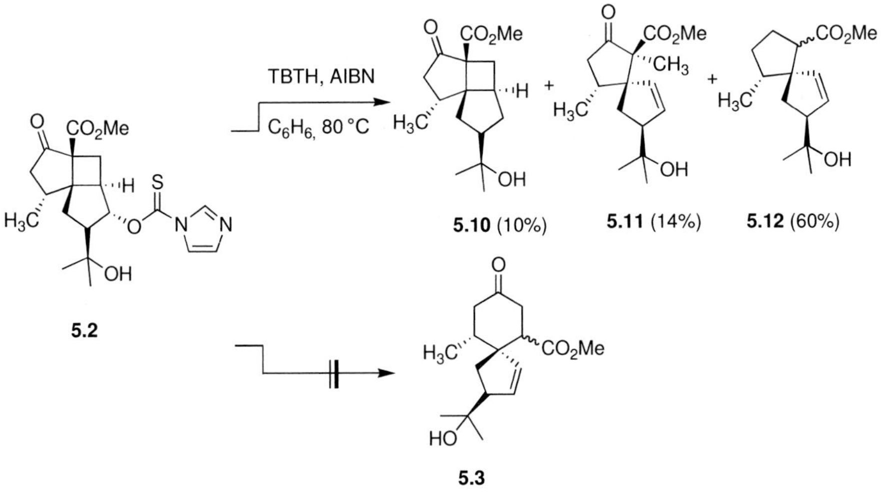

Scheme 5.4

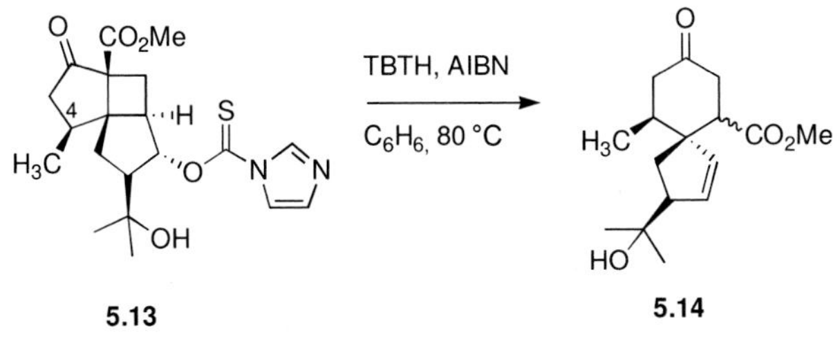

Scheme 5.5

In accordance with these results, the methyl group at C4 had to be introduced once the radical rearrangement had occurred. This forced a major tactical change in the planned synthesis. Now, the primary photoadduct **5.5** (Scheme 5.3) which lacks the troublesome Me-group, was converted to thiocarbamate **5.15**, which reacted with TBTH/AIBN in benzene to yield, as predicted, the mixture of epimeric esters **5.16** in 92% yield (Scheme 5.6).

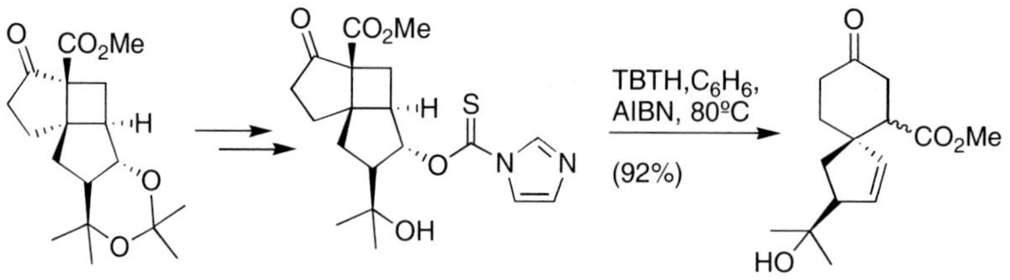

Scheme 5.6

After the successful implementation of the radical fragmentation/Dowd–Beckwith ring expansion step, it is necessary to install the C4-Me group to complete the synthesis of lubiminol from intermediate **5.16**.

Step 3. Introduction of the C4 Me-group and completion of the synthesis (Scheme 5.7):

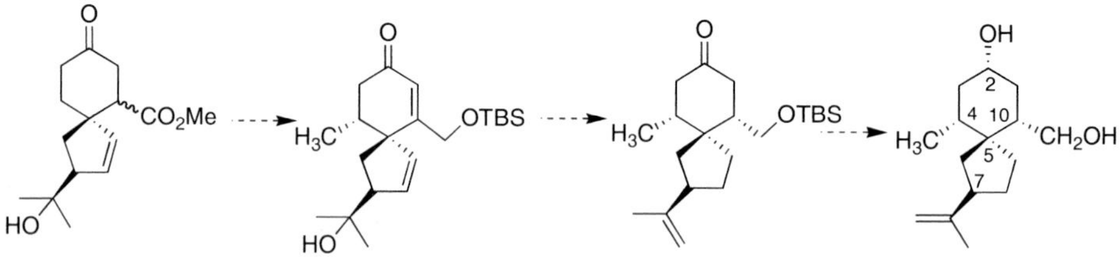

Scheme 5.7

To accomplish the synthesis of lubiminol, three main tasks remain. First, the placement of the C4 Me-group with the right stereochemistry, second the stereoselective reduction of the carbonyl group, and finally the regioselective dehydratation of the tertiary alcohol. The mixture of epimers **5.16** was reduced to the corresponding mixture of alcohols **5.19** with LiAlH$_4$, after protection of the ketone as acetonide. The acetonide was cleaved and the primary alcohol was blocked as TBS-ether yielding **5.20**. Compound **5.20** was transformed into dienone **5.21** by sequential treatment with NaH/methyl benzene-sulfinate, LDA/PhSeCl and final oxidation. Dienone **5.21** will be used as a key inter-mediate to introduce the C4 Me-group and to place the correct stereochemistry on the hydroxymethyl group. In this regard, the C4 Me-group was introduced by addition of Me$_2$CuLi to dienone **5.21**. The addition of the cuprate proceeded with complete regio- and stereoselectity, to yield exclusively **5.17** (Scheme 5.8).

Having the C4 Me-group in place with the correct stereochemistry, the reduction of the two alquenes and the carbonyl group of bicycle **5.17** was undertaken. Hydrogenation of **5.17** [Pd(C)] occurred with high stereoselectivity yielding **5.22**.

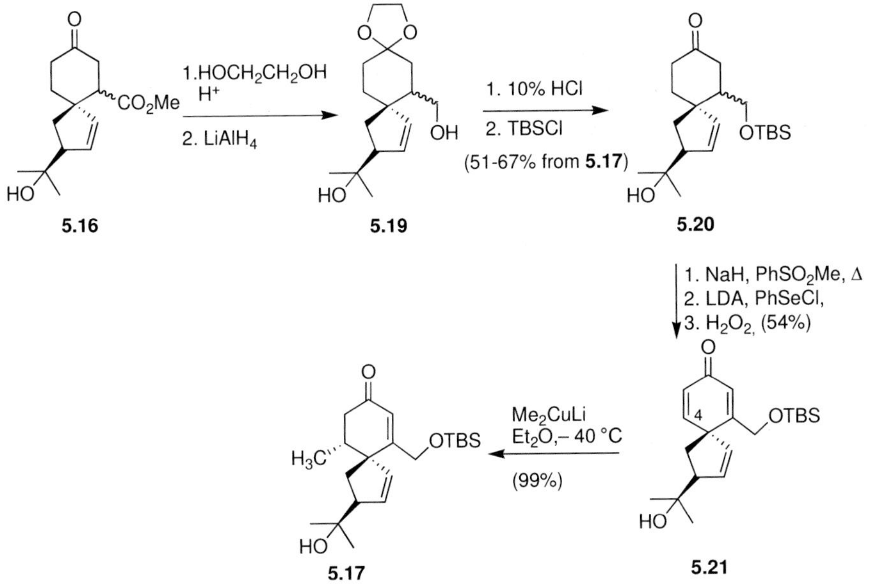

Scheme 5.8

Dehydration was performed by heating **5.22** with Al$_2$O$_3$/py, producing the isopropyli-dene derivative **5.18**. Treatment of **5.18** with LiAlH$_4$ provided a 9:1 mixture of alcohols favoring the isomer having the correct stereochemistry at C2. Removal of the sylil-group with TBAF provided (±)-lubiminol **5.1** (Scheme 5.9).

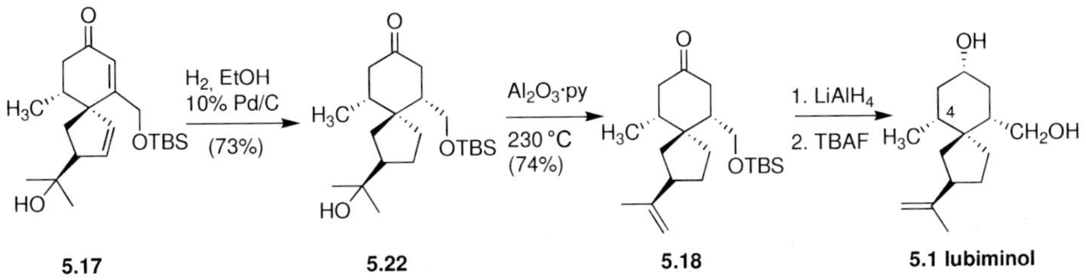

Scheme 5.9

Evaluation

The preparation of (±)-lubiminol is a nice example of how the unexpected appearance of an inert group, which inhibits a key step in a rearrangement, can thwart a planned route. However, in this case the dead-end is solved by using a new synthetic approach that is essentially identical to that originally designed, except for the order of the steps, but the synthesis had to be restarted almost from the beginning.

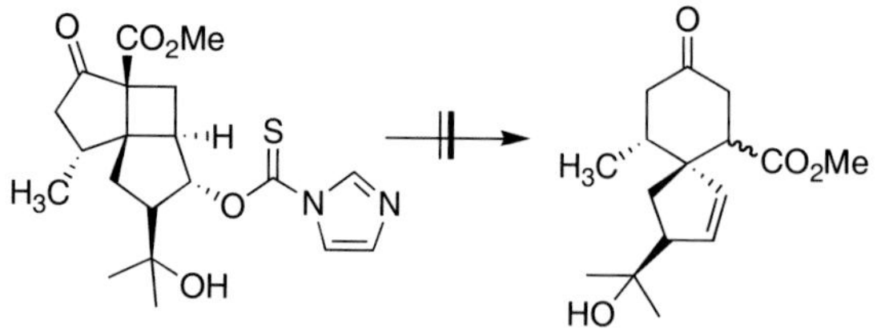

5.2   5.3

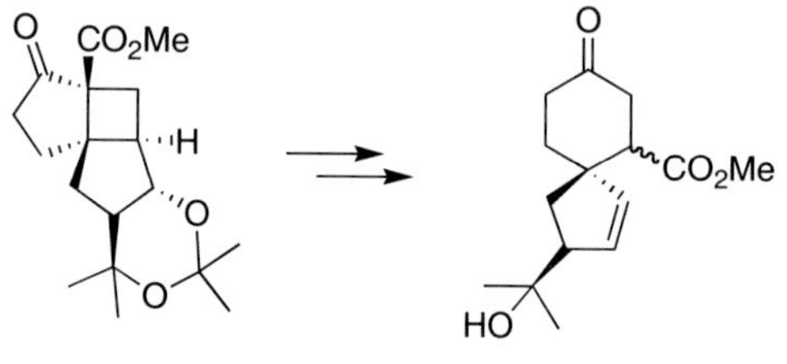

5.5   5.16

Key synthetic reaction

Dowd–Beckwith rearrangement: The Dowd–Beckwith rearrangement used as a key step in Lubiminol synthesis is a general procedure used to effect a radical ring-expansion [3]. The generation of a radical in pendant group β to a carbonyl group promotes the rearrangement. For example, the fragmentation of the cyclobutane ring of **5.15**, occurred after generation of the radical **5.23** from the thiocarbamate moiety. This fragmentation placed the new radical **5.24** in a suitable position to effect the *Dowd–Beckwith rearrangement*. The net result is a ketone **5.16** having a CH_2 more than the initial ring (Scheme 5.10).

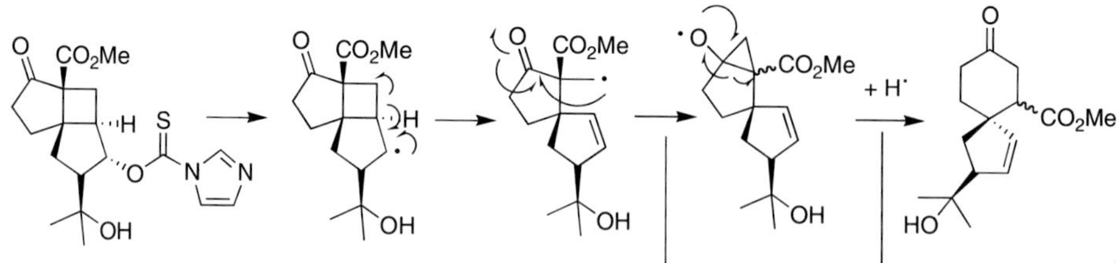

5.15   5.23   5.24   5.25   5.16

Dowd–Beckwith rearrangement

Scheme 5.10

In this context, the remote C4 Me-group of **5.2** inhibits the addition of the Me-radical to the carbonyl group, and hence the Dowd–Beckwith rearrangement.

5.1.2   (±)-Myrocin C [4]

5.26 (±) myrocin C

Target relevance

(±)-Myrocin C **5.26** is a pentacyclic pimarane diterpenoid, isolated from the fungus *Myrothecium verrucaria* [5]. It showed a broad spectrum of antimicrobial activity against Gram-positive bacteria.

Synthetic plan

The planned synthesis of (±)-myrocin C **5.26** features an intramolecular Diels–Alder reaction on a substrate like **5.27**, to construct the bicyclic template **5.28** from which the remaining structure will be elaborated. The angular methyl group at C4 would be contributed by the diene fragment. Adduct **5.28** has to be elaborated to intermediate **5.29**, from which the cyclopropane fragment will be built by an intramolecular alkylation leading to an intermediate like **5.30**. The final ring of myrocin will be constructed on intermediate **5.31** through an intramolecular Diels–Alder process (Scheme 5.11).

Predictable problems

- According to the precedents the behavior of *p*-benzoquinones in Diels-Alder cycloadditions is sluggish [6].
- The primary Diels-Alder adducts **5.28** may undergo aromatization under the strong thermal conditions required to force *p*-benzoquinone reaction.

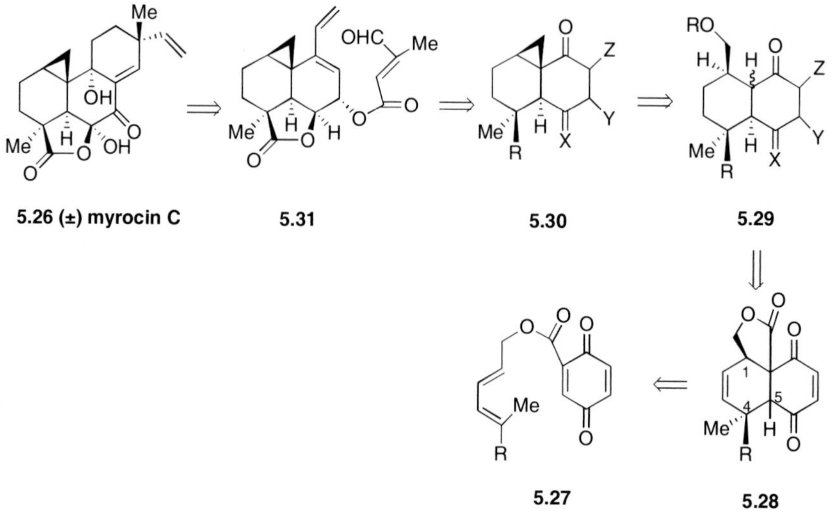

5.26 (±) myrocin C   **5.31**   **5.30**   **5.29**

5.27   **5.28**

Scheme 5.11

Synthesis

Step 1. Initial intramolecular Diels–Alder cycloadditions of 5.27 and the synthesis of adducts 5.28 (Scheme 5.12):

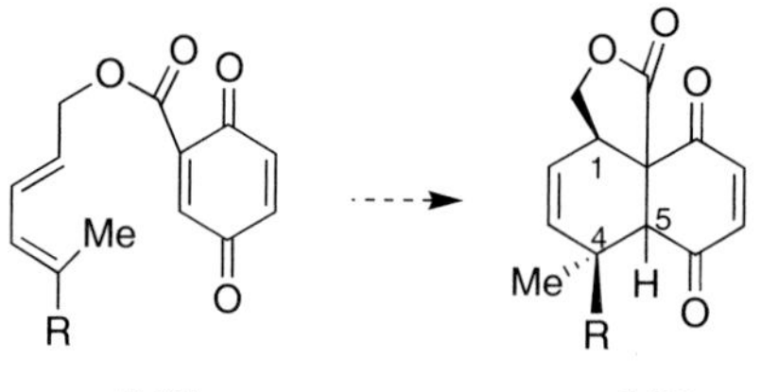

5.27   **5.28**

Scheme 5.12

Attempts to realize the intramolecular Diels–Alder reaction of compounds **5.27** were unsuccessful. Thermolysis of these quinones resulted either in no reaction or, by forcing the reaction conditions, in the destruction of the starting material. The failure of this cycloaddition reaction is attributable to the presence of the Me-group in the diene fragment. In fact, the same reaction on model compound **5.32** produced the tricyclic intermediate **5.33** in good yield.

Albeit that it is a minor detour, the possibility of using **5.33** as an intermediate in the synthesis of (±)-myrocin C **5.26** was evaluated. Intermediate **5.33** was elaborated to bromide **5.34,** which, under basic conditions, yielded a mixture of *cis-* and *trans*-fused cyclopropanic decalines **5.35**. The preference for the *cis*-fused isomer (the undesired isomer) may be explained considering that in the natural *trans*-fused isomer there would be a 1,3-diaxial steric interaction between the cyclopropane and the carbomethoxy at C4 position. This interaction is not present in the major *cis*-fused isomer. To become truly useful, intermediate **5.35** required the incorporation of the C4-Me group. However, precedents concerning the carbon–carbon bond formation at C4 position in related compounds [7], either *via* enolate esters or *via* conjugate addition to Δ^4-6-keto-systems, indicated the preferential formation of an axial bond. This consideration, together with the unwanted stereochemical result of the cyclopropanation reaction, motivated the reconsideration of the Diels–Alder strategy, in order to incorporate the Me-group from the beginning, instead of taking the chance of a risky operation in latter stages of the synthesis. A strategic change, concerning the nature of the initial Diels–Alder reaction was made.

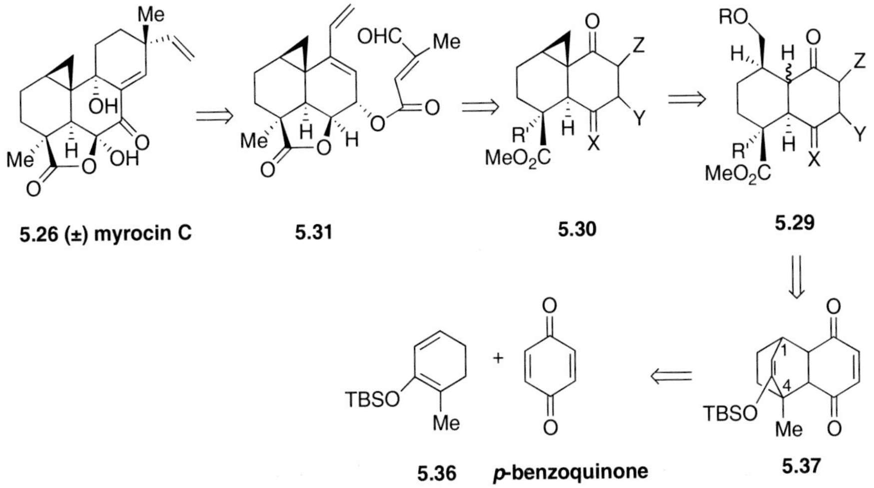

5.26 (±) myrocin C **5.31** **5.30** **5.29**

5.36 ***p*-benzoquinone** **5.37**

Scheme 5.13

Step 2. A revised synthetic plan (Scheme 5.14): The revised plan to myrocin **5.26** begins with a drastically different approach to the first bicyclic intermediate. Now an intermediate like **5.29** will be prepared by the intermolecular Diels–Alder reaction of the cyclic diene **5.36** and *p*-benzoquinone. It was anticipated that the electronic nature and the s-*cis* configuration of diene **5.36** should favor the production of intermediate **5.37**, from which the necessary substituents at C1 and C4 positions would be generated by oxidative cleavage (Scheme 5.14). The remaining synthetic design can be referred to the original plan.

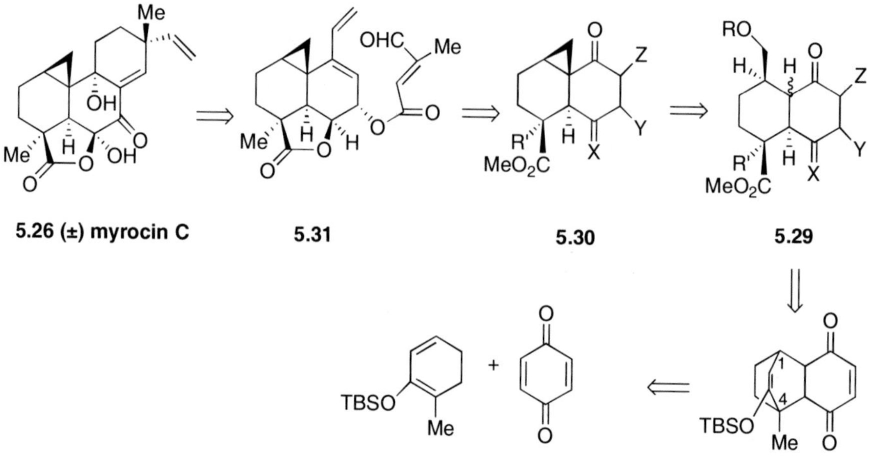

5.26 (±) myrocin C   **5.31**   **5.30**   **5.29**

5.36   *p*-benzoquinone   **5.37**

Scheme 5.14

Step 3. Intermolecular Diels–Alder and synthesis of the precursor for the installation of the cyclopropane moiety (Scheme 5.15):

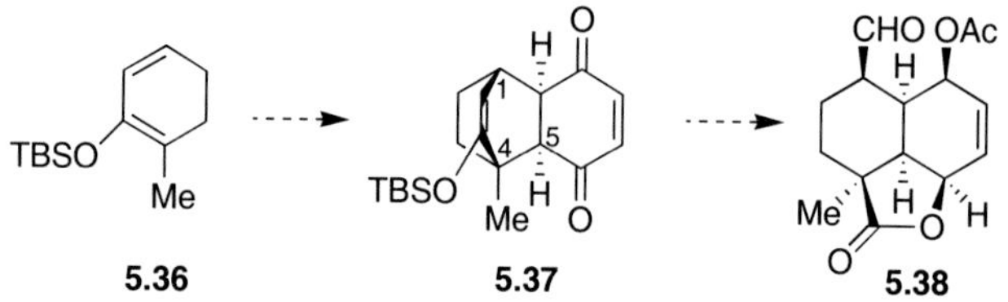

5.36   **5.37**   **5.38**

Scheme 5.15

In the event and following the plan depicted in Scheme 5.14, diene **5.36** was reacted with *p*-benzoquinone for five days in THF at room temperature producing cycloadduct **5.37** in 94% yield. The correct relative stereochemistry at carbons C1, C4 and C5 was ensured at this stage. The oxidative fragmentation of the silyl ether on **5.37**, would yield an intermediate related to **5.29** (see above), with the correct relative stereochemistry at C1 and C4.

Oxidation of compound **5.37** was not an easy task. After extensive experimentation, siloxy ketone **5.39** was obtained using 3,3-dimethyldioxirane as reagent. Reduction of **5.39**, by using the Luche conditions, afforded hemiketal **5.40** in 62% overall yield. The reduction of the two ketone groups took place by addition of hydride from the convex face of the bicyclic system. Intramolecular trapping of the emerging alcohol at carbon C6 by the third ketone group formed ketal **5.40**. This process was apparently faster than its reduction. Thus, the topology of the substrate **5.40** has flattered the correct installment of the chiral centers at carbons C1, C4, C6 and C10, as well as having facilitated the production of an intermediate with the four oxygenated functional groups fully differentiated. The oxidative fragmentation was performed on the pseudodiol **5.41**, obtained from **5.40** after acetylation and subsequent treatment with TBAF. Submission of **5.41** to the action of sodium metaperiodate (NaIO₄) produced the aldehyde lactone **5.38**. With the β-aldehyde substituent at carbon C1 and the α-Me in place at C4, the doors for the synthesis of the cyclopropane ring are open (Scheme 5.16).

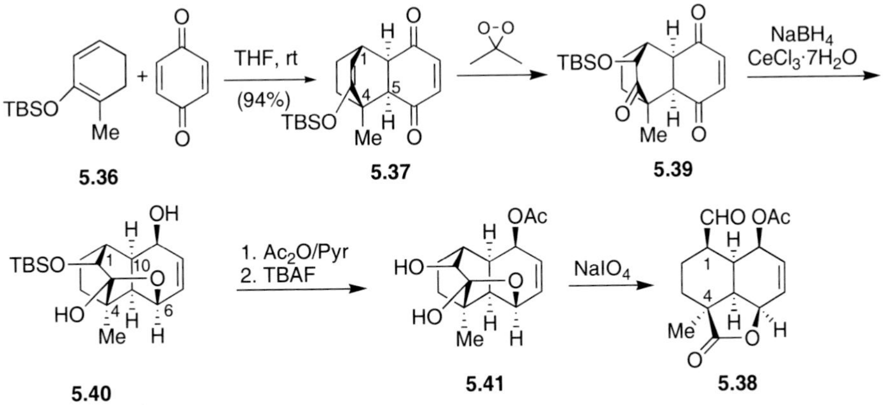

Scheme 5.16

Step 4. Cyclopropane installation (Scheme 5.17):

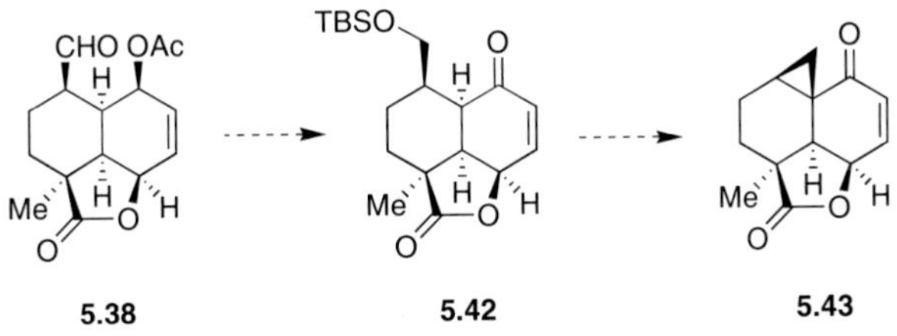

Scheme 5.17

According to the synthetic plan, the installation of the cyclopropane fragment will be achieved by intramolecular alkylation. With this aim, the aldehyde group at C1 on intermediate **5.38** has to become a leaving group. This was achieved by treatment with NaBH₄ yielding **5.44**. The hydroxyl group at C20 was protected as silylether, and the ketone group on C9 was retrieved producing intermediate **5.42**, on which the intramolecular alkylation was undertaken. Treatment of **5.42** with Ph₃PBr₂ produced an unexpected result, since the dihydrobenzofurane **5.45** was obtained instead of the desired bromide **5.46**. The change in reactivity has been motivated by the proximity of the silyloxymethyl group to the ketone at C9, which has participated in the reaction, thwarting the planned synthetic route (Scheme 5.18).

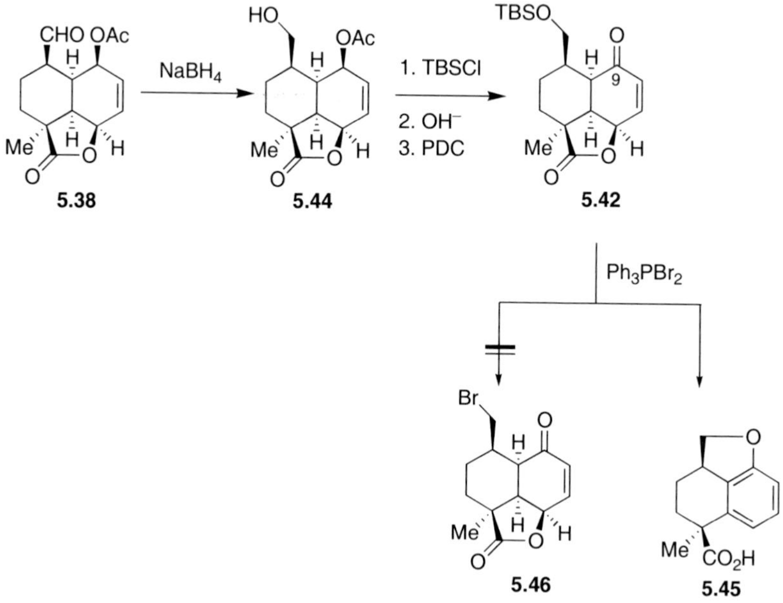

Scheme 5.18

It was reasoned at this point that removal of the double bond could avoid interference of the ketone in the intramolecular alkylation, since now the driving force proportionated by the aromatization of the system will not be present. Therefore, intermediate **5.42** was epoxidated with H_2O_2 in alkaline methanol and the epoxide **5.47** was transformed into mesylate **5.48** in two steps by desilylation followed by reaction of the primary alcohol with mesyl chloride/TEA/DMAP. Again, submission of **5.48** to LiMeO yielded the dihydro-epoxyfurane **5.49**. The intramolecular alkylation reaction was also tested in bromide **5.51**, obtained by hydrogenation of **5.42** forming **5.50** and bromination with Ph_3PBr_2. Again, involvement of the ketone in the reaction was observed producing the dihydrofurane **5.52** (Scheme 5.19).

Failures in the cyclopropanation reaction may be due to a common stereoelectronic effect. In the enolates derived from ketones **5.42**, **5.48** and **5.52**, the C20 leaving group is in the equatorial position. In such a configuration, the emerging backside orbital at C20 cannot overlap with the π-system of the C9–C10 enolate. A tactical change for the cyclopropane formation was needed. In the authors' words, *a less traditional cyclization modality had to be considered.*

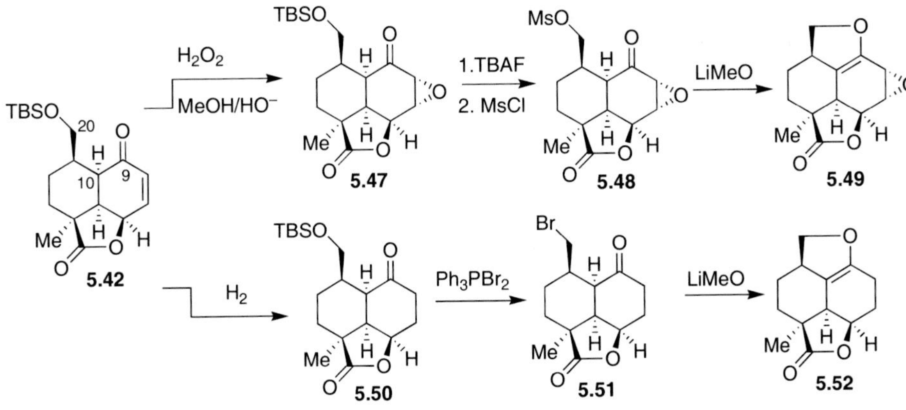

Scheme 5.19

Step 5. Second attempt at cyclopropane installation (Scheme 5.20): On a substrate like **5.53**, a carbanion can be generated at C20 that would interact with the C9–C10 double bond forming the cyclopropane **5.54**, with concomitant opening of the C7–C8 epoxide (Scheme 5.20). This new approach is a drastic change in tactics to effect the cyclopropane ring closure.

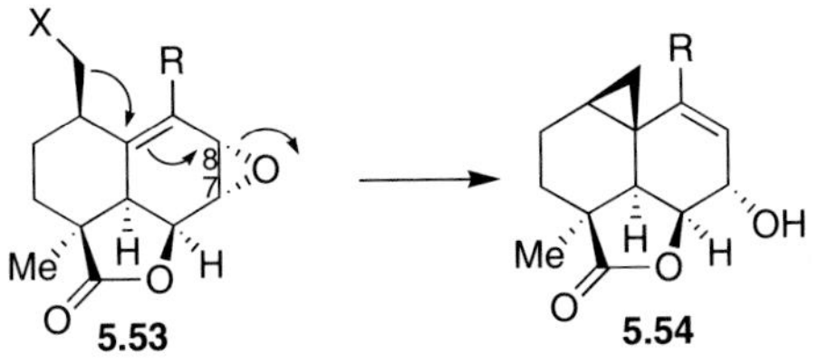

Scheme 5.20

To implement this new approach for cyclopropane ring formation, bromide **5.55** was prepared from *cis*-fused epoxide **5.47**. This compound was converted to the vinyl triflate **5.56** via the corresponding enol-phosphate. Diene **5.57** was obtained in 54% yield via Stille-like coupling with vinyltributyltin in the presence of Pd(II) and excess of LiCl. Buffered TBAF treatment of **5.57** liberated the C20 hydroxyl group affording alcohol **5.58**, which, upon reaction with triphenylphosphine-carbon tetrabromide, yielded **5.55** (Scheme 5.21).

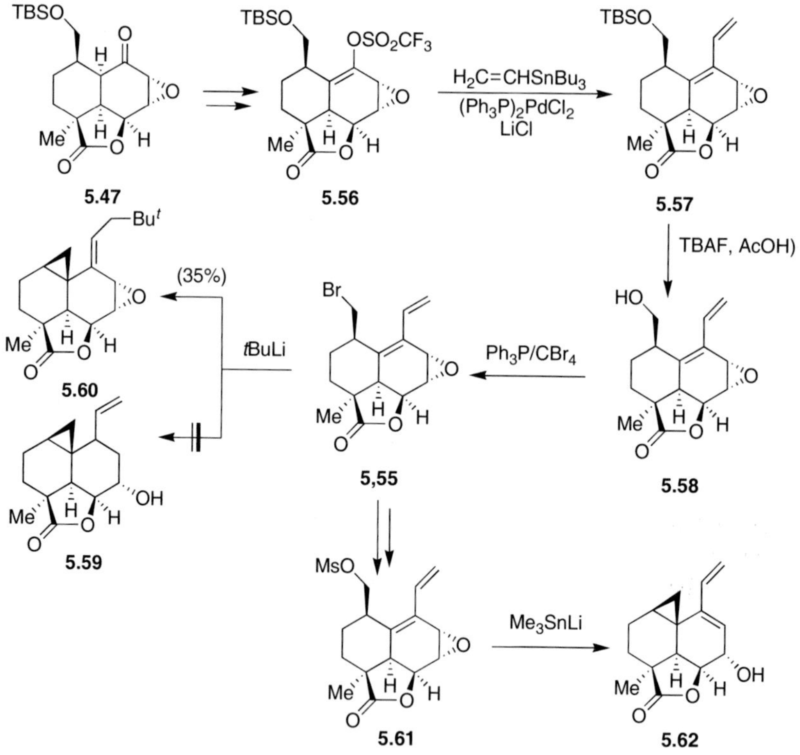

Scheme 5.21

Bromide **5.55** was treated with *t*BuLi in order to obtain **5.59** by lithiation of C20 followed by cyclization. Interestingly, under these conditions, vinylcyclopropane **5.60** was obtained in 35% yield, instead. Although intermediate **5.60** is not useful for the synthesis of myrocin C (**5.26**) its formation opens up the possibility of using a nucleophile that may serve as a substrate for the reductive opening of the epoxide. The reaction of mesylate **5.61**, obtained from bromide **5.55**, with (trimethylstannyl)lithium gave a 66% yield of cyclopropane **5.62**.

Thus, after many deviations we have arrived at the precursor of the originally designed intermediate **5.31** that was devised as the template to effect the full elaboration of the C-ring (Diels–Alder) and oxidation of carbon C6 to deliver, finally, myrocin C.

Step 6. Annulation of the C-ring by intramolecular Diels–Alder reaction and completion of the synthesis (Scheme 5.22):

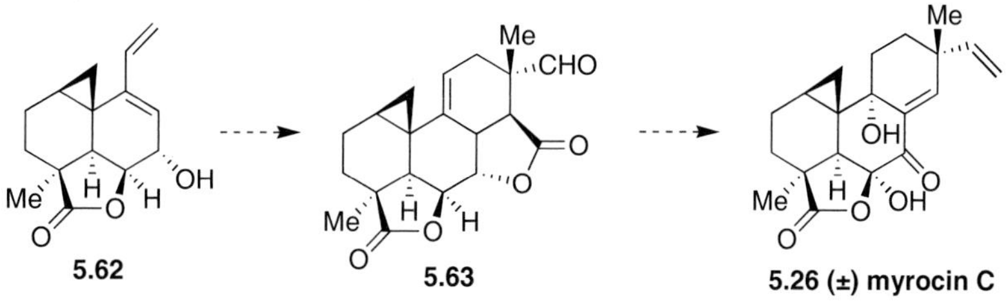

Scheme 5.22

Ring C would be assembled using an intramolecular Diels–Alder reaction to achieve good facial selectivity. Acid **5.64** was coupled with **5.62** in the presence of DCC/DMAP to form the ester **5.31** in excellent yield. Thermolysis of **5.31** yielded adduct **5.63** as a single stereoisomer. The facial selectivity has been imposed by the stereochemistry of C7 and by the predominance of endo control. The carboxylic carbon on **5.63**, introduced with the dienophile, was removed once its function was accomplished, to form **5.65**. The mirocyne exo-vinyl group was introduced prior to the removal of the carboxylic acid (Scheme 5.23).

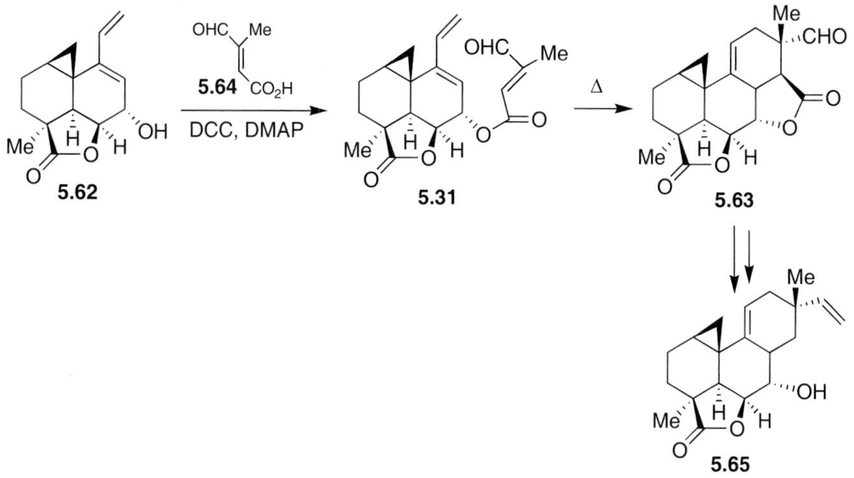

Scheme 5.23

Introduction of the C8–C14 double bond and hydroxylation at the C6 and C9 positions, are all that remains in order to access myrocin. In this regard, alcohol **5.65** was oxidized to the corresponding ketone **5.66**. Epoxidation with alkaline hydrogen peroxide yielded directly the α,β-epoxiketone **5.67**. Attempts to accomplish oxygenation of the enolate derived from **5.67** were completely unsuccessful. Surprisingly, reaction of **5.67** with Nozaki reagent [(*p*-MeOPhS)AlMe₃Li] yielded **5.68**. Finally, oxidation of **5.68** with 3,3-dimethyldioxirane formed 6-desoxymyrocin **5.69**, which upon exposure to potassium tert-butoxide in the presence of oxygen, followed by reduction of the hydroperoxide intermediate with triethyl phosphite, afforded racemic myrocyn C **5.26** (Scheme 5.24).

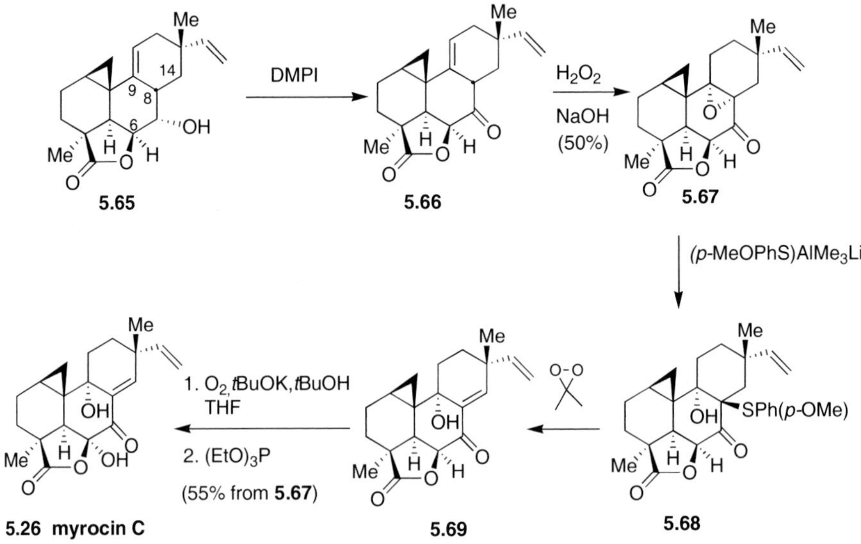

Scheme 5.24

Evaluation

The synthesis of myrocin C involves two major deviations from the original planning.

- The failure of the intramolecular Diels–Alder designed to place the C4 Me-group led, fortunately at an early stage of the synthesis, to a strategic change making use of an intermolecular Diels–Alder reaction. While compound **5.32**, lacking the key Me-group, reacts nicely, the analogous reaction failed in the case of compound **5.27**, with the desired methyl group placed at the double bond. Attempts to promote the reaction, by replacing the original carbomethoxy by a cyano group to sterically outperform the carbomethoxy function, were in vain. Reduction of the ester followed by protection of the resulting alcohol to downgrade its electron-withdrawing power in the Diels–Alder reaction, did not produce the desired reaction either. These experiments resulted in the abandonment of this approach, which introduces the C4 quaternary group at the early stages of the synthesis.

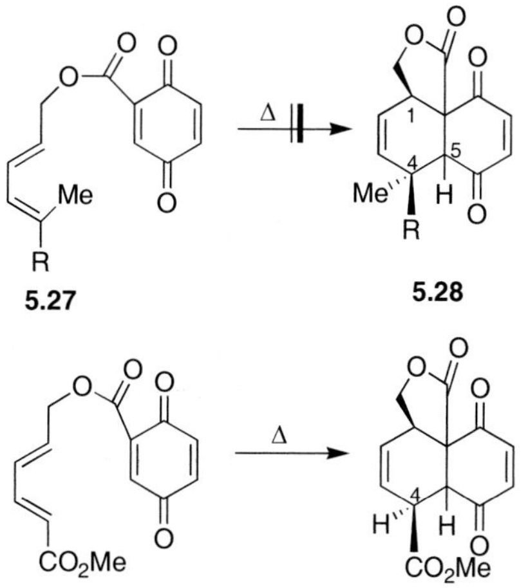

5.27   **5.28**

5.32   **5.33**

- Later in the synthesis the introduction of the fused cyclopropane ring presents serious problems. The first approach attempted encountered a live reactive ketone group (initially in compound **5.42**) that thwarted the planned route. All the modifications used to achieve the desired transformation: avoiding the interference of the carbonyl group; including the hydrogenation of the double bond (compound **5.51**), which may cause further problems since it is necessary to continue with the synthesis; or its epoxidation (compound **5.48**) were fruitless. The live ketone remote group killed this approach. A major tactical detour was undertaken to construct the troublesome cyclopropane.

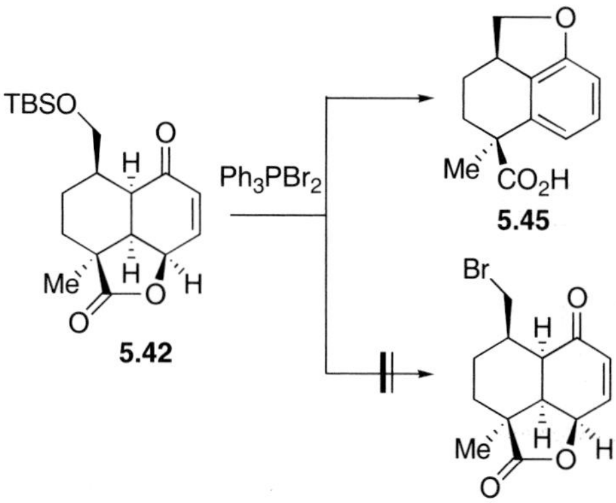

5.42   **5.45**

5.46

(±)-Myrocin C synthesis shows two of the problems with remote groups that we are discussing in this chapter. The methyl group is acting as "alive" dead-steric volume, while the ketone moiety of compounds **5.42, 5.48** and **5.51** is acting as a reactive remote group. The following section shows two more examples of this last very common situation.

5.2   The Reactive Remote Group

We have introduced this situation during the cyclopropane formation in the synthesis of (±)-myrocin (5.1.2). It is very frequent to encounter unexpected reactivity due to functional groups that, in principle, should not participate in the desired transformation. However, the densely functionalized and topologically complex intermediates in total synthesis made this scenario one of the most conflictive. The following examples teem in this topic.

5.2.1   (–)-Chaparrinone [8]

5.70 (–)chaparrinone

Target relevance
(–)-Chaparrinone **5.70** was first reported by Polonsky [9] and belongs to the quassinoid family of natural products, that were reported for the first time in 1961 from *Simarubaceae* plants [10], and some of them exhibit a moderate activity against P-388 lymphocytic leukemia.

Synthetic plan
The synthesis of (–)-chaparrinone **5.70** is based in the construction of the tricyclic fragment **5.71** containing the ABC core of the quassinoid by a Diels–Alder reaction between diene **5.72** and (*E*)-4-methyl-3,5-hexadienoate **5.73** as dienophile. The intermediate **5.71** is further elaborated to (–)-chaparrinone through lactone **5.74** (Scheme 5.25).

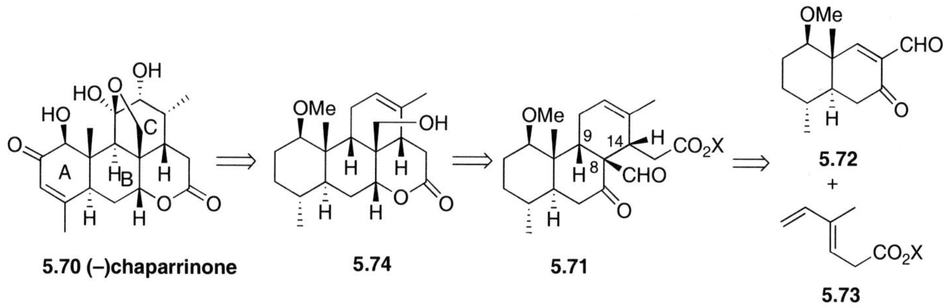

Scheme 5.25

Predictable problems

The main predictable problem of the synthetic planning depicted in Scheme 5.29 is the stereochemical outcome of the Diels–Alder reaction used to construct the tricyclic core present on intermediate **5.71**. Among the four possible [4+2] adducts derived from *endo*-transition states, those obtained by the approach of the diene **5.73** from the β-face of **5.72** (**5.75** and **5.76**) should be disfavored, due to the presence of the C19 Me-group. Therefore, the anticipated adducts **5.77** and **5.78** will have the undesired stereochemistry at C9. Subsequent inversion of the configuration at this center has to be considered (Scheme 5.26).

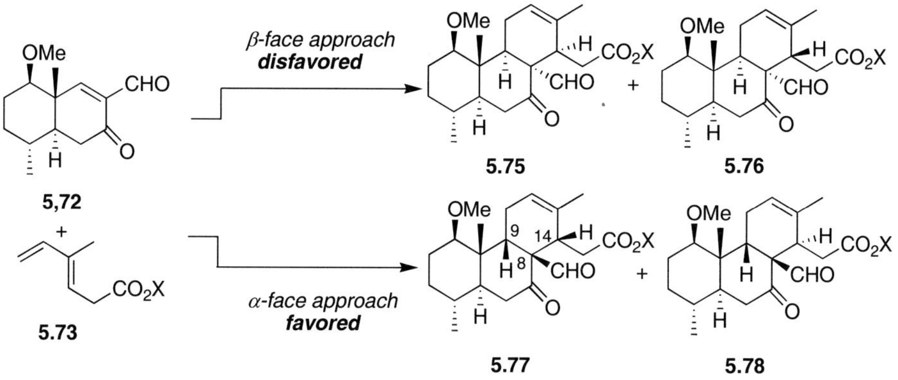

Scheme 5.26

Synthesis

*Step 1. Synthesis of the tetracyclic intermediate **5.74** (Scheme 5.27):*

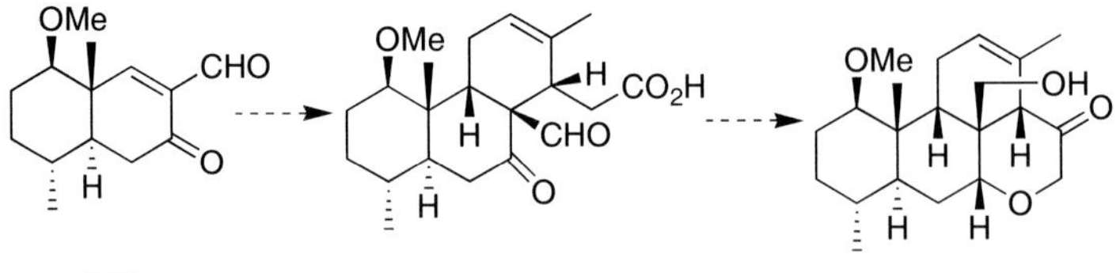

Scheme 5.27

The synthesis of (−)-chaparrinone **5.70** was initiated by the Diels–Alder reaction of **5.72** and the sodium salt **5.80** to yield the adduct **5.79**. The required selectivity in the Diels–Alder reaction was obtained by using this sodium salt in water. The use of the ester derived from acid **5.80** gave a mixture of diastereomeric adducts. It should be noted here that effectively compound **5.79** has the wrong stereochemistry at C9. Having accessed to tricyclic intermediate **5.79**, the next step is the installment of the lactone ring. This task was accomplished by submitting ketoaldehyde **5.79** to the action of sodium borohydride in aqueous THF. These conditions produce the reduction of the aldehyde and the stereoselective reduction of the C7 ketone group from the convex β-face, which yields a mixture of the corresponding lactones **5.81** and **5.74**. This mixture was quantitatively equilibrated to the desired, and more stable, lactone **5.74** by acid treatment (Scheme 5.28).

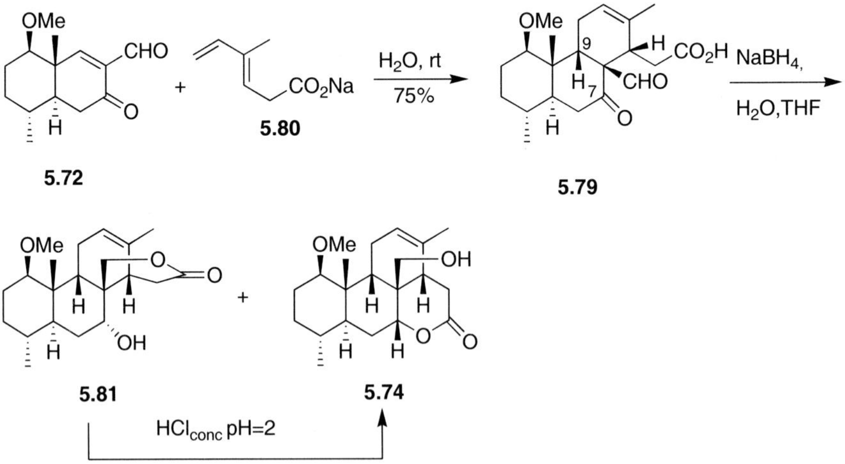

Scheme 5.28

*Step 2. Inversion of the configuration at C9 and functionalization of ring C. Synthesis of intermediate **5.82** (Scheme 5.29):*

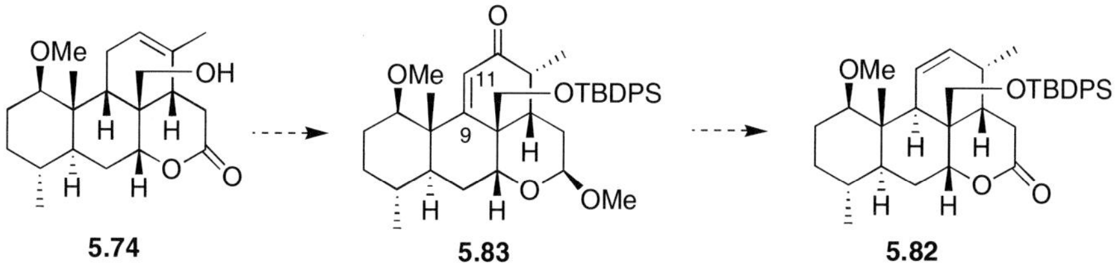

Scheme 5.29

Since the configuration at C9 was epimeric to chaparrinone **5.70**, the inversion of this center was undertaken before proceeding with the functionalization of the C ring.

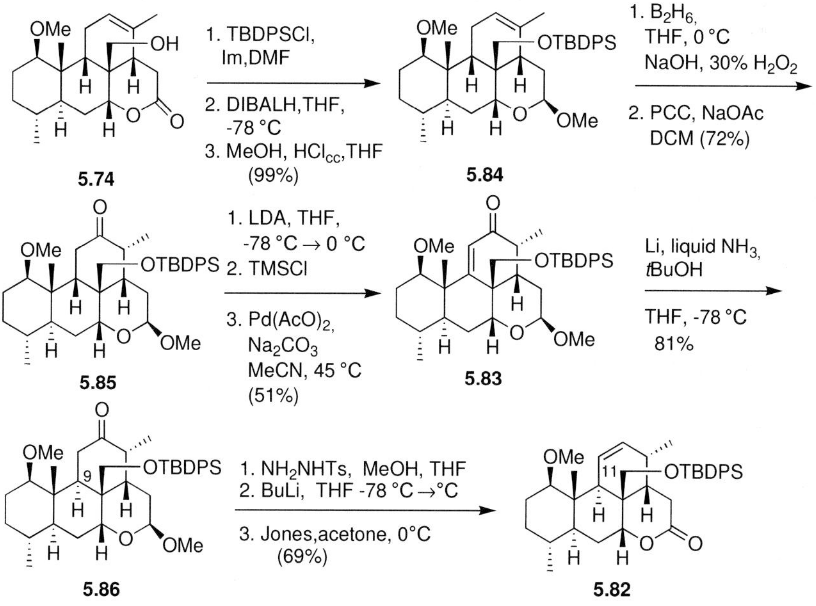

Scheme 5.30

Inversion of the configuration at C9 was achieved by metal ammonia reduction of the $\Delta^{9,11}$-C12 ketone **5.83**. To obtain **5.83** from lactone **5.74,** the C20 alcohol group was protected as silyl ether, and the lactone was reacted with DIBALH followed by treatment with hydrochloric acid in MeOH-THF. These reactions produced intermediate **5.84**, which was hydroborated and oxidized to yield the ketone **5.85**. Ketone **5.85** upon reaction with LDA, trapping of the kinetic enolate as silyl enol ether, and treatment of this Δ^{11}-silyl enol ether with $Na_2CO_3/Pd(OAc)_2$, afforded enone **5.83**. As expected, metal-ammonia reduction of **5.83** produced **5.86** having the correct stereochemistry at C9 (Scheme 5.34). At this stage, functionalization of ring C began by the placing of a Δ^{11} double bond, dihydroxylation of which would render the 11,12-trans dihydroxy grouping of chaparrinone. Reaction of ketone **5.86** with $TsNHNH_2$ and treatment of the hydrazone with BuLi yielded alkene **5.82** after Jones oxidation (Scheme 5.30).

Step 3: Functionalization of ring A and failed completion of the synthesis (Scheme 5.31):

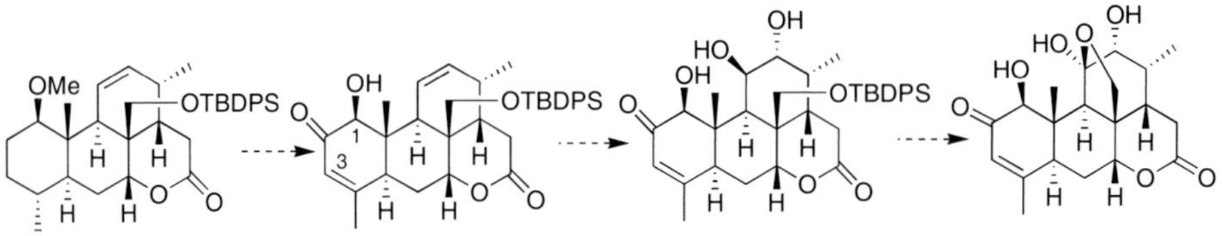

Scheme 5.31

To end the synthesis of chaparrinone, rings A and C have to be functionalized before effecting the final lactol ring closure. The instalment of the 1β-hydroxy-2-oxo-$\Delta^{3,4}$ olefin of derivative **5.87** was accomplished starting with oxidation of the C1 alcohol from which derived the unmasking **5.82**. Ketone **5.89** was elaborated to enone **5.90** using the standard sequence: enolization → seleniation → oxidation. The Δ^1-olefin of key enone **5.90** was selectively oxidized via its silyl enol ether, and the resulting hydroxy ketone **5.91** isomerized to **5.87** by base treatment (Scheme 5.32).

To complete the synthesis, in the words of the authors, *the remaining task of installing the ring C hemiketal array appeared to be a straightforward exercise.* It was reasoned that intermediate **5.87** could be transformed in chaparrinone **5.70** by selective espoxidation of the Δ^{11} double bond, followed by opening of the epoxide and selective protection of the C1 and C12 hydroxyl groups. Oxidation of the remaining C11 hydroxyl group and exhaustive removal of the protecting groups would render chaparrinone.

According to this synthetic plan, sketched in Scheme 5.35, intermediate **5.87** was treated with MCPBA affording epoxide **5.92** in 60%yield. However, reaction of this epoxide **5.92** with $HClO_4$ produced exclusively deoxychaparrinone **5.93**. The required *trans*-diol **5.88** was not detected. Deoxychaparrinone **5.93** may arise from the acid-catalyzed intramolecular epoxide opening by the hydroxymethyl group at C8 (Scheme 5.36). In this case the remote C8 hydroxymethyl substituent is not quite so remote (see **5.94**) taking part in the reaction and driving the synthesis to a real dead-end (Scheme 5.36). Changing the protecting group on the C8 hydroxymethyl did not modify the result of the reaction. Equally, it was impossible access to the β-epoxide from olefin **5.87**. Therefore a new strategy to prepare (–)-chaparrinone **5.70** was designed.

The situation depicted in this example is a real nightmare for the synthetic chemist. There is no other alternative than a major re-evaluation of the synthetic strategy.

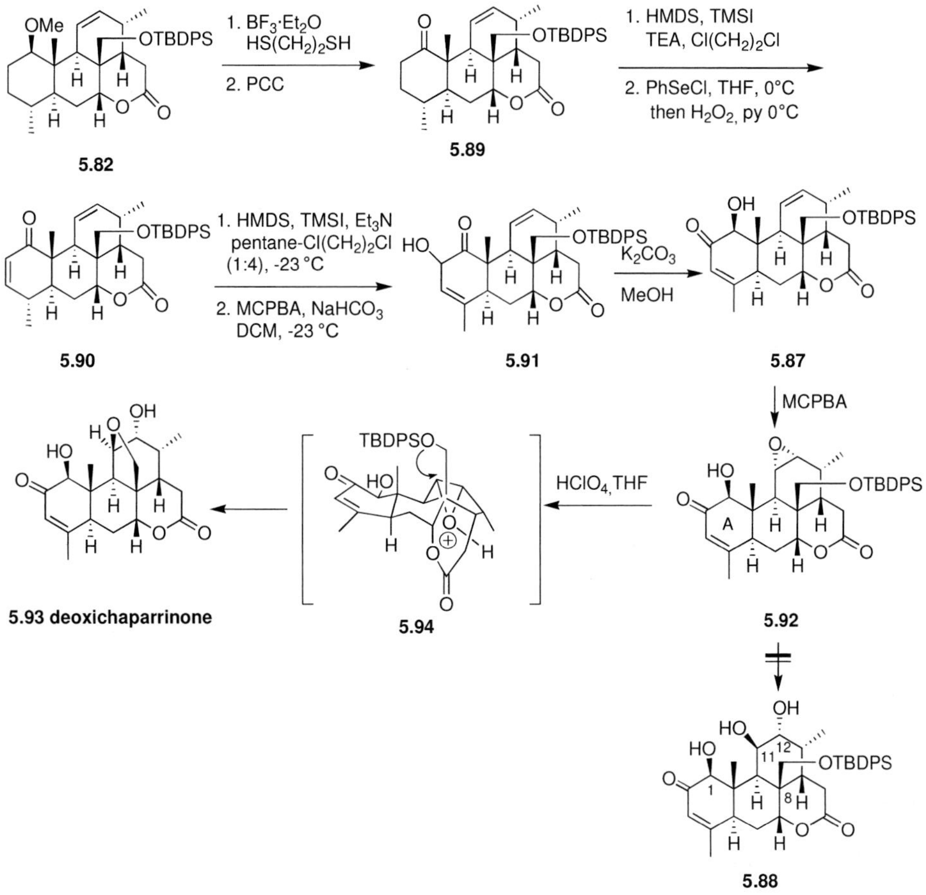

Scheme 5.32

Step 4. Development of a new strategy to complete the synthesis of (−)-chaparrinone 5.70 (Scheme 5.33): Owing to the inability to obtain **5.88**, the initial plan to access chaparrinone was abandoned. The new plan used, as starting material, intermediate **5.95** obtained from methoxy removal of **5.82**. The new plan will elaborate ring C to reach intermediate **5.96** from which elaboration of the A ring will be pursued to obtain **5.97**. From **5.97** (−)-chaparrinone **5.70** should be immediately available.

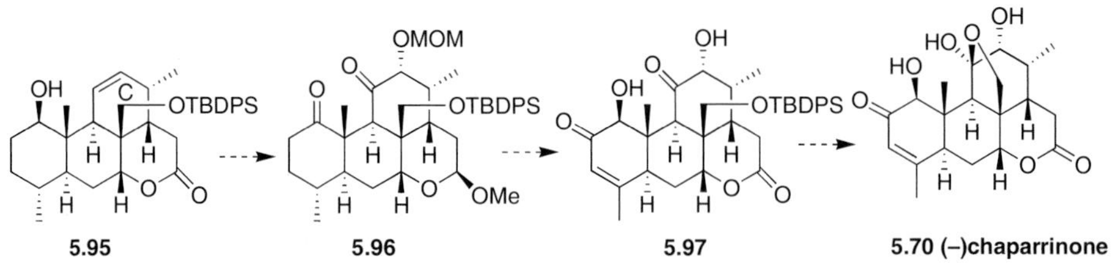

Scheme 5.33

To implement this sequence, alcohol **5.95** was transformed into the epimeric mixture of protected lactols **5.98**. Acetylation of secondary alcohol at C1 was followed by reaction with OsO₄ · py: and subsequent reduction with sodium bisulfite, affording diol **5.99** in overall 86% yield from **5.95**. Besides dihydroxylation, a probable intramolecular transesterification, exclusively to the C12, position, has taken place. The completely unexpected outcome of this reaction provided a suitable intermediate for simultaneous oxidation of the C1 and C11-hydroxyl groups and hydrolysis of the C12-acetate yielding diketone **5.100.** The fully functionalized A ring of chaparrinone was elaborated on MOMO-protected **5.101** to afford, in a multistep sequence, compound **5.97** in which the previously unsuccessful final lactol ring closure was attempted. Submission of inter-mediate **5.97** to TBAF produced chaparrinone **5.70** (Scheme 5.34).

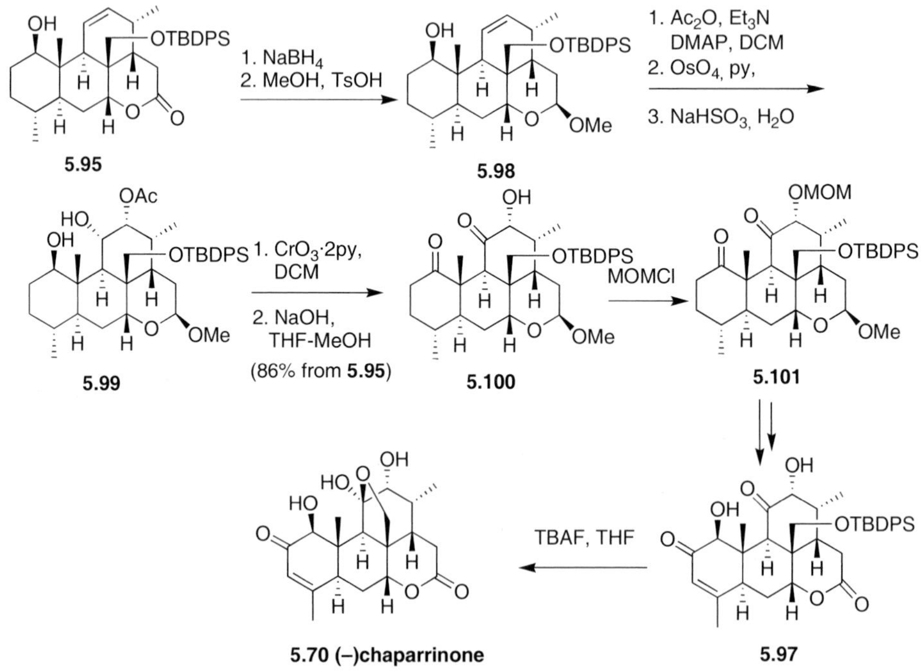

Scheme 5.34

Evaluation

The enormous disturbance caused by the inability of compound **5.92** to undergo the acid promoting epoxide aperture to form **5.88**, complicated by the exclusive formation of deoxychaparrinone **5.93**, was breathtaking. This is especially surprising when considering that this key step had been developed during the synthesis of klaineanone **5.102.** The key difference between chaparrinone **5.70** and klaineanone **5.102** that provokes a breakdown of the synthesis of compound **5.70**, was produced by the presence of an innocent remote hydroxyl substituent on the C20-Me-group.

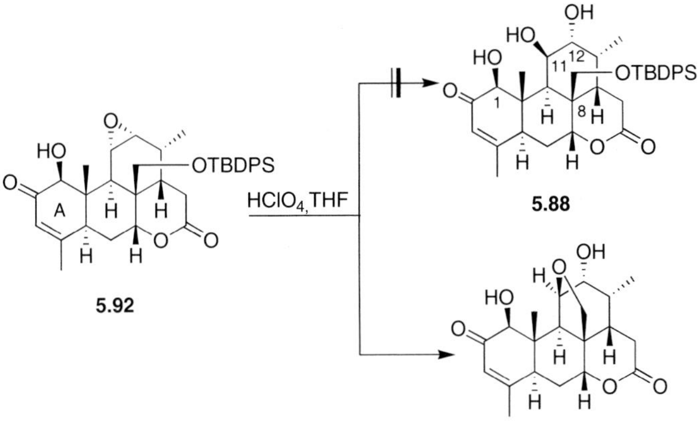

5.88

5.92

HClO$_4$, THF

5.93 deoxichaparrinona

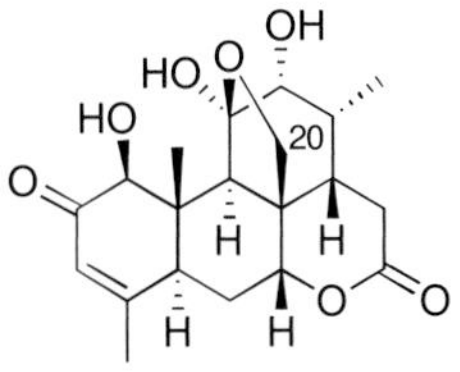

5.70 (–)chaparrinone

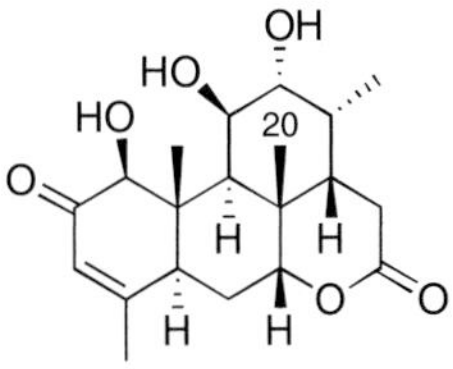

5.102 klaineanone

In the next example, the *distal substituent* is a protecting group that does not take part in the reaction, but modifies the reactivity of the substrate. The success of the synthesis in this case requires a careful choice of the protecting group.

5.2.2   (±)-Akuammicine [11]

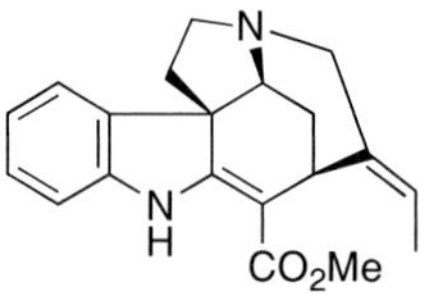

5.103 (±)-akuammicine

Target relevance
Akuammicine **5.103**, first isolated from the seeds of *Picralima klaineana* [12], has been found in different genera of *Apocinaceae*, in both optically active and racemic forms. Its structure was proposed by Robinson [13] and later confirmed by Smith [14].

Synthetic plan
The synthetic planning to prepare (±)-akuammicine is based in the disconnection of the dihydroindol ring to form an intermediate like **5.104**. The access to **5.104** was envisioned from intermediate **5.105** through the powerful and elegant aza-Cope–Mannich rearrangement [15] (Scheme 5.35).

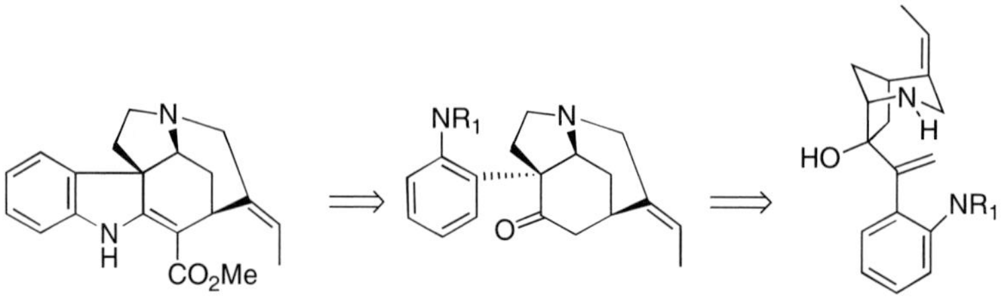

5.103 (±)-akuammicine   **5.104**   **5.105**

Scheme 5.35

Predictable problems
No problems were anticipated since the central aza-Cope–Mannich rearrangement had been successfully used in the synthesis of indol derivatives such as **5.106** from aminoalcohols **5.107** (Scheme 5.36).

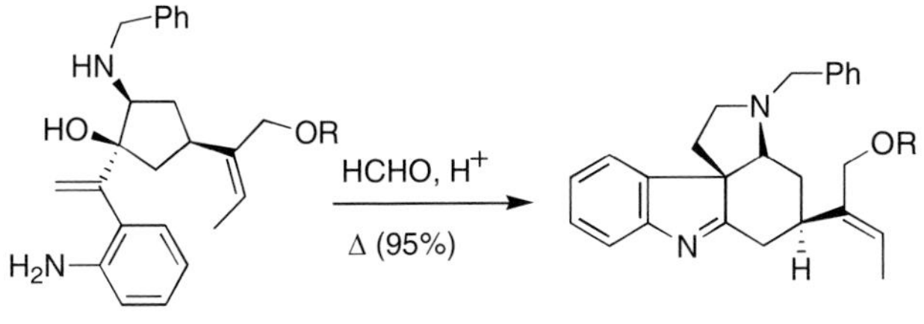

5.107 **5.106**

Scheme 5.36

Synthesis

*Step 1. The aza-Cope–Mannich rearrangement of **5.109**. Synthesis of (±)-dehydrotubifoline **5.111** (Scheme 5.37):* The use of KOH in refluxing Et_2O-H_2O was necessary to release the amino group of **5.108** and to produce the amino alcohol **5.109**. The harsh conditions were required as a result of the steric hindrance around the ring nitrogen. The submission of **5.109** to the standard mild conditions in which the aza-Cope–Mannich takes place (paraformaldehyde, CSA and Na_2SO_4 in refluxing MeCN), caused the rearrangement, providing compound **5.110** in 88% yield. Hydrolysis of **5.110** with KOH/EtOH yielded (±)-dehydrotubifoline **5.111** (Scheme 5.37).

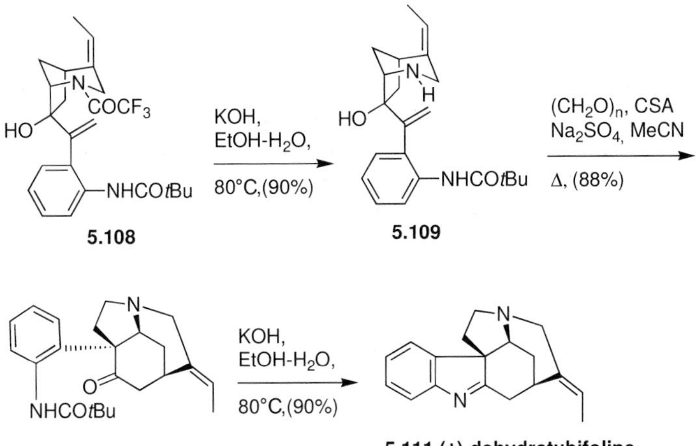

5.108 **5.109**

5.110 **5.111 (±)-dehydrotubifoline**

Scheme 5.37

*Step 2. Acylation of (±)-dehydrotubifoline **5.111**:* Transformation of (±)-dehydrotubifoline **5.111** into (±)-akuammicine **5.103** required only acylation at the C16 position. This is an apparently simple and straightforward operation. However, all attempts to directly acylate the C16 position were unfruitful. Thus, trapping of the lithium enolate, generated from (±)-dehydrotubifoline, resulted in complex reaction mixtures. Equally, formylation under different Vilsmeier–Haack [16] conditions provided only the *N*-formyl derivative **5.112** (Scheme 5.38).

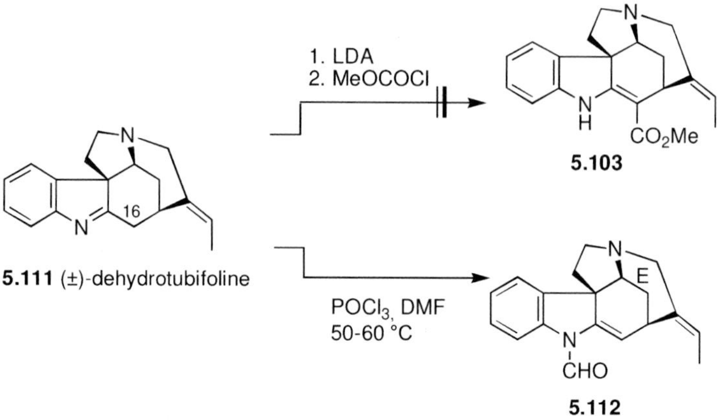

Scheme 5.38

The inability to acylate the C16 position of (±)-dehydrotubifoline **5.111** reflects the severe congestion caused by the adjacent bridged ring E. Therefore a new tactic consisting of the introduction of the carbomethoxy substituent prior to ring closure, had to be considered.

Step 3. Introduction of the troublesome methoxycarbonyl group before effecting the final ring closure: The acylation of intermediate **5.110** provided the ketoester **5.113**. Unfortunately, efforts to produce ring closure on **5.113** were unsuccessful, leading to unreacted material or to the formation of (±)-dehydrotubifoline **5.111**. The robustness of the pivaloyl protecting group hampers the ring closure to form the dihydroindol (Scheme 5.39).

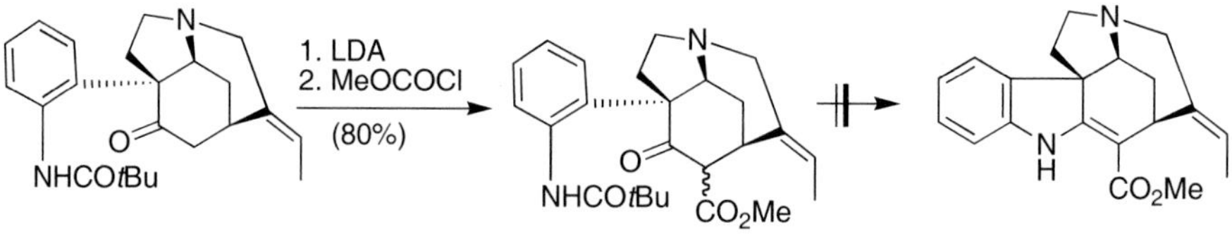

5.110 **5.113** **5.103 (±)-akuammicine**

Scheme 5.39

Step 4. The search for the suitable protecting group in order to achieve the ring closure: As depicted in Scheme 5.39 the *tert*-butoxycarbonyl (BOC) protecting group was chosen first, since the BOC group is removable under mild acidic conditions if access to akuammicine is gained. Preparation of the needed BOC derivative **5.114** is by no means direct since it required repeating the synthesis from a very early step, from **5.115**. In fact, intermediate **5.114**, the new substrate for the aza-Cope–Mannich rearrangement, was obtained after major difficulties (Scheme 5.40). In the words of the authors: *Although the general route (to the aza-Cope–Mannich substrates) is similar to that developed for the pivaloyl series, this seemingly small change in the nitrogen protecting group necessitates a number of critical modifications in the experimental sequence.*

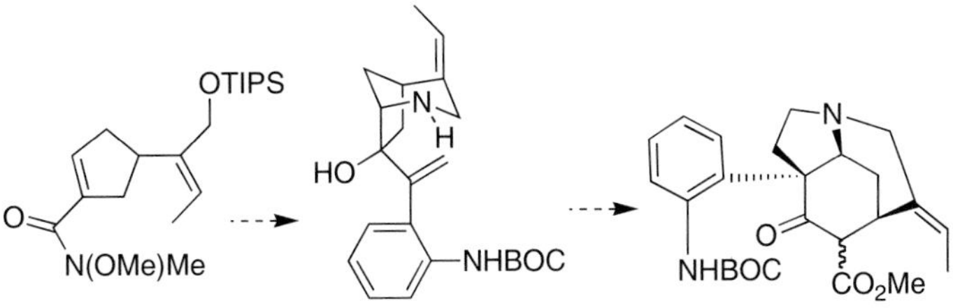

Scheme 5.40

Intermediate **5.114** was submitted to the standard aza-Cope–Mannich reaction conditions, yielding the dihydroindol **5.118** that could not be converted into akuammicine. In no case was the isolation of ketone **5.117** possible. The authors presume that the greater nucleophilicity of the BOC-protecting group ensures the cyclization to **5.118** (Scheme 5.41). It is clear that the BOC-protecting group is not suitable for the planned synthesis of akuammicine. The marked different reactivity between the pivaloyl and the BOC-protecting aniline series could not be known in advance.

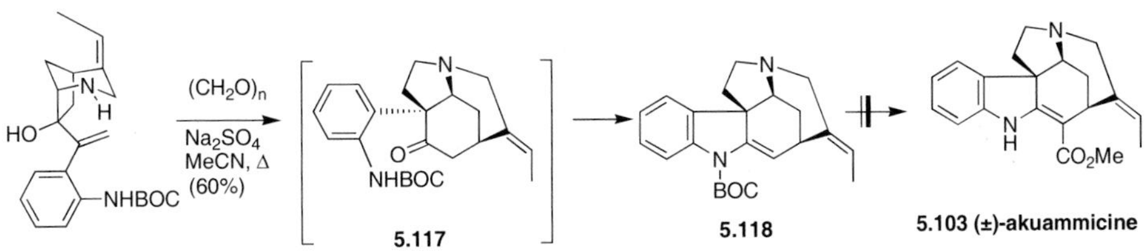

Scheme 5.41

According to the above results, the right protecting group must survive to the aza-Cope–Mannich rearrangement conditions and be easily removed and also must lock both hydrogens of the aniline nitrogen group. The 1,3-dimethylhexahydro2-oxo-1,3,5-triazine ("triazone") group, seemed to fulfill all the requirements, and the synthesis was started again from the very beginning. Following a sequence of similar reactions to those used for the first development of the pivaloyl series, the aza-Cope–Mannich rearrangement precursor **5.119** was accessed. Submission of **5.119** to the standard rearrangement conditions produced the havoc of the protecting group, but this time the result was not unexpected. In fact, driving the rearrangement in the absence of acid allowed isolation of ketone **5.120** in excellent yield. Acylation of **5.120** and subsequent acidic treatment of the resulting ketoester **5.121** furnished (±)-akuammicine **5.103** (Scheme 5.42).

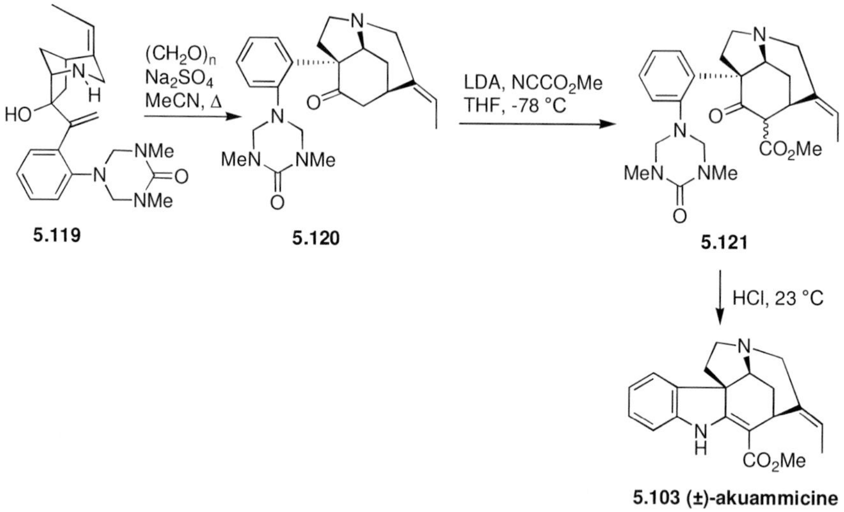

Scheme 5.42

Evaluation

- In the author's words the striking difference in reactivity of BOC- and pivaloyl-protected anilines in these series *underscores the profound effects on reaction outcome that can arise from small changes in electron density*. The choice of the right group was made after two unsuccessful approaches. The question is whether these troubles could be predicted *a priori* with our actual knowledge. The answer is, at this point, no. The robustness of the pivaloyl group and the reactivity of the BOC group encountered in this example, are not general but specific to this synthetic approach.

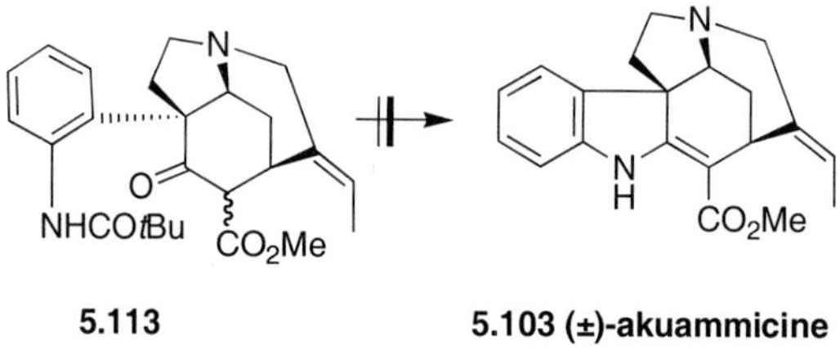

- The reactions on Scheme 5.42 elicit one additional intriguing question: why couldn't dehydrotubifoline **5.113** be transformed to (±)-akuammicine **5.103**? Apparently, the C16 position is fully shielded by the adjacent bridged E-ring thus precluding any transformation at this carbon. In agreement with this is the fact that acylation of compound **5.122** to yield **5.123** occurred in the Aspidosperma alkaloid series, lacking the bridged E-ring of the Strychnos alkaloids [17].

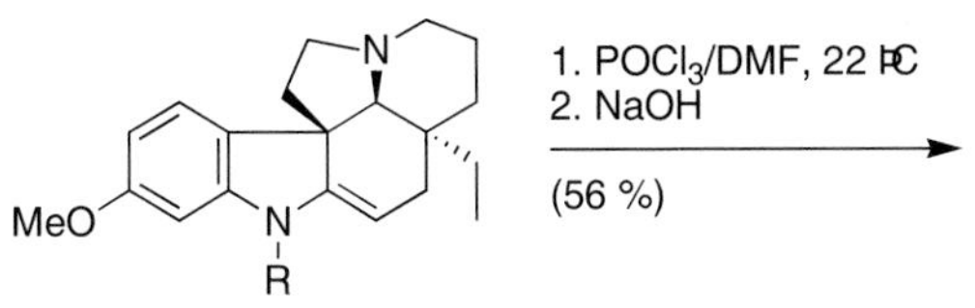

1. POCl$_3$/DMF, 22 °C
2. NaOH

(56 %)

5.122

5.123

R = COOMe

Key synthetic reactions

Aza-Cope–Mannich rearrangement: The one-pot formation of a Mannich salt, by reaction of a tertiary homoallyl amine with formaldehyde followed by the aza-Cope rearrangement [15].

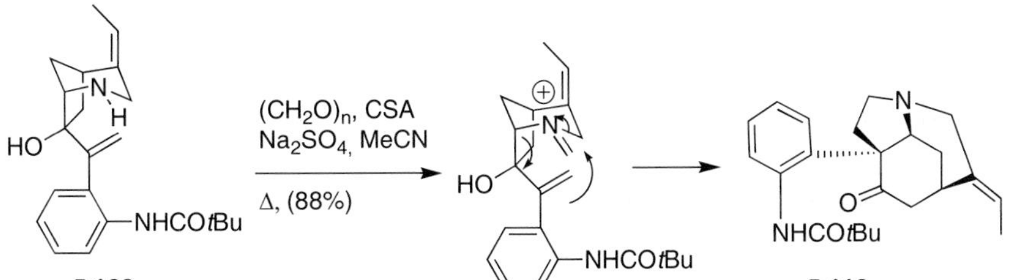

(CH$_2$O)$_n$, CSA
Na$_2$SO$_4$, MeCN

Δ, (88%)

5.109

5.110

To end this chapter we discuss the enormous influence in the outcome of well-planned synthesis that may be caused by a stereochemically badly placed remote functional group.

5.2.3   Mylbemycin D [18]

5.124 mylbemycin D

Target relevance

(+)-Milbemycin D **5.124** is one of the members of the family of natural products commonly known as milbemycins, the first example of which was reported in 1975 [19]. Their properties as antiparasitic and insecticidal agents, together with their low toxicity, have made possible their use for the prevention and treatment of human and animal parasitic diseases. The complexity and rarity of the structures of the milbemycin family have encouraged numerous synthetic efforts.

Synthesis plan

The synthesis developed by Crimmins is based in previous observations on the chemistry of these compounds [20]. Thus, it was known that the Δ^3 double bond of the hexahydrobenzofuran tends to migrate to the C2 position, causing C2 epimerization or aromatization by dehydration. At the same time, oxidation at C1 must be performed prior to installment of C3 and C8 double bonds, since it is not possible to oxidize C1 if both C3 and C8 are sp^2-hybridized. According to this, **5.125** was noted as a likely advanced intermediate in the synthesis of (+)-milbemycin D **5.124**. Rearrangement of an allylic sulfoxide at C3 would eventually solve the problem of constructing the C3 unsaturation. Intermediate **5.125** should be accessed by connecting spiroketal **5.126** and hexahydrobenzofuran **5.127** moieties through a Wittig olefination to form the C10–C11 double bond (Scheme 5.43).

Predictable problems

The background for the coupling of fragments related to **5.126** and **5.127** through a Julia coupling or a Wittig olefination, have resulted in only modest yields and in the use of up to three-fold excess of one of the two components [21]. The expense of such an amount of **5.126** or **5.127**, is at this point, unthinkable. Therefore, maintaining the stoichiometric reaction is indispensable and reaction conditions have to be carefully controlled to achieve the maximum yield and the maximum stereoisomeric purity in this coupling step.

Scheme 5.43

Synthesis

Step 1. The Wittig coupling of **5.128** *and* **5.129** *and the elaboration of substrate for the sigmatropic coupling:* A careful study was undertaken to join fragments **5.128** and **5.129** using a Wittig olefination reaction. Treatment of aldehyde **5.129** with the phosphorane derived from **5.128** by deprotonation with BuLi gave compound **5.130** as a 2:1 mixture of the *E* and *Z*-isomers across the newly formed double bond in only 33% yield. Clearly, these conditions are not appropriate at this advanced stage of the synthesis. A dramatic improvement in the yield of the coupling was obtained by preparing the phosphorane from phosphonium salt **5.128** using the MeLi.LiBr complex. In these conditions the coupled product **5.130** was a 10:1 mixture of *E:Z* isomers in 84% yield. Furthermore, the quantitative isomerization to the *E* isomer was effected by exposure of the mixture to traces of iodine in benzene (Scheme 5.44).

The next step requires oxidation at C1 followed by macrolactonization, prior to undertaking the rearrangement of the C3 sulfoxide. Oxidation of C1 was attempted after removal (K_2CO_3/MeOH) of both the C1-TBS and C7-TMS groups, to yield diol **5.131**. The oxidation of **5.131** to the carboxylic acid **5.132** proved to be impossible. Therefore, both C1 and C7-hydroxyls were reprotected as TMS-derivatives. Slow silica gel chromatography produced selective cleavage of the primary TMS-group, affording **5.133**. Reaction of **5.133** with Pr_4NRuO_4 [22] and consecutive oxidation of the corresponding aldehyde with $NaClO_2$, produced acid **5.134** in 43% overall yield from **5.131**. Again, the failure of a simple alcohol oxidation resulted in a significant detour to effect the same transformation. Release of the C19-hydroxyl by reaction of **5.134** with TBAF followed by submission of the seco acid to Keck's conditions [23] and caused macrolactonization rendering lactone **5.135** in 47% yield. At this point of the synthesis, since only rearrangement of the allyl sulfide and inversion of the resultant C5-estereocenter remained to be done, in the author's words *the completion of the synthesis seemed imminent* (Scheme 5.44).

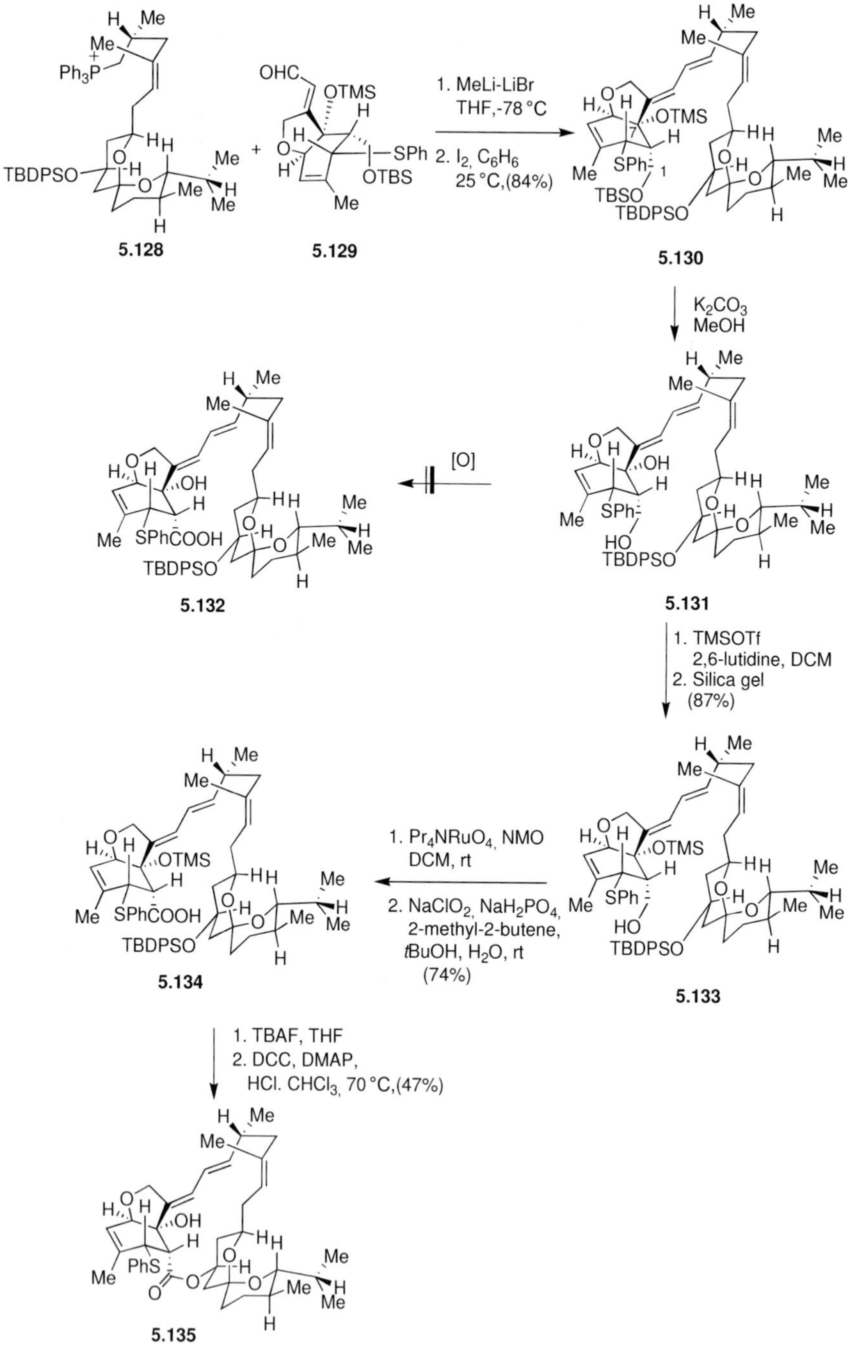
5.128
5.129
1. MeLi-LiBr THF,-78 °C
2. I2, C6H6 25 °C,(84%)
5.130
K2CO3 MeOH
5.131
[O]
5.132
1. TMSOTf 2,6-lutidine, DCM
2. Silica gel (87%)
5.133
1. Pr4NRuO4, NMO DCM, rt
2. NaClO2, NaH2PO4, 2-methyl-2-butene, tBuOH, H2O, rt (74%)
5.134
1. TBAF, THF
2. DCC, DMAP, HCl. CHCl3, 70 °C,(47%)
5.135

Scheme 5.44

Step 2. Sigmatropic rearrangement of the C3 sulfoxide on **5.136***:* It was anticipated that sulfoxide **5.136**, obtained by selective MCPBA oxidation of **5.135**, would render product **5.137** having the correct structure of (+)-milbemycin D. However, no rearranged product **5.137** was observed upon treatment of sulfoxide **5.136** with (MeO)₃P. The product of sulphoxide elimination, diene **5.138**, was produced instead (Scheme 5.45). Epimerization of either C2 or C3 favorably competes with the [2,3]-sigmatroic rearrangement, placing the C2 hydrogen and the sulfoxide group in a syn relationship suited for easy elimination. This fact causes the synthetic planning to reach a dead-end. Now, the reason for the failed approach is an active hydrogen capable of deprotonation and therefore of stereocenter inversion.

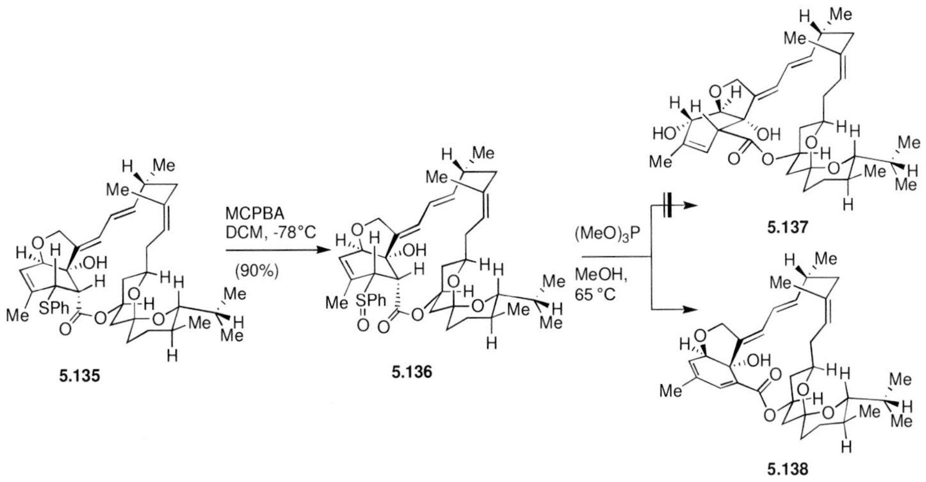

Scheme 5.45

Step 3. Revised plan of synthesis. Introduction of the troublesome C3–C4 double bond at the later stages of the synthesis: The new synthetic plan considered the introduction of the C3–C4 double bond in the later stages of the synthesis, using the aldehyde **5.139** in the coupling with the spiranic fragment **5.128**. The new strategy avoids the compromising sigmatropic rearrangement that is effected early during the preparation of **5.139**. Since the new fragment **5.139** lacks the double bond, oxidation problems at C1 will be avoided. Evidently, additional stages have to be included to rebuild the missing double bond.

    Joining of the aldehyde **5.139** with the spirane fragment **5.128** under the conditions described above (Scheme 5.44) afforded intermediate **5.140**, after double bond isomerization by I₂ treatment. Acid treatment of **5.140** caused deprotection of the primary C1 and C7 silyl ethers. Oxidation of the corresponding diol by a two-step sequence, yielded acid **5.141**. Selective release of the C19 over C5 hydroxyl proved to be necessary prior to macrolactonization. Therefore, **5.141** was treated with TBAF under controlled conditions and the alcohol submitted to Keck macrolactonization [23] conditions to produced macrolatone **5.142** in 53% yield. Transformation of **5.142** into milbemycin D **5.124** requires introduction of the C3–C4 double bond by means of selenylation and elimination on the ketone at C5. For this, the silyl ether at the secondary C5 was removed and the resulting hydroxyl was oxidized with Pr_4NRuO_4 yielding ketone **5.143**, the silyl enol ether of which was converted to the (1:1) mixture of selenides **5.144**. Taking into account Ley's observations, namely that the best endo:exo selectivity for the selenoxide elimination reaction [24] could be achieved if the α-isomer is used and the C(7) hydroxyl group is free, the β-isomer was recycled to ketone **5.143** to increase the yield of the α-selenide. Deprotection of the C7 hydroxyl group on the α-selenide isomer **5.144**, by the action of HF-pyridine, followed by reaction with sodium periodate and immediate reduction of the ketone, delivered a 6:1 mixture of milbemycin D **5.124** and the exo-olefin isomer (Scheme 5.46).

Evaluation

- A major strategic change has to be effected during synthesis of (±)-milbemycin **5.124**. The original synthetic planning was designed to achieve a [2, 3] sigmatropic rearrangement on macrolactone sulfoxide **5.136**. However, the epimerization of C2 or C3 results in an intermediate having a *syn* relationship between the sulfoxide and the C2 hydrogen, thereby allowing thermal elimination to occur. This fact was surprisingly overlooked. The alternative, and finally successful route to (+)-milbemycin D involved the removal of the alkene present in the six-membered ring of the furane fragment **5.139**, to avoid the elimination problems. Obviously, most of the work had to be restarted from the beginning. The C3–C4 olefin was then introduced at the end of the synthesis, through selenylation and elimination.

5.137   5.136   5.138

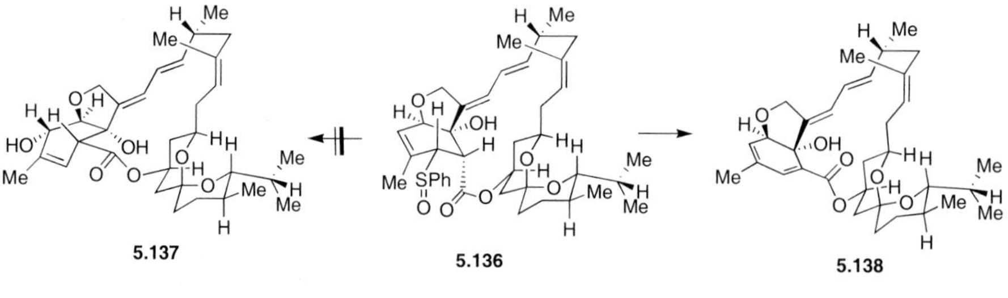

- There is another interesting point in the synthesis of (±)-milbemycin **5.124**: the impossibility of oxidizing **5.131** to acid **5.132**. This fact is quite intriguing and should derive from the lability of the tertiary free hydroxyl group, since the protection of this alcohol (which assumes a few more synthetic steps) solves the problem.

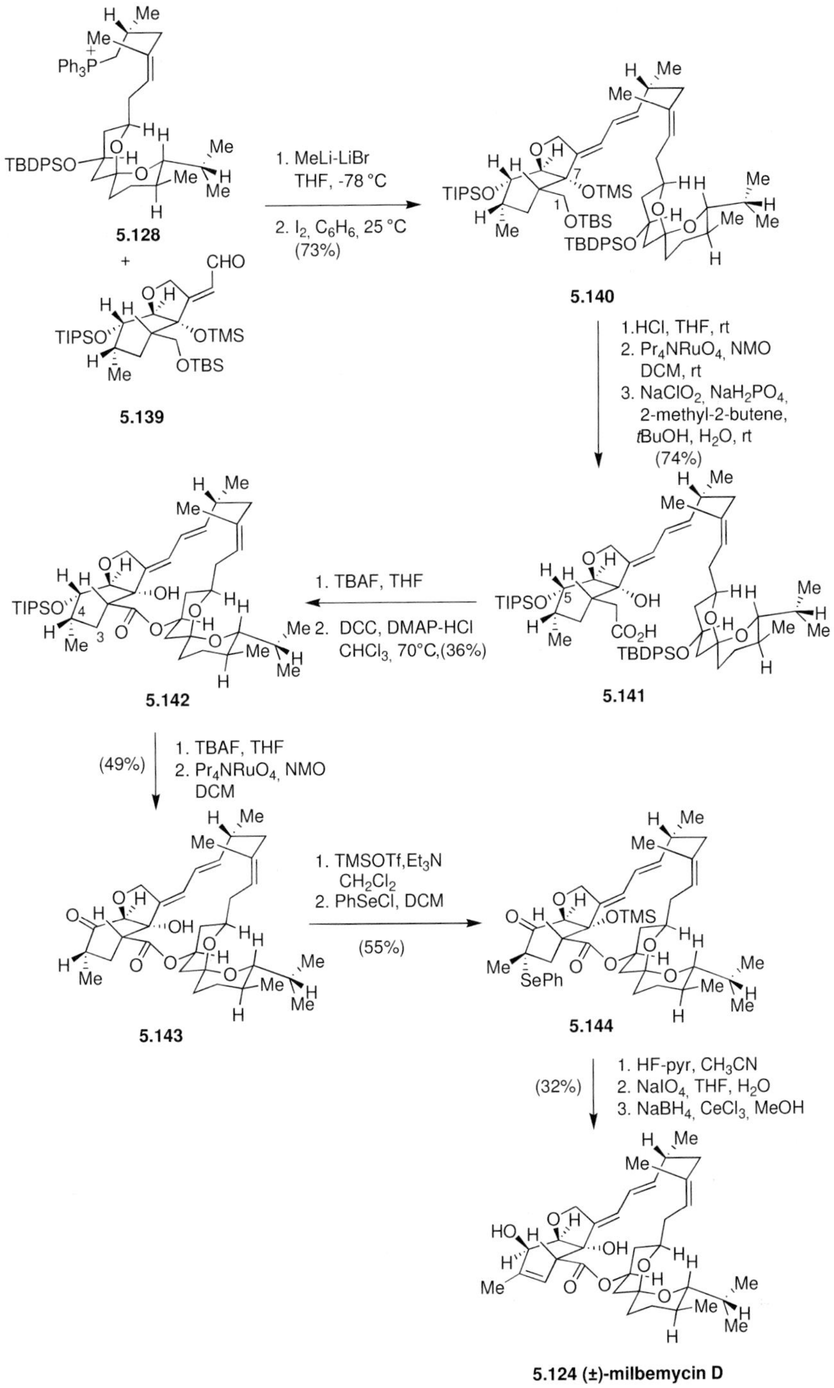
5.128
5.139
1. MeLi-LiBr THF, -78 °C
2. I_2, C_6H_6, 25 °C (73%)
5.140
1. HCl, THF, rt
2. Pr_4NRuO_4, NMO DCM, rt
3. $NaClO_2$, NaH_2PO_4, 2-methyl-2-butene, tBuOH, H_2O, rt (74%)
5.141
1. TBAF, THF
2. DCC, DMAP-HCl $CHCl_3$, 70°C,(36%)
5.142
1. TBAF, THF
2. Pr_4NRuO_4, NMO DCM
(49%)
5.143
1. TMSOTf,Et_3N CH_2Cl_2
2. PhSeCl, DCM
(55%)
5.144
1. HF-pyr, CH_3CN
2. $NaIO_4$, THF, H_2O
3. $NaBH_4$, $CeCl_3$, MeOH
(32%)
5.124 (±)-milbemycin D

Scheme 5. 46

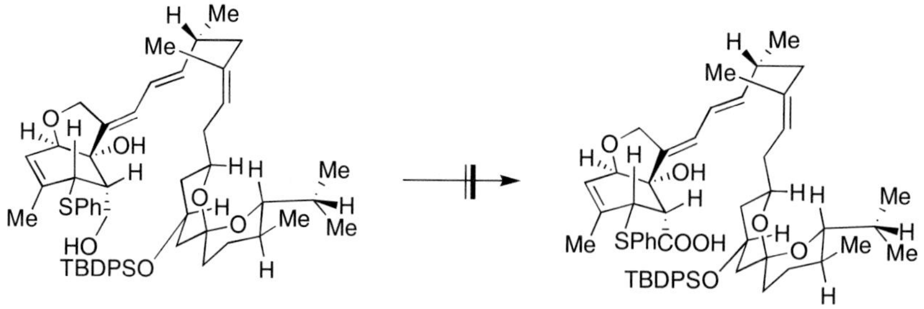

5.131 → **5.132**

Particularly interesting reagents

Ley oxidation reagent (aka Ley–Griffith Reagent) [22], (Pr₄NRuO₄/NMO/4Å-Ms):
Pr₄NRuO₄ is abbreviated TPAP and used in catalytic amounts. An extremely efficient
reagent for the oxidation of alcohols, to the corresponding carbonyl groups at rt, in the
presence of sensitive functional groups. The reagent is used in organic solvents and the
presence of molecular sieves usually favors the reaction.

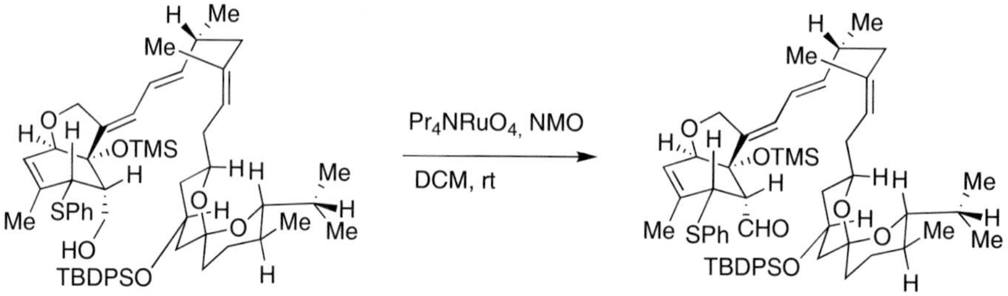

Pr₄NRuO₄, NMO

DCM, rt

5.133   **5.145**

References

1. Crimmins, M. T.; Wang, Z.; McKerlie, L. A. *J. Am. Chem. Soc.* **1998**, *120*, 1747.
2. a) Iwata, C.; Takemoto, Y.; Kuboto, H.; Yamada, M.; Uchida, S.; Tanaka, T.; Imanashi, T. *Chem. Pharm. Bull.* **1989**, *37*, 866; b) Iwata, C.; Takemoto, Y.; Kuboto, H.; Yamada, M.; Uchida, S.; Tanaka, T.; Imanashi, T. *Chem. Pharm. Bull.* **1988**, *36*, 4581; c) Iwata, C.; Kuboto, H.; Yamada, M.; Takemoto, Y.; Uchida, S.; Tanaka, T.; Imanashi, T. *Tetrahedron Lett.* **1984**, *25*, 3339.
3. Dowd, P.; Zhang, W. *Chem. Rev.* **1993**, *93*, 2091.
4. Chu-Moyer, M. Y.; Danishefsky, S. J.; Schulte, G. K. *J. Am. Chem. Soc.* **1994**, *116*, 11213,
5. Nakagawa, M.; Hsu, Y.-H.; Hirota, A.; Shima, S.; Nakayama, M. *J. Antibiot.* **1989**, *42*, 2128.
6. a) Woodward, R. B.; Bader, F. E.; Bickel, H.; Frey, A. J.; Kierstead, R. W. *Tetrahedron* **1958**, *2*, 1; b) Bohlmann, F.; Mathar, W.; Schwarz, H. *Chem. Ber.* **1977**, *110*, 2028; c) Hayakawa, K.; Ueyama, K.; Kanematsu, K. *J. Chem. Soc., Chem. Commun.* **1984**, 71.
7. a) Burke, S. D.; Powner, T. H.; Kageyama, M. *Tetrahedron Lett.* **1983**, *24*, 4529.
8. Grieco, P. A.; Collins, J. L.; Moher, E. D.; Fleck, T. J.; Gross, R. S. *J. Am. Chem. Soc.* **1993**, *115*, 6078.
9. Polonsky, J.; Bourguignon-Zylber, N. *Bull. Soc. Chim. Fr.* **1965**, 2793.
10. Polonsky, J. *Prog. Chem. Org. Nat. prod.* **1985**, *47*, 22.
11. Angle, S. R.; Fevig, J. M.; Knight, S. D.; Marquis Jr, R. W.; Overman, L. E. *J. Am. Chem. Soc.* **1993**, *115*, 3966.
12. a) For a comprehensive review see: Husson, H. P. *Indoles: Monoterpene Alkaloids*, Saxon, J. E., Ed.; Wiley: New York, 1983; Chapter 7; b) Henry, T. A.; Sharp, T. M. *J. Chem. Soc.* **1927**, 1950; c) Edwards, P. N.; Smith, G. F. *Proc. Chem. Soc., London* **1960**, 215.
13. a) Millson, P.; Robinson, R.; Thomas, A. F. *Experientia* **1953**, *9*, 89; b) Robinson, R.; Thomas, A. F. *J. Chem. Soc.* **1955**, 2049; c) Robinson R.; Aghoramurthy, K. *Tetrahedron* **1957**, *1*, 172.
14. Edwards, P. N.; Smith, G. F. *J. Chem. Soc.* **1961**, 152.
15. For reviews see: a) Overman, L. E.; Ricca, D. J. *Comp. Org. Synth.* **1991**, *2*, 1007; b) Overman, L. E. *Acc. Chem. Res.* **1992**, *25*, 352.
16. Jones, G.; Stanforth, S. P. *Org. React.* **1997**, *49*, 1.
17. Cardwell, K.; Hewitt, B.; Ladlow, M.; Magnus, P. *J. Am. Chem. Soc.* **1988**, *110*, 2242.
18. Crimmins, T. M.; Al-awar, R. S.; Vallin, I. M.; Hollis Jr, W. G.; O'Mahony, R.; Lever, J. G.; Bankaitis-Davis, D. M. *J. Am. Chem. Soc.* **1996**, *118*, 7513.
19. a) Mishima, H.; Kurabayashi, M.; Tamura, C.; Sato, S.; Kuwano, H.; Saito, A. *Tetrahedron Lett.* **1975**, 711; b) Takiguchi, Y.; Ono, M.; Muramatsu, S.; Ide, J.; Mishima, H.; Terao, M. *I. Antibiot.* **1983**, *36*, 502; c) Mishima, H.; Junya, I.; Muramatsu, S.; Ono, M.; *J. Antibiot.* **1983**, *36*, 980; d) Ono, M.; Mishima, H.; Takugichi, Y.; Terao, M. *J. Antibiot.* **1983**, *36*, 991.

20. a) Fraser-Reid, B. O.; Wolleb, H.; Faghih, R.; Barchi, J., Jr. *J. Am. Chem. Soc.* **1997**, *109*, 933; b) Pivnichny, J. V.; Arison, B. H.; Preiser, F. A.; Shim, J.-S. K.; Mrozik, H. *J. Agric. Food Chem.* **1988**, *36*, 826; c) Hirama, M.; Noda, T.; Yasuda, S.; Ito, S. *J. Am. Chem. Soc.* **1991**, *113*, 1830.
21. a) Hanessian, S.; Dube, D.; Hodges, P. J. *J. Am. Chem. Soc.* **1987**, *109*, 7063; b) Hanessian, S.; Ugolini, A.; Hodges, P. J.; Beaulieu, P.; Dube, D.; Andre, C. *Pure Appl. Chem.* **1987**, *59*, 299; c) White, J. D.; Bolton, G. L. *J. Am. Chem. Soc.* **1990**, *112*, 1626; d) White, J. D.; Bolton, D. L.; Dantanarayana, A. P.; Fox, C. M. J.; Hiner, R. N.; Jackson, R. W.; Sakuma, K., Warrier, U. S. *J. Am. Chem. Soc.* **1995**, *117*, 1908; e) Hirama, M.; Noda, T.; Yasuda, S.; Ito, S. *J. Am. Chem. Soc.* **1991**, *113*, 1830.
22. Griffith, W. P.; Ley, S. V. *Aldrichim. Acta* **1990**, *23*, 13.
23. Boden, E. P.; Keck, G. *J. Org. Chem.* **1985**, *50*, 2394.
24. a) Armstrong, A.; Ley, S. V. *Synlett* **1990**, 323; b) Diez-Martin, D.; Grice, P.; Kolb, H.C.; Ley, S. V.; Madin, A. *Synlett* **1990** 326; c) Ley, S. V.; Armstrong, A.; Díez-Martín, D.; Ford, M. J.; Grice, P.; Knight, J. G.; Kolb, H. C.; Madin, A.; Marby, C. A.; Mukherjee, S.; Shaw, A. N.; Slawin, A. M. Z.; Vile, S.; White, A. D.; Williams, D. J.; Woods M. *J. Chem. Soc., Perkin Trans. 1* **1991**, 667.

Chapter 6
The Elusive Side Chain

Many synthetic routes are designed to build the fully functionalized, fully protected main carbon skeleton of a target product. To complete the synthesis a few carbons (as few as one) in a side chain may remain to be included at a late stage. How easy is it to join these "harmless" side chains to the main body of the target molecule? Apparently, it should be quite easy, as very few comments about this question are found in the papers that deal with total synthesis. Nothing could be further from the truth. The problem of introducing a side chain is even more daunting since it occurs in the last steps of the synthesis, a fact that often involves making a complete strategic change or at least a significant detour. Let us discuss some of these situations. We will begin with synthesis of (±)-scopadulcic acid B as a good example of how difficult it may be to join a methyl group to a very elaborate molecular skeleton.

6.1   (±)-Scopadulcic Acid B [1]

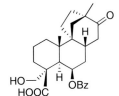

| 6.1 (±)-scopadulcic acid B | 6.2 (±)-scopadulcic acid A | 6.3 (±)-scopadulciol |

Dead Ends and Detours: Direct Ways to Successful Total Synthesis
Miguel A. Sierra and María C. de la Torre
Copyright © 2004 WILEY-VCH Verlag GmbH & Co. KGaA, Weinheim
ISBN: 3-527-30644-7

Target relevance
Scopadulcic acid B **6.1** is a tetracyclic diterpene that contains a bicyclo[3.2.1]octane substructure which is also characteristic of many other diterpenes. Scopadulcic acid B **6.1** was isolated from *Scoparia dulcis* L. and, together with scopadulcic acid A **6.2** and scopadulciol **6.3**, it is responsible for the biological activity associated with this plant [2]. Scopadulcic acid B **6.1** shows activity as H^+,K^+-adenosine triphosphate and against herpes simplex virus type-1 (HSV-1).

Synthetic plan
The tetracyclic skeleton of scopadulcic acid B **6.1** will derive from a tandem intramolecular Heck/cyclization on the dienyl aryl iodide **6.4**, which would produce the tetracyclic ring system **6.5** in a single step. The key intermediate **6.4** would derive from the enoxysilane **6.6** (Scheme 6.1).

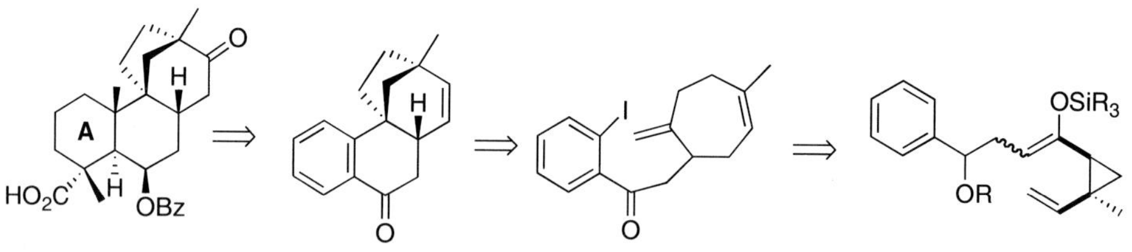

Scheme 6.1

Predictable problems
- *A risky tandem intramolecular Heck/cyclization.* By the time that this strategy was designed (1988), only one precedent of a tandem intramolecular Heck/cyclization had been reported in the literature [3]. Therefore, this was considered a quite intrepid strategy. Although the Heck reaction [4] has proved to be extremely useful for the synthesis of complex molecules and for the construction of quaternary centers, this synthesis relied on an unprecedented tandem reaction [5].
- *Functionalization of ring A.* The ring A of the target molecule is introduced as an aryl ring that may be not adequate for the complete development of the full functionality of a scopadulan A ring.

Synthesis

Step 1. Preparation of dienyl aryl iodide 6.4 (Scheme 6.2):

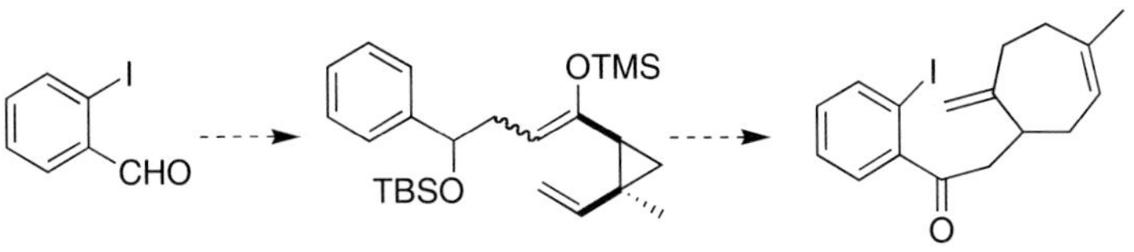

Scheme 6.2

The key intermediate **6.4**, on which the Heck cyclization was to be effected, was planned to be accessed by using a divinylcyclopropane rearrangement [6] on the cyclopropane **6.8**. Compound **6.8** would be prepared from commercially available 2-iodobenzaldehyde **6.7**.

The synthesis of **6.4** began with the reaction of **6.7** with allylmagnesium bromide, followed by protection of the resulting alcohol as TBS-derivative. Hydroboration of **6.9** followed by Swern oxidation [7] provided aldehyde **6.10**. Addition of the lithium reagent derived from the *cis/trans* mixture of cyclopropylbromides **6.11/6.12** yielded alcohols **6.13–6.14** in low yield, though, due to the competing reaction with the aryl-iodide group. The result on the formation of alcohols **6.13–6.14** was not satisfactory in terms of yield. It was found that the organomagnesium derivatives formed from **6.11–6.12**, produced an increment in the yield of alcohols **6.13/6.14** which, without further purification, were submitted to oxidation with PCC affording the required *cis* and *trans*-cyclopropyl ketones **6.15/6.16** (Scheme 6.3).

The mixture of isomeric ketones was treated with TMSOTf and the enoxysilane derivatives **6.17/6.18** were heated in refluxing benzene. As expected, only the enoxy-silane **6.17** derived from the *cis*-isomer underwent the divinylcyclopropane rearrangement, yielding, after selective cleavage of the enol ethers, cycloheptenone **6.19** in 51% yield together with 26% of unreacted *trans*-cyclopropyl ketone **6.16**. Hydrolysis of **6.19** formed compound **6.20**, which was transformed into the desired **6.7** by methylenation with Wittig methyleneylide followed by TBAF-removal of the silyl ether group and oxidation of the resulting alcohol with PCC (Scheme 6.4).

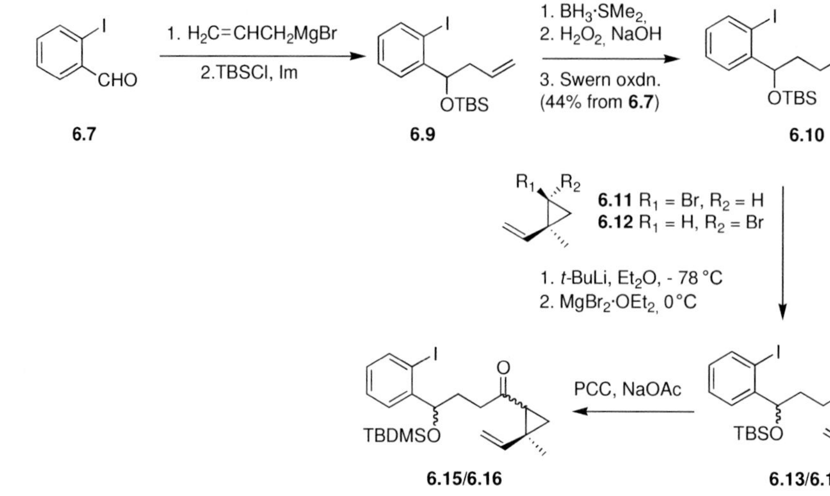

Scheme 6.3

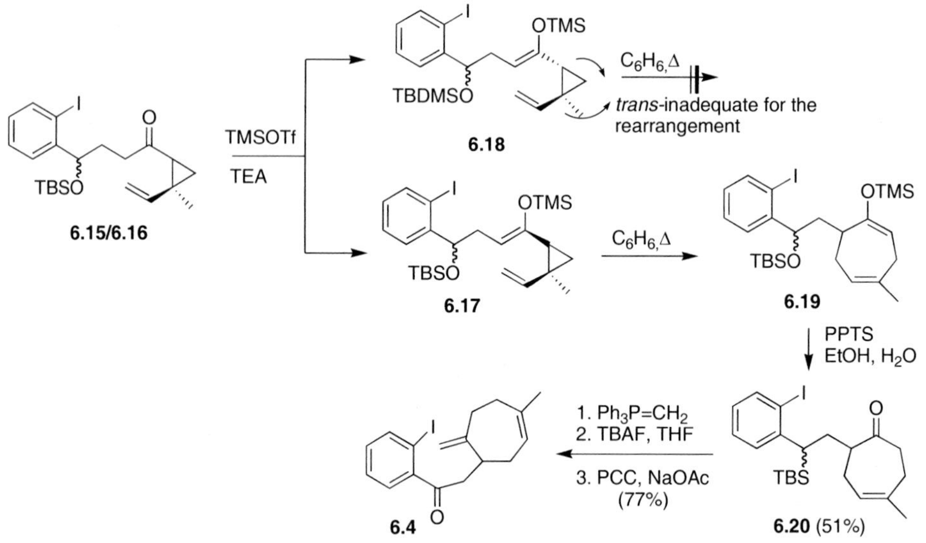

Scheme 6.4

Step 2. The intramolecular tandem Heck/cyclization and the synthesis of the tetracyclic core of scopadulcic acid B (Scheme 6.5):

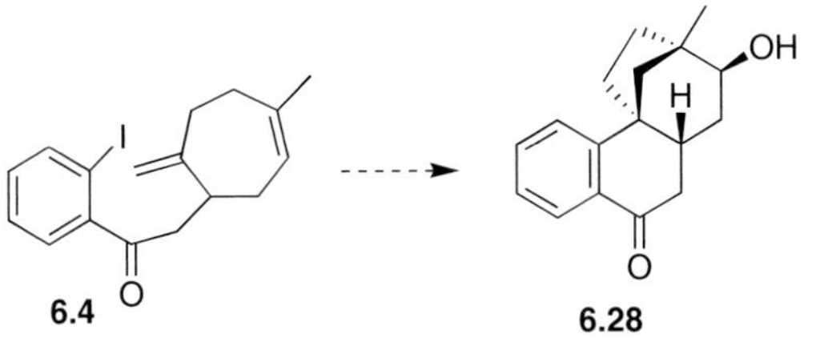

Scheme 6.5

The Heck coupling of **6.4** was achieved by using 10% Pd(OAc)₂, 20% of Ph₃P and an excess of TEA in refluxing MeCN. These conditions ensure reproducible and high yields of enones **6.24** and its conjugated isomer in 3.1:1 ratio. Having successfully accessed to the tetracyclic intermediate **6.24**, introduction of the oxidation at C13, the elaboration of ring A and finally, the introduction of the quaternary C10 Me-group, are all that remain for the completion of the synthesis. The functionalization of ring C was undertaken first. The mixture of ketones was treated with DDQ producing dienone **6.25**. Reaction with MCPBA formed exclusively the epoxide product **6.26** arising from oxidation of the γ,δ double bond from the less hindered β-face. Reduction of the epoxide with NaTeH yielded the C13 β-alcohol **6.27** in 42% overall yield from the mixture of the ketones obtained during the Heck cyclization. The saturation of the Δ⁷ double bond was achieved by reduction of the enone **6.27** with H₄AlLi in THF. One single diastereomer (the one having the desired stereochemistry) **6.28** was obtained in 73% yield (Scheme 6.6).

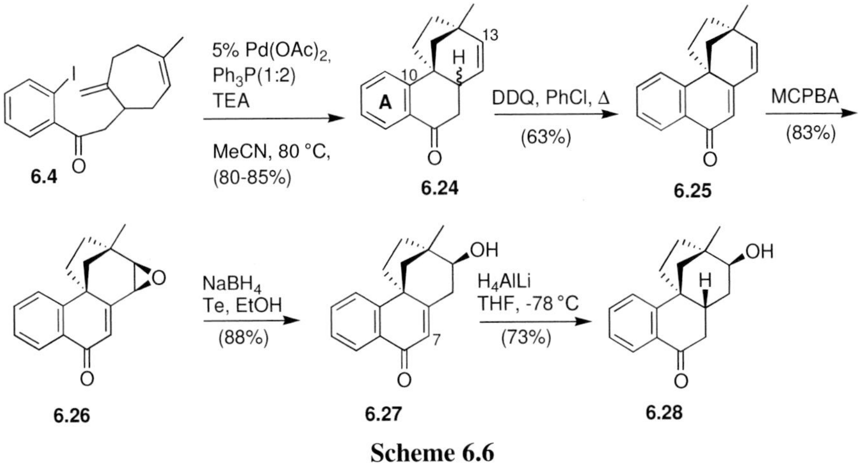

Scheme 6.6

Step 3. Functionalization of ring A and elaboration of the substrate for the final introduction of the C10 Me-group (Scheme 6.7):

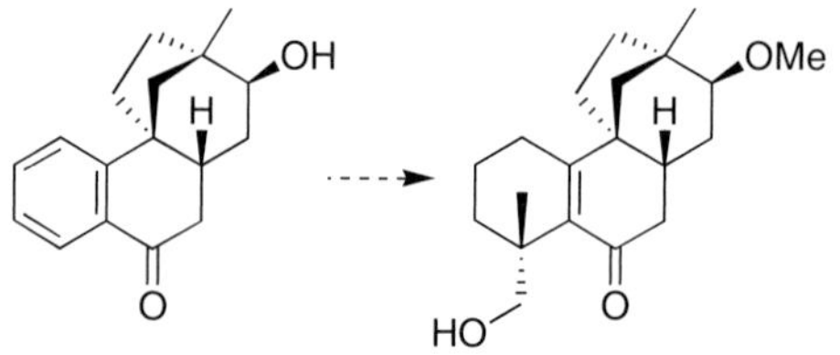

6.28   **6.29**

Scheme 6.7

Functionalization of ring A and the introduction of the angular methyl group are the tasks remaining to end the synthesis of (±)-scopadulcic acid B **6.1**. To functionalize the aromatic ring was used a Birch reduction [8]. This process was attempted on intermediate **6.30** delivered after protection of the C13 alcohol and the benzylic ketone as methyl ether and imidazoline, respectively. Reaction of **6.30** with excess Li in NH_3-THF containing *t*BuOH, followed by cleavage of the diamine protecting group, and further hydrogenation of the disubstituted double bond delivered a low yield (15–25%) of **6.31** (Scheme 6.8).

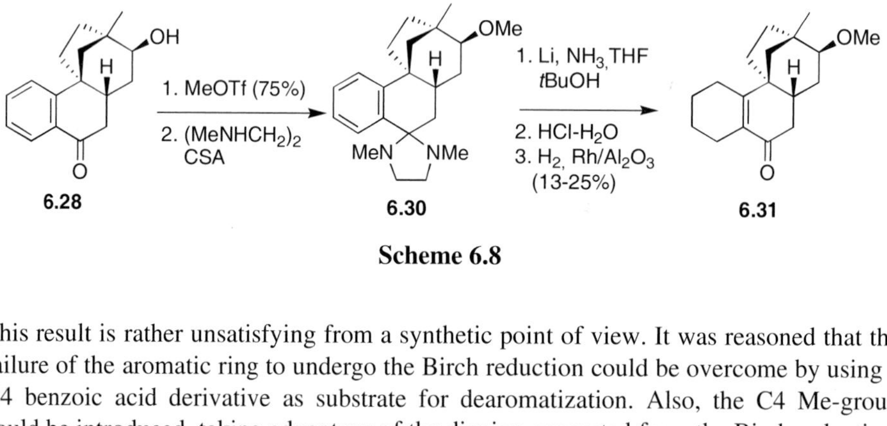

6.28   **6.30**   **6.31**

Scheme 6.8

This result is rather unsatisfying from a synthetic point of view. It was reasoned that the failure of the aromatic ring to undergo the Birch reduction could be overcome by using a C4 benzoic acid derivative as substrate for dearomatization. Also, the C4 Me-group could be introduced, taking advantage of the dianion generated from the Birch reduction. The desired acid **6.32** was accessed from alcohol **6.33**, obtained by reduction of ketone **6.34**, by reaction of the dianion generated by an excess of BuLi in refluxing TMDA-pentane with solid CO_2. A 10% yield of lactone **6.35**, which could be converted in acid **6.32**, was also obtained in this reaction. Birch reduction of **6.32** and *in situ* methylation was followed by hydrogenation over Rh/Al_2O_3 producing a 65% of lactone **6.36**, on which the C4 methyl group has been introduced exclusively from the β-face. The reduction of lactone **6.36** with H₄AlLi and the oxidation of the C6 hydroxyl group rendered intermediate **6.37** (Scheme 6.9).

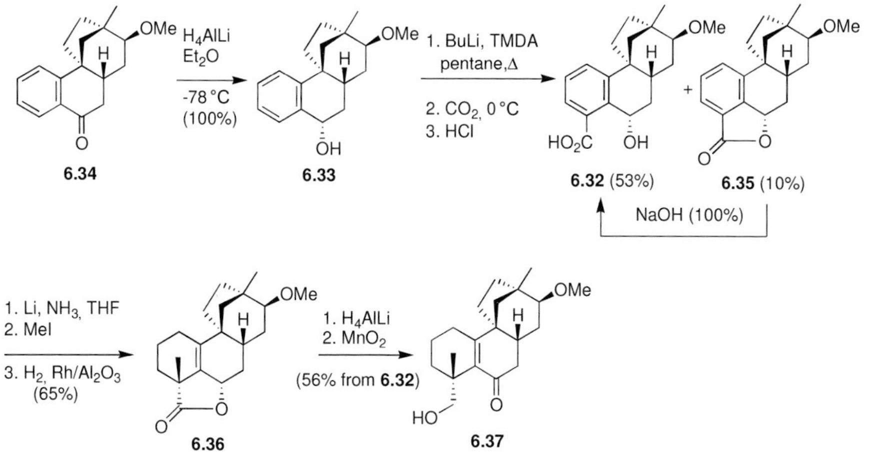

Scheme 6.9

Step 4. Introduction of the quaternary C10-Me group: From the beginning, it was anticipated that introduction of the quaternary methyl at C10 would be a difficult task, since C10 was adjacent to a quaternary center of the bicyclo[3.2.1]octane of **6.37**. Nevertheless, reaction of enone **6.37** with Me₂CuLi was attempted, since the 1,4-adduct **6.40** was produced in good yield in the reaction of β,β-disubstituted enone **6.39** with Me₂CuLi (Scheme 6.10).

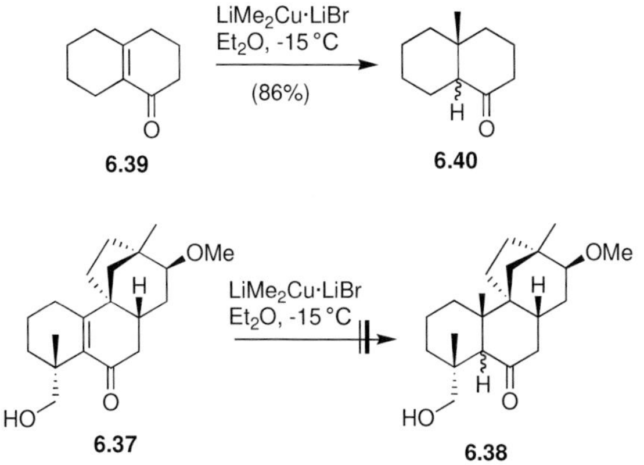

Scheme 6.10

Disappointingly, compound **6.37** did not react at all, under similar reaction conditions. Enone **6.37** failed to produce the desired **6.38** in the many different reagents and reaction conditions used, including Me$_2$CuLi/TMSCL, MeCu,Bu$_3$P, "higher order" cyano methyl cuprates, BF$_3$·Et$_2$O-catalysed cuprate addition, and nickel acetylacetonate-catalysed addition of dimethylzinc. Therefore, the introduction of the C20-Me group was unfeasible by the planned route. The introduction of the C20-Me group required a different tactic, and a detour had to be taken.

Step 5. Alternate procedure for the introduction of the C20-Me group and completion of the synthesis: It was decided to introduce the C20-Me group of the scopadulane skeleton as cyanide and reduce it afterwards. With this aim, Et$_2$AlCN was chosen as reagent since it is very effective for conjugate addition to highly congested centers [9]. Reaction of **6.37** with Et$_2$AlCN in fact produced the 1,4-adduct **6.41** in 48% yield. At this point, the stereochemistry of the addition was not unambiguously established but tentatively assigned according to the, well known, axial bias of the addition. Reaction of intermediate **6.41** with LiAlH$_4$ in refluxing THF, produced the β-alcohol at C6 and, unexpectedly, the C20 was blocked as aminal giving product **6.42**. The resulting **6.42** was submitted to Wolff–Kishner reduction providing **6.43** in 74% yield from **6.41**. Selective protection of the primary alcohol as silylether and benzoylation of the secondary axial hydroxyl group, afforded **6.44**. Cleavage of the silylether followed by oxidation of the primary alcohol with RuO$_4$, produced (±)-scopadulcic acid B **6.1** (Scheme 6.11).

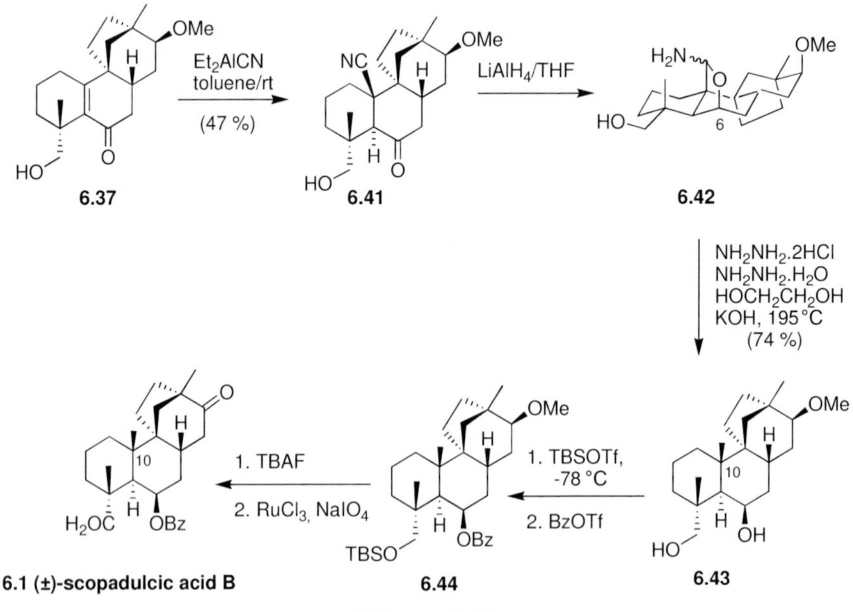

Scheme 6.11

Evaluation

In this case, the impossibility of introducing the remaining C10-Me group in the final stages of the synthesis has been circumvented by changing the Me group by a cyanide group, which is a better nucleophile. This modification entails the need for elaboration of the cyanide group to the Me group, adding steps to the synthesis. This is not a general solution because the required reduction must be compatible with the remaining functionalities of the molecule. In this particular case, the fortunate unexpected result of the reduction of the cyano-group, leading to aminal **6.42**, facilitates the final installment of the angular Me group.

6.37 → (LiMe₂Cu·LiBr, Et₂O, -15 °C) → 6.38

Key synthetic reactions

Divinylcyclopropyl rearrangement: A class of the general [3,3]-sigmatropic rearrangements closely related to the Cope and Claisen rearrangement. The final product is a 1,4-cycloheptadiene and the requisite for this concerted rearrangement is that both vinyl groups should be in a *cis*-disposition [6].

6.18 — C₆H₆, Δ — *trans*-inadequate for the rearrangement

6.17 — C₆H₆, Δ → **6.19**

The problem of placing a pendant chain in a preformed cyclic system is quite general. Many of the standard nucleophilic additions failed, probably due to the topology of the molecule or to the high steric hindrance of the center in which the chain is to be placed. The synthesis of (±)-scopadulcic acid can be solved with a significant detour using different nucleophiles. Other times, as in the following case, the pendant chain has to be built stepwise, adding many steps to the overall synthetic scheme.

6.2   Dysidiolide [10]

6.45 dysidiolide

Target relevance

Dysidiolide **6.45** was isolated from the sponge *Dysidea ethereal* by Gunasekara and Clardy [11]. It shows *in vitro* activity against A-549 human lung carcinoma and P338 murine leukemia, probably due to inhibition of cdc25A protein phosphatase. Dysidiolide belongs to the sesterterpene class of natural products and it was the first member of a new structural type not previously encountered in nature.

Synthetic plan

The synthesis utilizes a linear strategy consisting of the introduction of the appendages at C1 and C5 on a performed [4.4.0] byciclic nucleus. The byciclic core of dysidiolide **6.47** arises from decaline **6.46** through a biomimetic rearrangement, which simultaneously generates the C1 quaternary centre and the endocyclic double bond. The bicyclic ketal enone **6.48** is the starting point of the synthesis (Scheme 6.12).

Predictable problems

The designed plan features a compromising biomimetic carbocation rearrangement [11] to create an unusual quaternary center at C1 as well as the endocyclic double bond. All these simultaneous transformations should occur within a highly substituted bicyclic core.

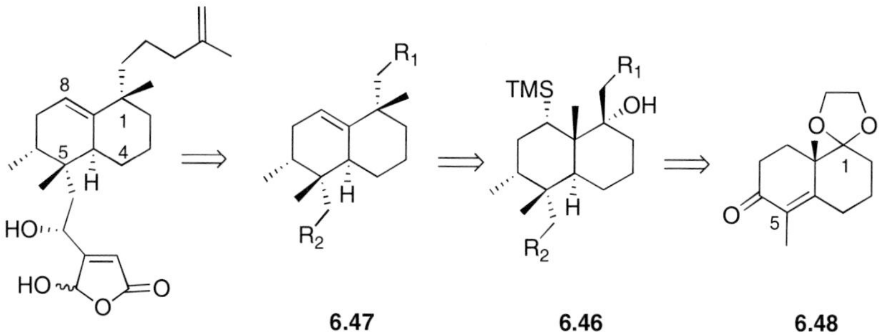

Scheme 6.12

*Step 1. Functionalization of C5 in the decaline core. Formation of intermediate **6.49** (Scheme 6.13):*

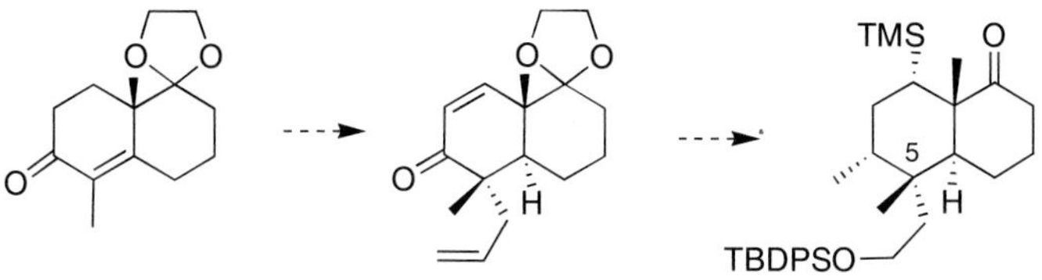

Scheme 6.13

The first step of the synthesis is the introduction of the chain at C5. Birch reduction of enone **6.48** and trapping of the resulting enolate with allylbromide afforded the *trans*-decaline **6.51** as a single diastereoisomer in 82% yield. Introduction of a silane group at the C8 position was required to trigger the rearrangement that would produce the quaternary centre at C1 as well as the Δ^8 double bond. For this, ketone **6.51** was transformed into the α,β-unsaturated derivative **6.50** by sequential deprotonation with LDA, phenylsulfenation with diphenyl disulfide, oxidation with MCPBA, and elimination of the resulting α-phenylsulfinyl ketone by heating in the presence of $(MeO)_3P$. Reaction of enone **6.50** with an excess of LiTMS produced conjugate addition affording exclusively the axial α -TMS ketone **6.52**. Prior to introduction of the chain at the C1 position, the C6-Me group was introduced by reaction of **6.52** with methylene-triphenylphosphorane. The allyl appendage was transformed into a 2-hydroxyethyl group by Sharpless dyhydroxylation of the vinyl group, followed by glycol cleavage with NaIO$_4$. Final reduction of the aldehyde thus formed, provided alcohol **6.53** in 91% overall yield. Release of the protected ketone and protection of the alcohol as silyl ether were followed by hydrogenation with the Wilkinson's catalyst providing intermediate **6.49** (Scheme 6.14).

Having accessed intermediate **6.49** the next goal was the introduction of the chain at position C1.

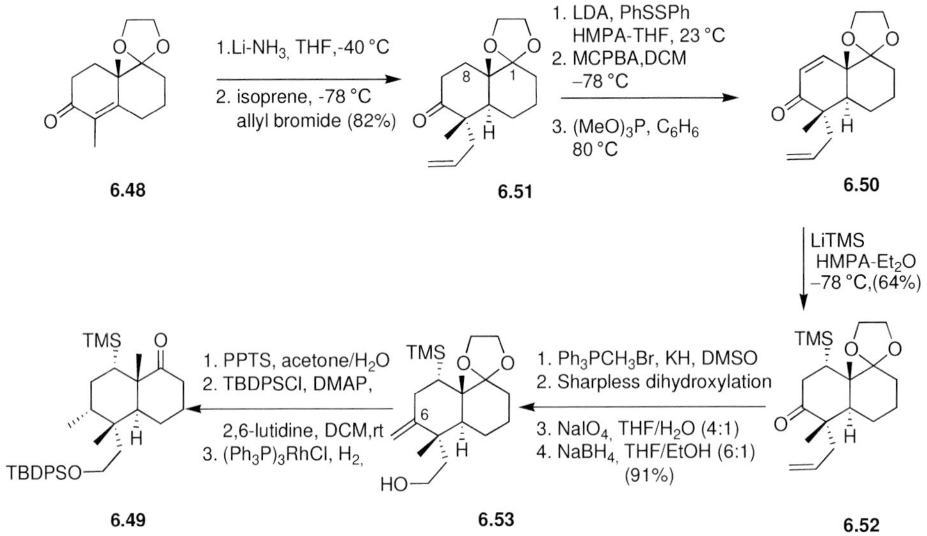

Scheme 6.14

Step 2. Placing the appendage at C1, an apparently innocuous task (Scheme 6.15):

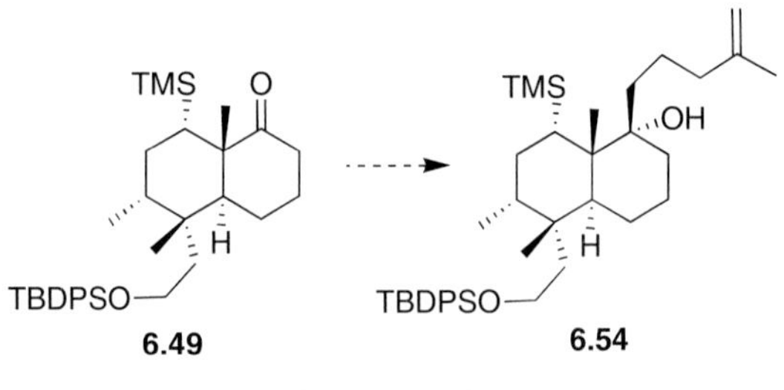

Scheme 6.15

The introduction of the side chain at C1 seems obvious. The straight nucleophilic addition of 4-methyl-4-pentenyl lithium to ketone **6.49** produced only the recovery of starting material. None of the desired product **6.54** was observed. The same result derived from the use of the corresponding magnesium or cerium reagents. This is one of the most unpleasant situations in a synthesis, namely the lack of reactivity of a given substrate towards a key transformation, since it represents a dead-end. The problem had to be solved by effecting a serious tactical detour.

In view of the impossibility of introducing the side chain in one step, ketone **6.49** was reacted with allylmagnesium bromide with the aim of completing the elaboration of the full chain later in the synthesis. Treatment of **6.49** with allylmagnesium bromide gave stereospecifically the tertiary alcohol **6.55** in excellent yield. Therefore, why cannot the 4-methyl-4-pentenyl fragment be introduced in the same way? (Scheme 6.16). No answer to this question is available.

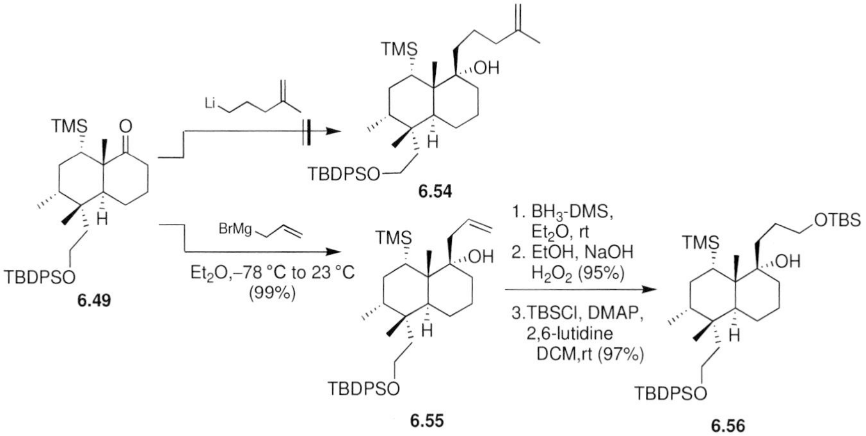

Scheme 6.16

The chain at C1 of **6.55** was transformed into the TBS-protected 3-hydroxy-propyl group through a hydroboration–oxidation sequence, followed by selective reaction of the primary alcohol with TBSCl. In this way, compound **6.56** was obtained in 92% overall yield. Intermediate **6.56** was used as the substrate for the biomimetic rearrangement that would yield the hydrocarbon core of dysidiolide. The full description of the side chains will be delayed to the later steps of the synthesis.

*Step 3. Biomimetic rearrangement of **6.56** and completion of the synthesis (Scheme 6.17):*

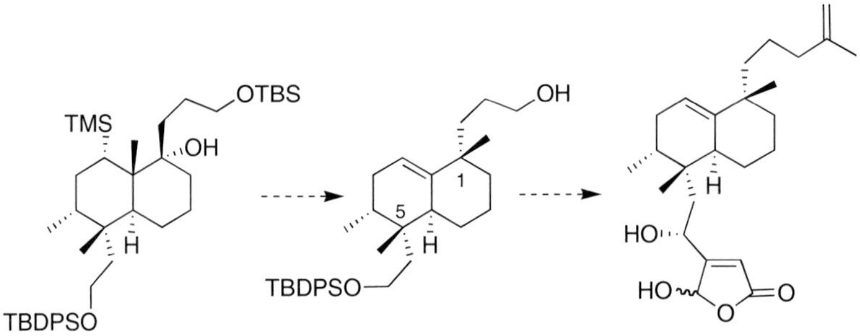

Scheme 6.17

The critical biomimetic rearrangement on which the synthetic planning rests, will be effected on intermediate **6.56**. The tertiary alcohol **6.56** was treated with an excess of BF_3 followed by reaction with PPTS to produce the cleavage of the silyl ether, yielding the desired **6.57** in 70% overall yield for the two steps. The TMS group at carbon C8 plays a double role in the rearrangement. First, the TMS group facilitates the migration of the Me group by stabilization of the emerging cationic center. In addition, the subsequent easy elimination of the TMS ensures the correct location of the double bond. The rearrangement only works with the TMS derivative. The analogue DMPS does not allow elimination as replacement of the DMPS group by $HOMe_2Si$ is the main reaction pathway. Having accessed to the carbocyclic core of dysidiolide, transformation of intermediate **6.57** into the target molecule required elaboration of the two side chains at carbons C1 and C5. The side chain at C1 was responsible for a major tactical drawback because its introduction thwarts the biomimetic rearrangement. This drawback is due to the competition between the cation–olefin cyclization and the rearrangement of the Me group. The C1 side chain was completed by coupling of the vinyl cuprate derived from 2-lithiopropene, with the iodide derived from alcohol **6.57**. Compound **6.58** has the full C1 side chain and was obtained in 94% overall yield from **6.57**.

The hydroxybutenolide moiety was introduced as a β-substituted furan ring. Cleavage of the TBDPS group was followed by oxidation of the resulting alcohol with Dess–Martin periodinane. The aldehyde reacted with 3-lithiofuran to afford a mixture (1:1) of epimeric alcohols **6.59** separable by silica gel chromatography. Finally, photochemical oxidation of furane **6.59** provided dysidiolide **6.45** in 98% yield (Scheme 6.18).

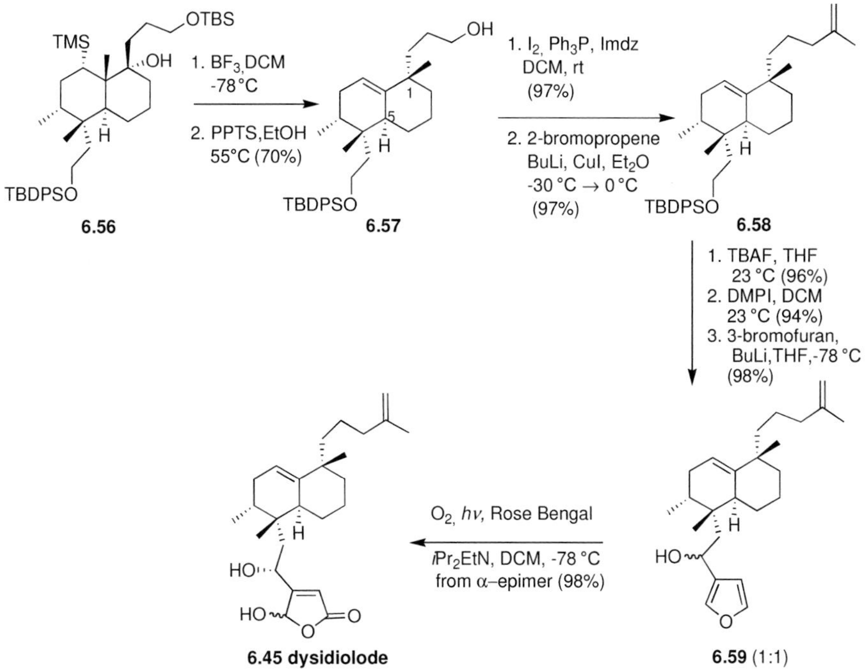

Scheme 6.18

Evaluation

The reluctance of the 4-methy-4-pentenyl lithium to react with ketone **6.49** forces the sequential introduction of the side chain at C1. This fact means that five more steps have to be added to the synthesis. Nevertheless, it is interesting to note that, even if the placement of the appendage succeeds and the major tactical drawback had not occurred, the side chain at C1 would have thwarted the rearrangement since cation–olefin cyclization onto the exocyclic olefin double bond, competes successfully with the movement of the Me group to form a spiro ring.

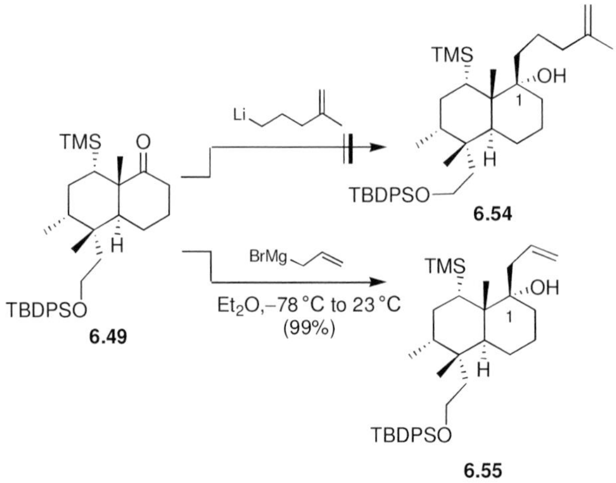

6.3   (+)-Epoxydictymene [13]

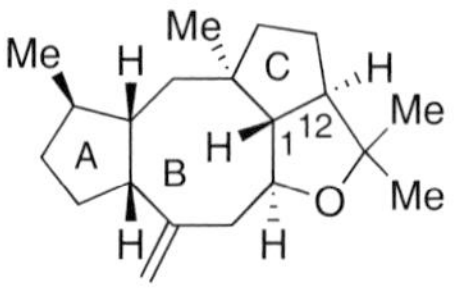

6.60 (+)-epoxydictymene

Target relevance

(+)-Epoxydictymene **6.60** belongs to the diterpene family of natural products. It is isolated from the brown algae *Dictyota dichotoma* [14]. From a structural point of view it has a tetracyclic fused ring system, part of which is an eigh-membered ring and a strained *trans*-fused 5–5 ring systems.

Synthetic plan

It was reasoned that (+)-epoxydictymene **6.60** could be obtained from enone **6.61** by a reductive methylation. Therefore, intermediate **6.61** became the primary objective of the synthesis. The authors, using an acid-promoted Nicholas reaction followed by an intramolecular Pauson–Khand reaction [15], had previously obtained cyclopentenone analogues to **6.61**. Accordingly, the designed precursor **6.61** of (+)-epoxydictymene could arise from the dicobalt cluster like **6.62**, obtained using an intramolecular acid-promoted Nicholas reaction on enyne **6.63**. The synthesis of **6.63** would be attempted by joining the allylsilane **6.65** and the propargylic acetal **6.64** using a nucleophilic displacement (Scheme 6.19).

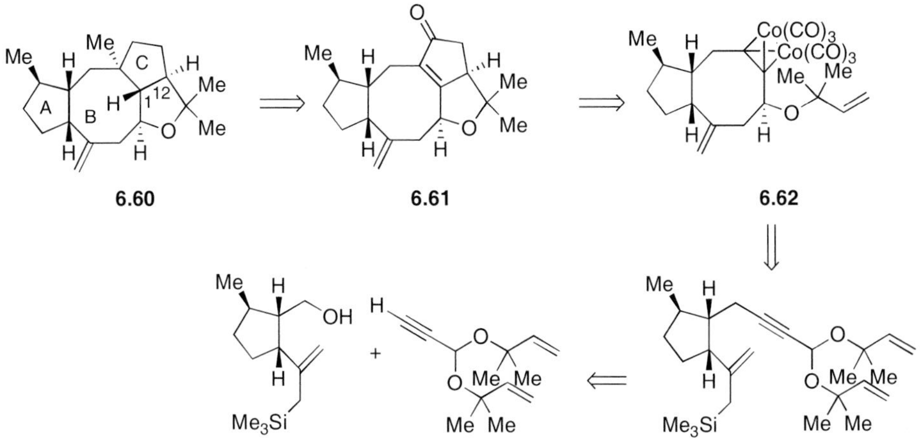

Scheme 6.19

Predictable problems

- The building of the trans 5–5 ring system was uncertain. The reductive methylation on the late intermediate **6.61** might install the quaternary carbon at C1, but it is highly speculative that this reaction would render the *trans* 5–5 ring system. In fact *trans*-fused 5–5 systems are less stable than the *cis* isomers (approximately 6 Kcal/mol) [16]. The calculated difference in the heats of formation of the *cis*- and *trans*-3-oxabicyclo[3.3.0]octanes is over 10 Kcal/mol [17]. Nevertheless, similar fused 5–5 enones had been obtained *via* intramolecular Pauson-Khand reactions [18] .
- Formation of eight-membered ring via cyclization of acyclic precursors in a Nicholas reaction.

Synthesis

Step 1. Synthesis and coupling of allylsilane 6.65 and propargylic acetal 6.64 (Scheme 6.20):

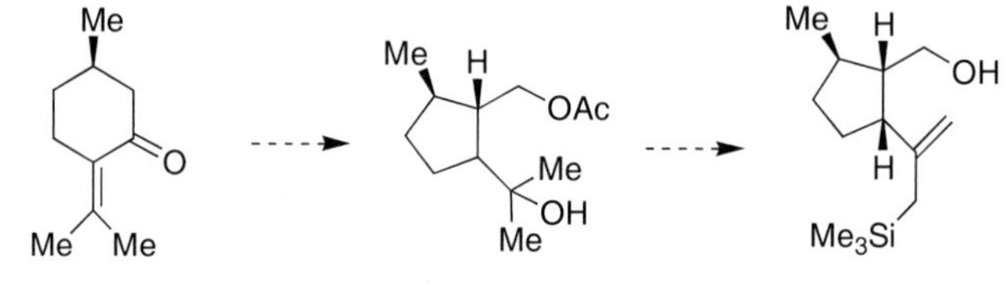

6.66 *(R)*-pulegone   6.67   6.65

Scheme 6.20

Allylsilane **6.65** was obtained from *(R)*-pulegone **6.66** as follows. Favorskii ring contraction [19] yielded exclusively the *trans*-isomer of the methylester **6.68**. Saponification and acid catalysed intramolecular cyclization of **6.68** afforded the *cis*-fused lactone **6.69** in 56% overall yield. Reduction of lactone **6.69** with LiAlH$_4$ followed by selective acetylation of the less-hindered primary alcohol, yielded the acetate **6.67**, which under boiling acetic anhydride and subsequent elimination of the tertiary acetate group rendered a 1:2 mixture of the two possible separable alkenes **6.70** and **6.71**. Reaction of **6.71** with Schlosser's base [20] produced an intermediate dianion that was reacted with TMSCl producing, after hydrolysis of the silyl ether, a 50% yield of allylsilane **6.65** (Scheme 6.21).

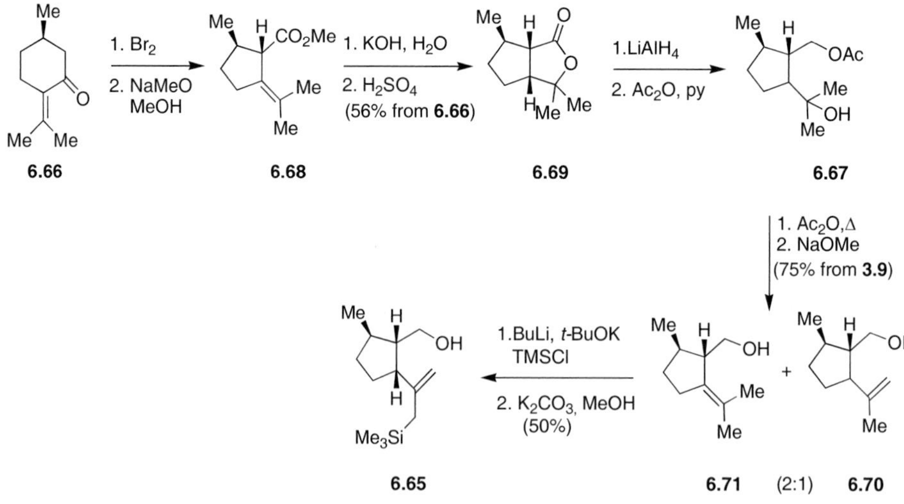

6.66   6.68   6.69   6.67

6.65   6.71   (2:1)   6.70

Scheme 6.21

The synthesis of the propargylic acetal **6.64** was attempted by acid-catalysed transacetalization of the de diethyl acetal **6.72**, but disappointingly, all efforts were unsuccessful. The mixed acetal **6.73** was obtained in 54% yield by treatment of **6.72** with bromodimethylborane followed by trapping of the bromomethyl ether with 2-methyl-3-buten-2-ol. Because the site-selectivity in the mixed acetal **6.73**, during the Nicholas reaction was uncertain this reagent was discarded as the starting material. 1,3-Dioxane **6.74** was considered a more suitable substrate for the acid-promoted Nicholas reaction, due to its reported complete site selectivity [21]. Compound **6.74** was obtained in 66% yield by reaction of **6.72** with 3-methyl-1,3-butanediol in the presence of TsOH (Scheme 6.22). The impossibility of preparing the designed starting material **6.64** will assume, as matter evolves, a change in the plan of synthesis that will increase the number of steps (see below).

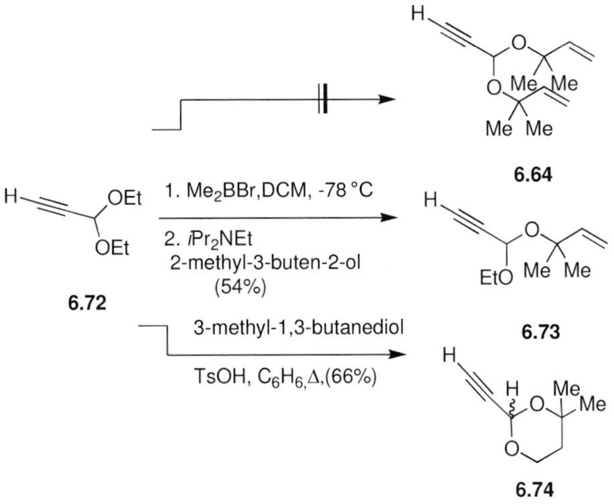

Scheme 6.22

Coupling of allylsilane **6.65** and cyclic acetal **6.74** required a rigorous control of the reaction conditions. The triflate generated from alcohol **6.65** was treated immediately with the lithium anion derived from **6.74**, yielding the coupled product **6.75** as a mixture (1:1) of diastereoisomers in the ketal center (Scheme 6.23).

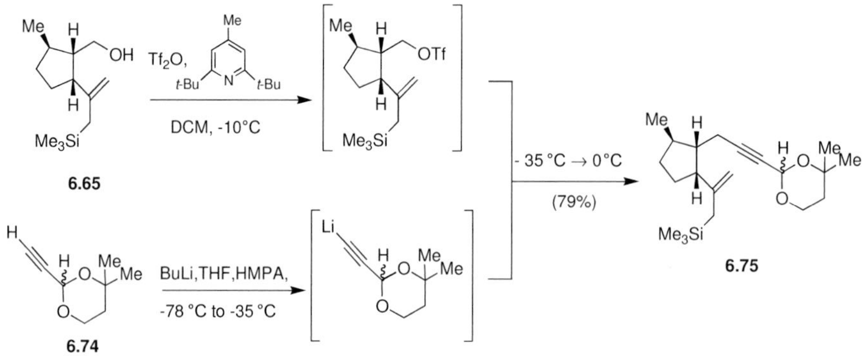

Scheme 6.23

Step 2. The intramolecular Nicholas cyclization-Pauson–Khand cyclocarbonylation (Scheme 6.24):

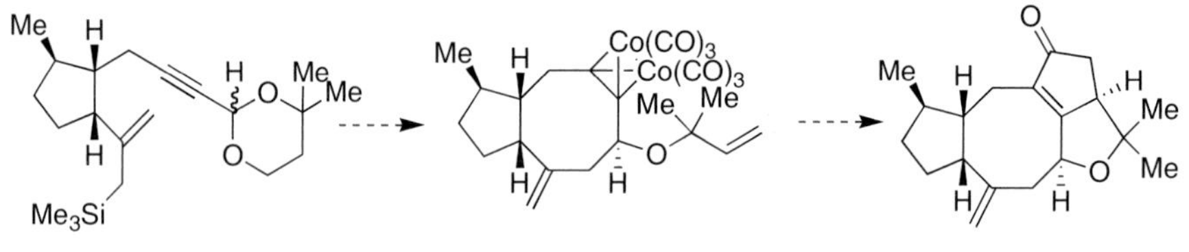

Scheme 6.24

The required alkyne–dicobalt cluster **6.76** in which the Nicholas reaction was to be carried out, was obtained in 90% yield by reaction of **6.75** with Co$_2$(CO)$_8$ in ether. Treatment of **6.76** with a stoichiometric amount of Et$_2$AlCl in DCM produced 91% of the fused 5–8 ring as a single diastereoisomer. It is worth noting the complete site-selectivity of the reaction. Having efficiently accessed the fused ring system, the side chain primary alcohol was converted to terminal olefin **6.62**. Since mild conditions compatible with the sensitivity of the dicobalt cluster to oxidation were required, the method of Grieco and Nicolaou [22] was used. The oxidation step was performed using the phenyloxaziridine of Davis [23] under basic conditions. Intermediate **6.62** was obtained in 61% yield from **6.77** (Scheme 6.25).

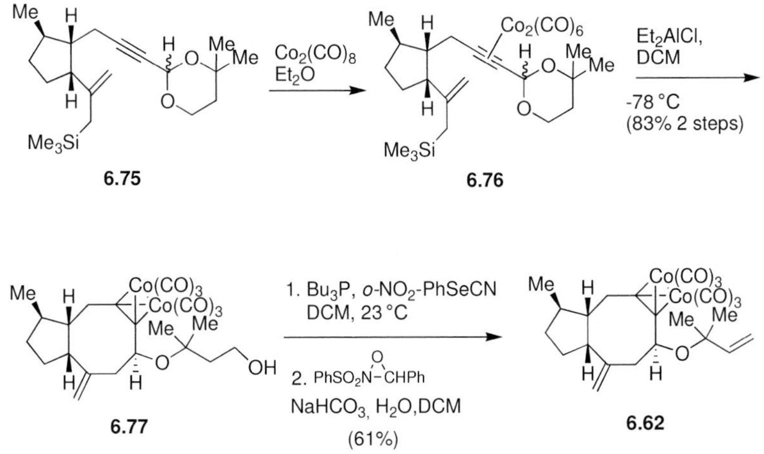

Scheme 6.25

It is necessary to note at this point that the sequence **6.77** → **6.62** was necessary because the planned substrate **6.63** for the Nicholas reaction was not accessible due to the impossibility of preparing acetal **6.64**. Lateral failures in trivial reactions resulted in increasing the number of steps of the overall synthetic scheme. The next step was the Pauson–Khand cyclization of **6.62**. Oxidative conditions were used for initiating the reaction. Treatment of **6.62** with NMO leads to the isolation of the tetracyclic enone **6.61** in a 70% yield as a (11:1) mixture of epimers at C12, favoring the undesired isomer. It has to be mentioned that the thermally initiating Pauson–Khand reaction in an atmosphere of air, yielded a (5:1) mixture of isomers at C12 also favoring the undesired isomer (Scheme 6.26). Attempts to isomerize the undesired epimer at C12 were fruitless. Therefore, to follow the planned synthesis, the authors had to choose to go ahead with the minor isomer, or to continue with the main isomer delaying the inversion at C12 to a further stage of the synthesis. This second option was the one chosen.

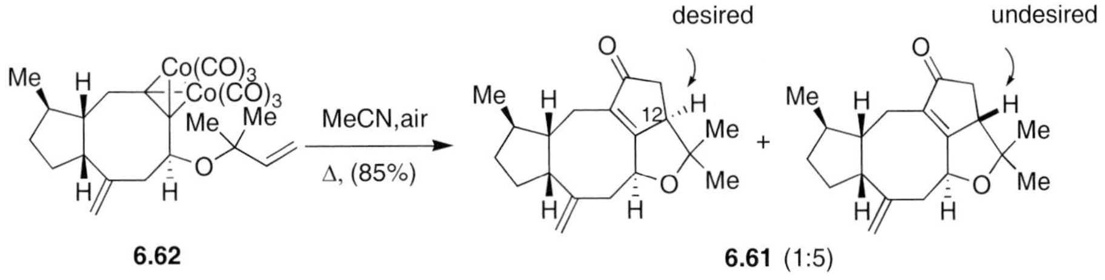

Scheme 6.26

Step 3. The introduction of the C4-Me group and the selective reduction of the enone double bond leading to a full dead-end (Scheme 6.27):

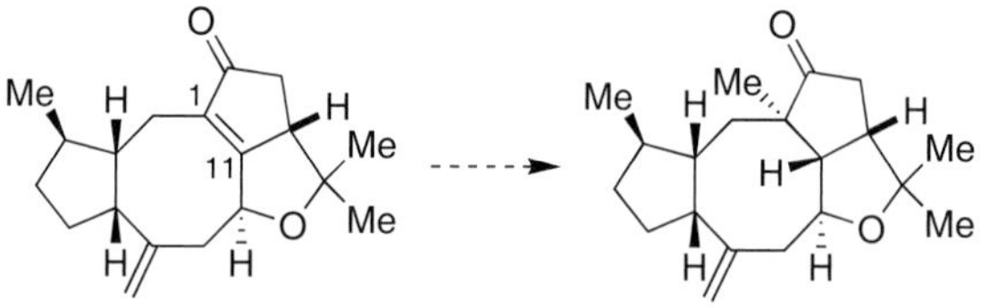

6.61 6.78

Scheme 6.27

The installation of the quaternary C1 α Me group as well as the β-selective reduction of enone **6.61** at C11, were attempted jointly. Dissolving metal reduction followed by reaction of the resulting enolate with iodometane produced ketone **6.79** in high yield. The *cis* fused 5–5 ring was obtained exclusively, as predicted on the basis of literature precedents. However, the methyl group of **6.79** has been installed with the undesired stereochemistry. It was recognized that methyl ketone **6.79** could not easily be transformed into epoxyditymene (Scheme 6.28). Therefore, this intermediate is a complete dead-end of the synthesis. A new strategy for introduction of the methyl group had to be devised.

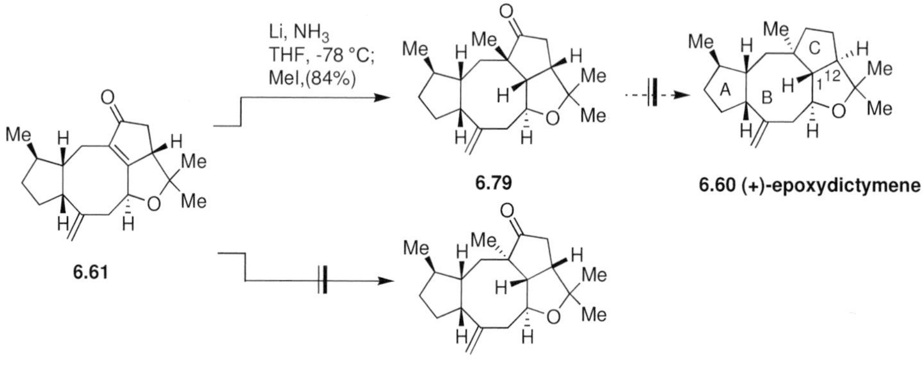

6.61 6.79 6.60 (+)-epoxydictymene 6.78

Scheme 6.28

*Step 4. A new strategy for the introduction of the C4-Me group and the introduction of
the right stereochemistry at C1 leading to the completion of the synthesis (Scheme 6.29):*

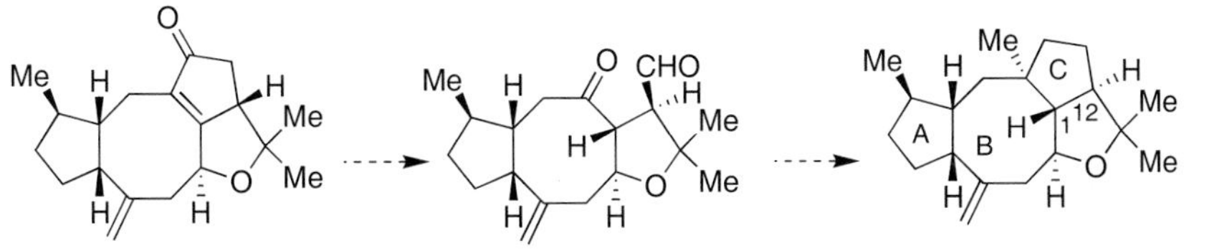

Scheme 6.29

Results depicted in Scheme 6.28 show that the synthesis of epoxydictymene by means of
a deconjugative methylation probed to be unfeasible. The impossibility of introducing a
methyl substituent on a performed hydrocarbon backbone was not predicted in advance
at all. All the above results suggest that the α-configuration of the methyl group will not
be achieved unless prior inversion of the stereochemistry at C12 is performed. Therefore,
inversion of the configuration at C12 became the prime task before going ahead with the
completion of the synthesis. On account of the failure to invert the configuration at C12
on the complete tetracyclic skeleton, an approach based on the fragmentation of ring C
was considered (Scheme 6.29).

   Enone **6.61** was reduced with a Li/NH$_3$ system followed by isoprene/NH$_4$Cl
quench, affording ketone **6.81**, which has the correct configuration at C11 in the 8-5-5
fused system. Double α-hydroxylation of **6.81** proceeded from the less-hindered β-face
yielding keto-diol **6.82**, which, upon directed reduction of the carbonyl with
NaHB(OAc)$_3$ afforded triol **6.83**. Pb(OAc)$_4$ oxidation of **6.83** followed by treatment with
DBU, and produced keto aldehyde **6.80** as a mixture (3:1) of diastereoisomers favoring
the isomer having the configuration at C12 inverted (Scheme 6.30).

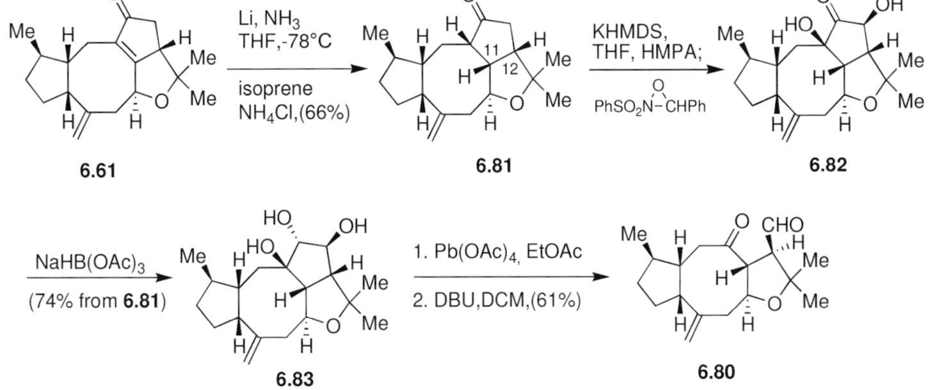

Scheme 6.30

The inversion of the configuration at C12 could be accomplished but only by paying the price of opening a five-membered ring that then had to be reclosed. This is by no means a trivial task. The methodology involved in building the 8-5-5 fused system was based on the procedure developed by Bailey [24] for the synthesis of *trans*-bicyclo[3.3.0]-octanes based on the intramolecular anionic cyclization of 5-hexenyl iodides. In this case intermediate **6.84** was prepared from ketoaldehyde **6.80** in a five-step sequence. The critical anionic cyclization step was attempted by exposure of the iodide **6.84** to metal–halogen exchange conditions. Surprisingly and contrary to the results obtained with simpler substrates, **6.84** yielded only the protodehalogenation product **6.85**. A radical cyclization was tested using (trimethylsilyl)silane and AIBN. In this case cyclization occurs but in a 6-*endo* way, producing a 5-8-6-5 fused system **6.86** (Scheme 6.31).

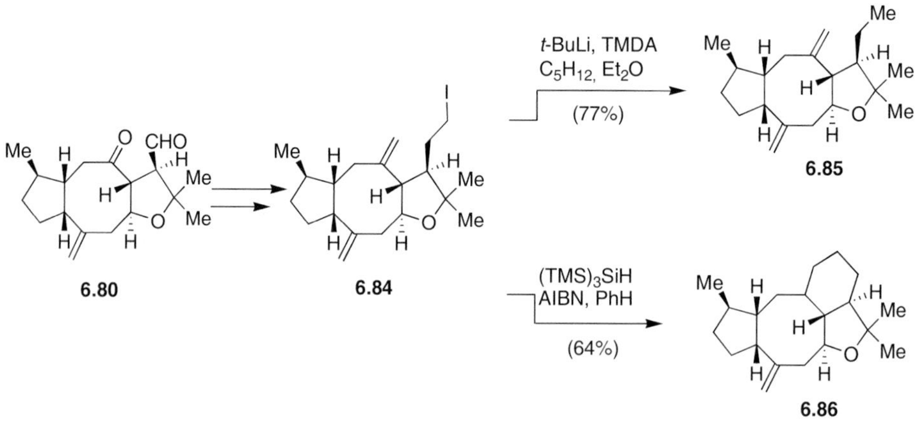

Scheme 6.31

The cyclization **6.84** → **6.86** was not surprising with hindsight, since it is known [25] that the radical cyclization of 5-alkyl-5-hexenyl iodides occurs in 6-*endo* mode preferably to the 5-*exo*-mode. Nevertheless, having a possible substrate for a given reaction even contrary to predictions is a temptation chemists cannot resist. Precisely because we are conscious of the many variables we do not count on obtaining the desired product but to do so against the odds is always a temptation few can resist. Nevertheless, epoxydictymene will not be accessed from iodide **6.84**, which has clearly closed another pathway in the synthesis, but the knowledge acquired in these reactions facilitates the design of a new substrate for radical cyclization. This new substrate has to present a clear bias for the 5-exo-mode of cyclization. The best way to solve this problem is to use a compound like **6.87** having a cyano group at C15. This group may facilitate the radical as well as the anionic cyclization alternate processes. In addition, the cyano grouping could be removed in a single step. The cyanomethylene group will thus finally become the solution to placing the troublesome C15-Me group.

The iodo acrylonitrile **6.87** was prepared from keto aldehyde **6.80**, through the methyl enol ether **6.88** by a double Wittig sequence. Submission of iodide **6.87** to radical cyclization conditions afforded again the 6-*endo* product **6.89**. On the contrary, the anionic process promoted by halogen–metal exchange on **6.87**, yielded 74% of a cyclization product **6.90** containing the complete 5-8-5-5 fused-ring system of epoxy-dictymene. Finally, reductive decyanation delivered synthetic (+)-epoxydictymene **6.60** (Scheme 6.32).

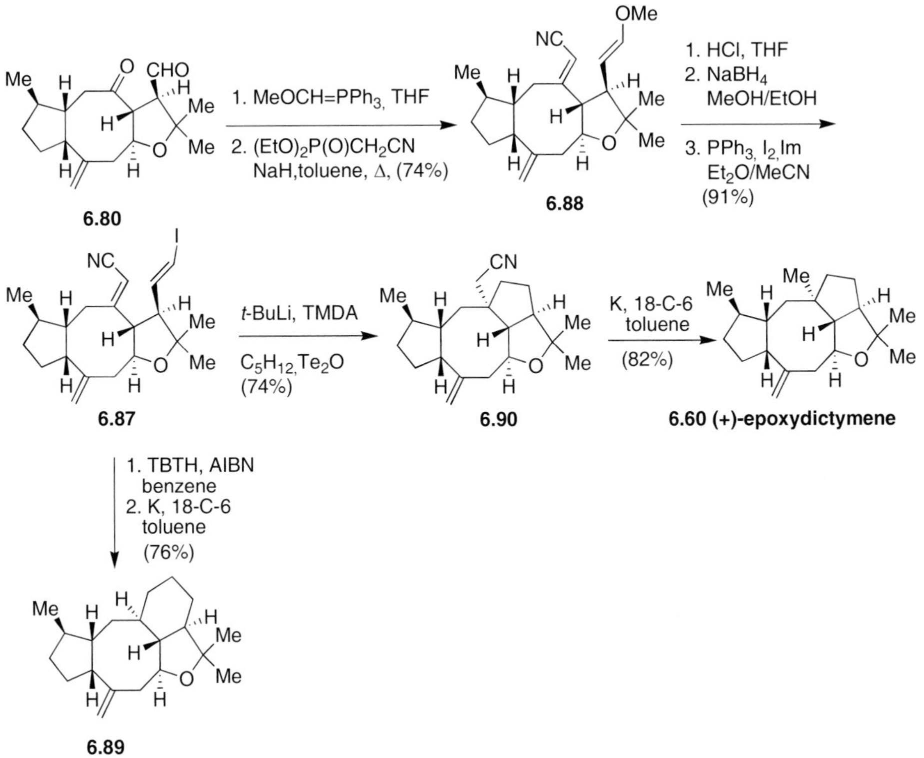

Scheme 6.32

Evaluation

- The construction of the tetracyclic core **6.61** of (±)-epoxydictimene is an impressive "tour de force" considering that the fused–ring system can be built, from the appropriate starting materials, in **three steps**. However, the cyclization mainly formed the undesired stereisomer. This intermediate **6.61** failed to attach the remaining Me group and to correct the stereochemistry at C11. Therefore, a tedious redesign of the synthesis had to be undertaken.

- Installation of the C4-Me group involves ring-C breakage, differentiation of two carbonyl groups after correcting the wrong troublesome stereochemistry, and final ring-C closure. This is an important strategic change since it extends the entry to (+)-epoxydictymene in many steps; especially considering that the fused ring system is constructed in an amazing way.

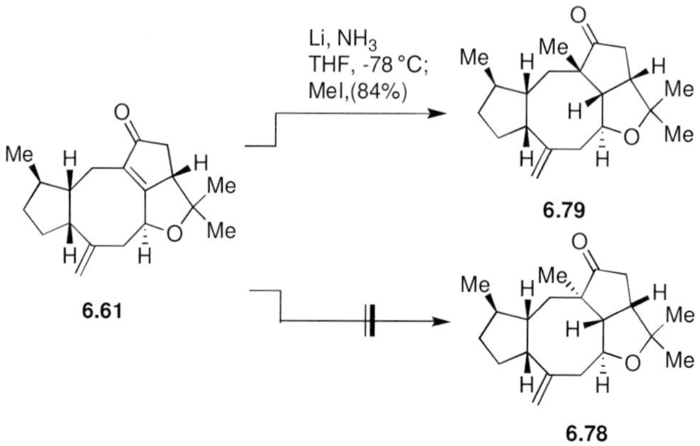

Li, NH₃
THF, -78 °C;
MeI,(84%)

6.61

6.79

6.78

Key synthetic reactions

Favorski rearrangement (*aka* Faworski–Wallach rearrangement) [19]: Ketone ring contraction promoted by α-halogenation followed by base treatment. The reaction produced an acid or acid derivative through an α-cyclopropanone intermediate.

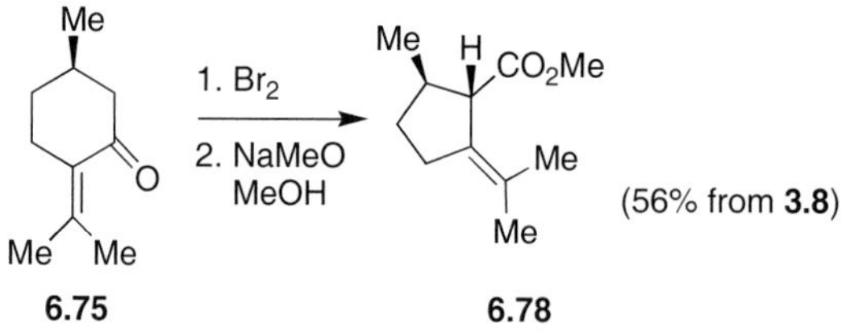

1. Br₂
2. NaMeO
   MeOH

(56% from **3.8**)

6.75   **6.78**

The Nicholas reaction: Generation by a protic or Lewis acid of a carbocation α- to an alkyne-Co₂(CO)₆ cluster and its subsequent inter or intramolecular reaction with a nucleophile [26].

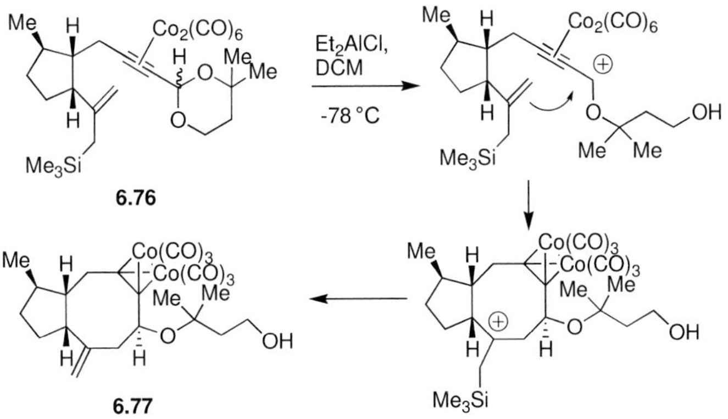

6.76

6.77

The Pauson–Khand reaction: The carbonylative cyclization of an alkyne-Co₂(CO)₆ and one olefin to yield a cyclopentenone. The process may be either inter or intramolecular [18].

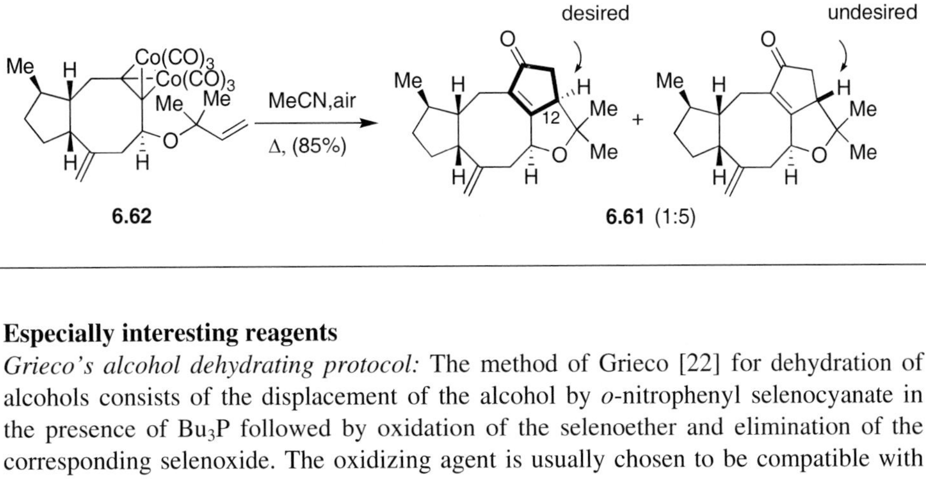

6.62   **6.61** (1:5)

Especially interesting reagents

Grieco's alcohol dehydrating protocol: The method of Grieco [22] for dehydration of alcohols consists of the displacement of the alcohol by *o*-nitrophenyl selenocyanate in the presence of Bu₃P followed by oxidation of the selenoether and elimination of the corresponding selenoxide. The oxidizing agent is usually chosen to be compatible with the remaining functional groups of the molecule. Even complexes, which are extremely sensitive to oxidation, such as Co-clusters **6.87**, can be dehydrated using this method.

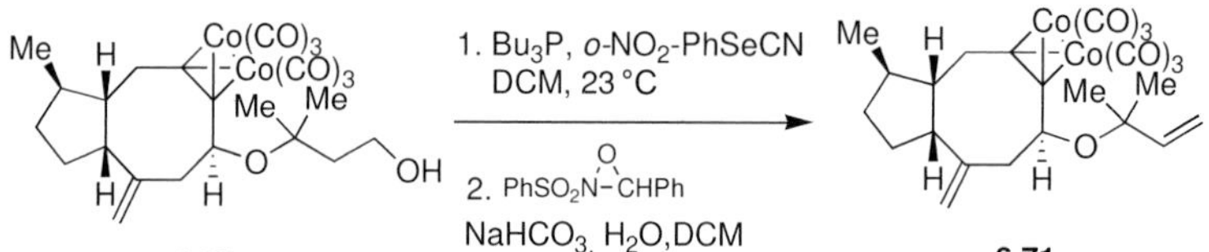

6.87 → **6.71**

1. Bu₃P, o-NO₂-PhSeCN
DCM, 23 °C

2. PhSO₂N-CHPh
NaHCO₃, H₂O, DCM

6.4   (±)-Lepadiformine [27]

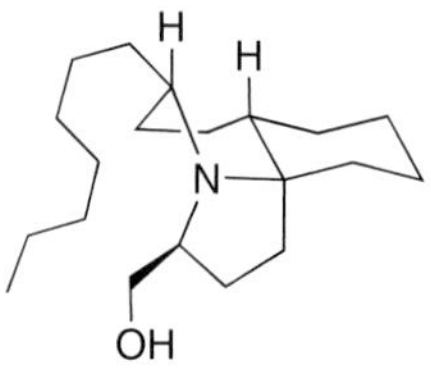

6.91(±)-lepadiformine

Target relevance

Lepadiformine **6.91** was isolated from the tunicate *Clavelina lepadiformis* in 1994 [28], but its structure was not unambiguously established until 2000 when Kibayashi and co-workers synthesized the hydrochloride salt of the natural product [29].

Synthetic plan

The planned synthesis is based on the construction of the *trans*-perhydroquinoline fragment of the natural product using a Diels–Alder reaction of the amidoacrolein **6.92** with diene **6.93**. The resulting tosylamide **6.94** could be elaborated to an activated oxaziridine **6.95**, on which the homoallylic appendage should be introduced by reaction with the corresponding organometallic reagent. Iminium ion **6.96** should be capable of stereochemically controlling the introduction of the hexyl side chain, forming **6.97**. Finally, electrophile-promoted cyclization of the amine, derived from tosylamide **6.97**, should deliver lepadiformine **6.91** (Scheme 6.33).

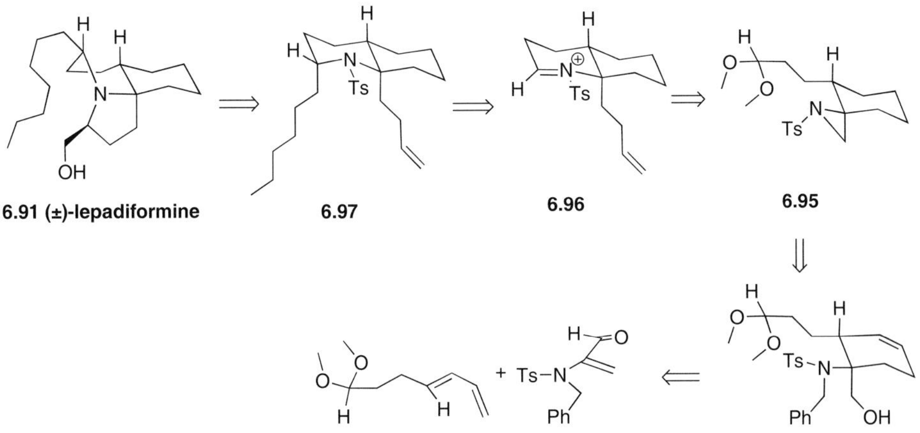

6.91 (±)-lepadiformine 6.97 6.96 6.95

6.93 6.92 6.94

Scheme 6.33

Predictable problems

- No problems were anticipated since the synthesis relies on the well known intermolecular Diels–Alder reaction of an amidoacrolein with a diene. These Diels–Alder reactions show complete regio and stereocontrol, in the sense required for the synthesis for the final construction of the *trans*-perhydroquinoline core.
- With respect to the introduction of the hexyl substituent, addition of an organometallic reagent to the iminium ion **6.96** should be stereoelectronically controlled proceeding *via* chair conformation from the α-face.

Synthesis

Step 1. Diels–Alder reaction and construction of aziridine intermediate ***6.95****:* Thermal Diels–Alder reaction of **6.92** with diene **6.93** produced mainly polymerization of dienophile. However, submission of the reagents to high pressure (12 Kbar) gave only the expected *endo*-cycloadduct **6.98**. The aldehyde was reacted with $NaBH_4$ to form **6.94**, which, in the presence of hydrogen and Pearlman's catalyst, simultaneously hydrogenated the double bond and removed the benzyl group. Cyclohexane **6.99** was obtained in 94% yield. Intramolecular Mitsunobu ring-closure of **6.99** was accomplished under standard conditions ($Ph_3P/I_2/Im$) yielding aziridine **6.95** (Scheme 6.34).

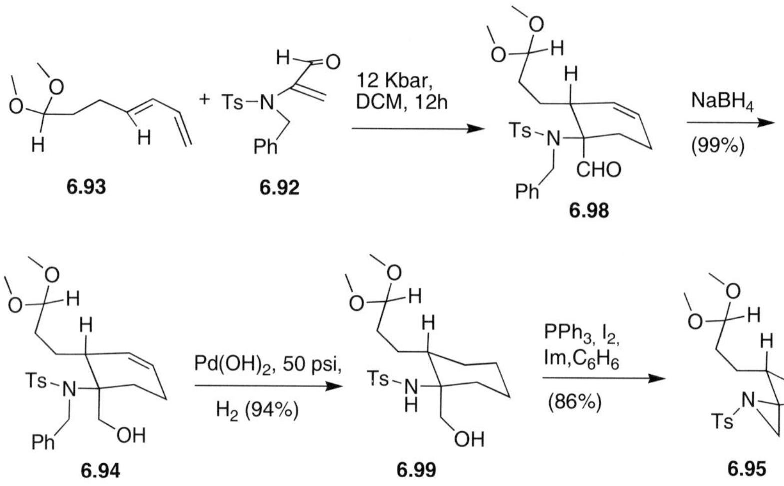

Scheme 6.34

Step 2. Introduction of the side chains. Failure in obtaining **6.97**: First, the butenyl side chain that would eventually form the pyrrolidine fragment of the lepadiformine **6.91** was introduced. Nucleophilic ring opening of the tosylaziridine **6.95** with excess allylmagnesium bromide yielded tosylamide **6.100**. At this point attempts to introduce the hexyl substituent were made. In any case the desired adduct **6.97** was obtained under a variety of acid catalysts and in the presence of hexylmagnesium bromide. Enamide **6.101** was obtained using BF_3 at 0 °C. Due to this serious failure in the planned synthesis, compound **6.100** was transformed to aldehyde **6.102**. Compound **6.102** is not in ring-chain tautomerism with the cyclic form **6.103** (the precursor of iminium salt **6.104**) in a variety of solvents. Therefore, the planned addition of a hexyl-nucleophile onto the iminium salt **6.104** failed (Scheme 6.35).

Step 3. Re–evaluation of the methodology to appendage placement. Completion of the synthesis: It was obvious that the introduction of the hexyl group should be effected independently of the formation of the second ring, contrary to the original synthetic planning. Aldehyde **6.102** was reacted with hexylmagnesium bromide in the presence of $Yb(OTf)_3$ yielding alcohol **6.105** as an inseparable (10.9:1) mixture with its C13 epimer. This increase in the stereoselectivity could be the consequence of a process controlled by quelation, on which the greater steric bulk of the organoytterbium reagent, coupled to the attenuated reactivity of the magnesium-chelated aldehyde, may favor the formation of the desired isomer. The *trans*-perhydroquinoline **6.106** was accessed by submission of the mixture of alcohols to Mitsunobu cyclization conditions, followed by removal of the tosyl group. Nucleophilic cyclization promoted by iodine yielded an (iodomethyl)pyrrolidinium salt intermediate **6.107**, which was directly treated with aqueous NaOH containing Bu_4NI, delivering racemic lepadiformine **6.91** in 77% yield (Scheme 6.36).

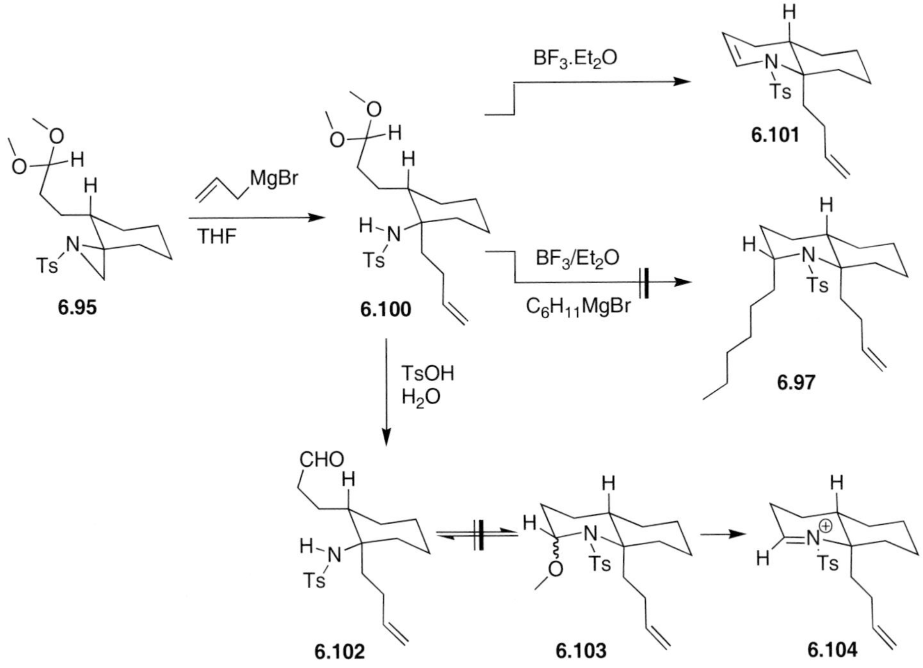

Scheme 6.35

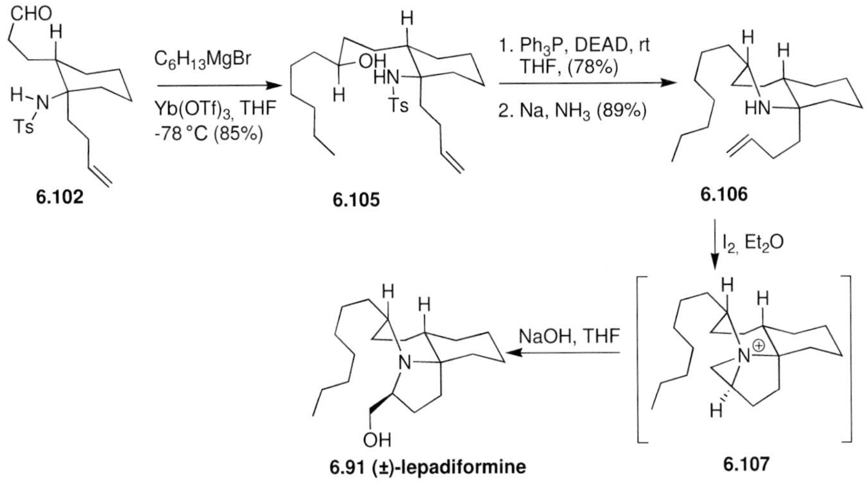

Scheme 6.36

Evaluation

The lepadiformine synthesis is a simple example of the difficulties encountered in appending a side chain in a preformed bicyclic molecule. In this case the key point of the planned process was the ability to generate an iminium salt from an, *in situ,* generated carbinol amine. Tosylamides are probably not sufficiently nucleophilic in these cases to effect this cyclization as demonstrated by the second attempt made in aldehyde **6.114**. The failure on generating the iminium salt thwarted the planned synthesis.

References

1. Overman, L. E.; Ricca, D. J.; Tran, V. D. *J. Am. Chem. Soc.* **1997**, *119*, 12032.
2. a) Hayashi, T.; Kishi, M.; Kawasaki, M.; Arisawa, M.; Shimizu, M.; Suzuki, S.; Yoshizaki. M.; Morita, N.; Tezuka, Y.; Kikuchi, T.; Berganza, L. H.; Ferro, E.; Basualda, I. *Tetrahedron Lett.* **1987**, *28*, 3693; b) Hayashi, T.; Asano, S.; Mizutani, M.; Takeguchi, N.; Kojima, T.; Okamura, K.; Morita, N. *J. Nat. Prod.* **1991**, *54*, 802.
3. Abelman, M. M.; Overman, L. E. *J. Am. Chem. Soc.* **1988**, *110*, 2328.
4. a) Gibson, S. E.; Middleton, R. J. *Contemp. Org. Synth.* **1996**, *3*, 447; b) de Meijere, A.; Meyer, F. E. *Angew. Chem. Int. Ed. Engl.* **1994**, *33*, 2379; c) Brase, S.; deMeijere, A. in *Metal-Catalyzed Cross-Coupling Reactions;* Diederich, F.; Stang, P. J. Eds.; Wiley-VCH, 1998; d) Tsuji, J. *Palladium Reagents and Catalysts;* John Wiley & Sons: Chichester, 1995; e) Heck, R. F. *Palladium Reagents in Organic Synthesis*; Academic Press: New York, 1985.
5. The intramolecular Heck reaction: Link, J. T. *Org. React.* **2002**, *60*, 157
6. a) Piers, E.; Nagakura, I. *Tetrahedron Lett.* **1976**, 3247; b) Marino, J. P.; Browne, L. J. *Tetrahedron Lett.* **1976**, *41*, 3245; c) Wender, P. A.; Filosa, M. P. *J. Org. Chem.* **1976**, *41*, 3490; d) Wender, P. A.; Eissenstat, M. A.; Filosa, M. P. *J. Am. Chem. Soc.* **1979**, *101*, 2196. e) Hudlicky, T.; Fan, R.; Reed, J. W.; Gadamasetti, K. G. *Org. React.* **1992**, *41*, 1.
7. a) Tidwell, T. T. *Synthesis* **1990**, 857; b) Tidwell, T. T. *Org. React.* **1990**, *39*, 297; c) Mancuso, A. J.; Huang, S.-L.; Swern, D. *J. Org. Chem.* **1979**, *44*, 4148. d) Swern, D.; Mancuso, A.; Huang, S. *J. Org. Chem.* **1978**, *43*, 2480.
8. a) Bennett, C. R.; Cambie, R. C. *Tetrahedron* **1966**, *22*, 2845; b) Fringuelli, F.; Mancini, V.; Tatticchi, A. *Tetrahedron* **1969**, *25*, 4249.

9.   Nagata, W.; Yoshioka, M. *Org. React.* **1977**, *25*, 255.

10. Corey, E. J.; Roberts, B. E. *J. Am. Chem. Soc.* **1997**, *119*, 12426.

11. Gunasekera, G. P.; McCarthy, P. J.; Kelly-Borges, M.; Lobkvosky, E.; Clardy, J. *J. Am. Chem. Soc.* **1996**, *118*, 8759.

12. de la Torre, M. C.; Sierra, M. A. *Angew. Chem. Int. Ed.* **2004**, *43*, 160.

13. Jamison, T. F.; Shambayati, S.; Crowe, W. E.; Schreiber, S. L. *J. Am. Chem. Soc.* **1997**, *119*, 4353.

14. Enoki, N.; Furusaki, A.; Suehiro, K.; Ishida, R.; Matsumoto, T. *Tetrahedron Lett.* **1983**, *24*, 4341.

15. a) Schreiber, S. L.; Sammakia, T.; Crowe, W. E. *J. Am. Chem. Soc.* **1986**, *108*, 3128; b) Schreiber, S. L.; Klimas, M. T.; Sammakia, T. *J. Am. Chem. Soc.* **1987**, *109*, 5749.

16. Chang, S.; McNally, D.; Shary-Tehrany, S.; Hickey, M. J.; Boyd, R. H. *J. Am. Chem. Soc.* **1970**, *92*, 3109.

17. Schoening, A.; Friedrichsen, W. *Z. Naturforsch.* **1989**, *44b*, 975.

18. a) Schore, N. E. *Org. React.* **1991**, *40*, 1; b) N. E. Schore in *Comprehensive Organic Synthesis*, Eds. Trost B. M. and Fleming, I. 1992, Elsevier, vol. 9, p. 1037; c) Schore, N. E. in *Comprehensive Organometallic Chemistry II*, ed. Abel, E. W.; Stone F. G. A. and Wilkinson, G. Elsevier, 1995, vol. 12, p. 703; e) Geisand O.; Schmalz, H.-G. *Angew. Chem., Int. Ed.*, **1998**, *37*, 7.

19. Wolinsky, J.; Gibson, T.; Chan, D.; Wolf, H. *Tetrahedron* **1965**, *21*, 1247.

20. Schlosser, M. *Angew. Chem. Inet. Ed. Engl.* **1974**, *13*, 701.

21. Ishihara, K.; Yamamoto, H.; Heathcock, C. H. *Tetrahedron Lett.* **1989**, *30*, 1825.

22. a) Grieco, P. A.; Gilman, S.; Nishizawa, M. *J. Org. Chem.* **1976**, *41*, 1485; b) Grieco, P. A.; Jaw, J. Y.; Claremon, J. A.; Nicolaou, K. C. *J. Org. Chem.* **1981**, *46*, 1215.

23. Davis, F. A.; Chen, B. C. *Chem. Rev.* **1992,** *92*, 919.

24. a) Bailey, W. F.; Khanolkar, A. D. *Tetrahedron Lett.* **1990**, *31*, 5993; b) Bailey, W. F.; Khanolkar, A. D.; Gavaskar, K. V. *J. Am. Chem. Soc.* **1992**, *114*, 8053.

25. Curran, D. P. *Radical Cyclization and sequential Radical Reactions;* Curran, D. P., Ed.; Pergamon Press: New York, 1991; Vol. 4.2, pp 779-831.

26. a) Melikyan, G. G.; Nicholas, K. M., in *Modern Acetylene Chemistry,* Stang, P. J.; Diederich, F. Eds. VCH, Weinheim, **1995**, pp. 118; b) Caffyn, A. J. M.; Nicholas, K. M. *Comprehensive Organometallic Chemistry II,* Abel, E. W., Stone, F. G. A., Wilkinson, G., Eds.; Pergamon: Oxford, 1995; Vol 12, pp. 685; c) Nicholas, K. M. *Acc. Chem. Res.* **1987**, *20*, 207.

27. Greshock, T. J.; Funk, R. L. *Org. Lett.* **2001**, *3*, 3511.

28. Biard, J. F.; Guyot, S.; Rossakis, C.; Verbist, J. F.; Vercauteren, J.; Weber, J. F.; Boukef, K. *Tetrahedron Lett.* **1994**, *35*, 2691.

29. Abe, H.; Aoyagi, S.; Kibayashi, D. *J. Am. Chem. Soc.* **2000**, *122*, 4583.

Chapter 7
The Unpredictable Stereochemistry

Up to what point is the stereochemistry of one synthetic step or one synthetic scheme controllable? Following the superb achievements in the development of methodologies to control the stereochemistry of many fundamental processes, it might be thought that this question has been answered. In fact, during the last twenty years the highly stereoselective methods developed to effect chiral control and chiral discrimination, with the help of internal or external chiral auxiliaries and catalysts, have produced exceptional results [1]. Nevertheless, the wrong stereochemistry obtained in one key synthetic step or the impossibility of controlling the stereochemistry, even in well established processes, is still a problem, causing several tactical and strategic modifications to planned routes. The following examples show the diverse situations in which the formation of the wrong isomer causes deviations or modifications in the synthetic plan.

7.1   Nocardiones A and B [2]

7.1 nocardione A   7.2 nocardione B

Dead Ends and Detours: Direct Ways to Successful Total Synthesis
Miguel A. Sierra and María C. de la Torre
Copyright © 2004 WILEY-VCH Verlag GmbH & Co. KGaA, Weinheim
ISBN: 3-527-30644-7

Target relevance

(–)-Nocardione A **7.1** and (–)-nocardione B **7.2** were isolated in minute amounts from the culture broth of *Nocardia* sp. TP-A0248 [3]. Both compounds are new tyrosine phosphatase inhibitors with moderate antifungal and cytotoxic activities. The absolute configurations of these two furano-*o*-naphtoquinones remained unknown and their synthesis was undertaken in order to solve this problem.

Synthetic plan

The planning devised to prepare compounds **7.1** and **7.2** delayed the construction of the unstable *o*-naptoquinone system to the final stages of the synthesis. Both compounds will be derived from a common intermediate **7.3** that will be accessed from tetralone **7.4** and (*R*)-propylene oxide **7.5**. The furan ring would be closed through a Mitsunobu reaction (Scheme 7.1).

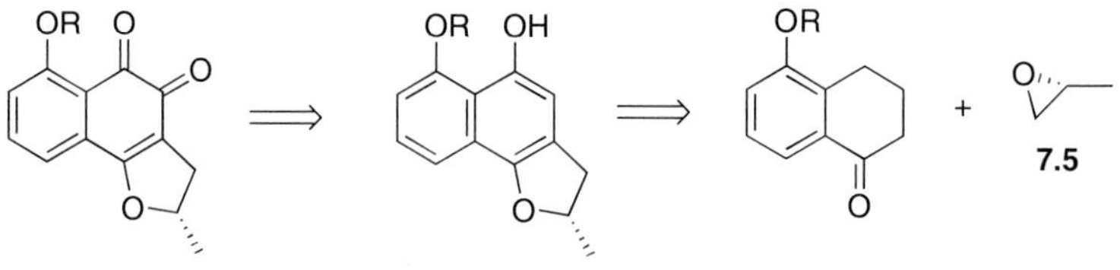

7.1 R = H
7.2 R = Me

7.3

7.4

7.5

Scheme 7.1

Predictable problems

The simple approach to nocardiones A and B sketched in Scheme 7.1 should not, in principle, present problems, since it is based on well tested reactions. There are no compromising transformations in the scheme.

*Step 1. Synthesis of nocardione A **7.1** (Scheme 7.2):*

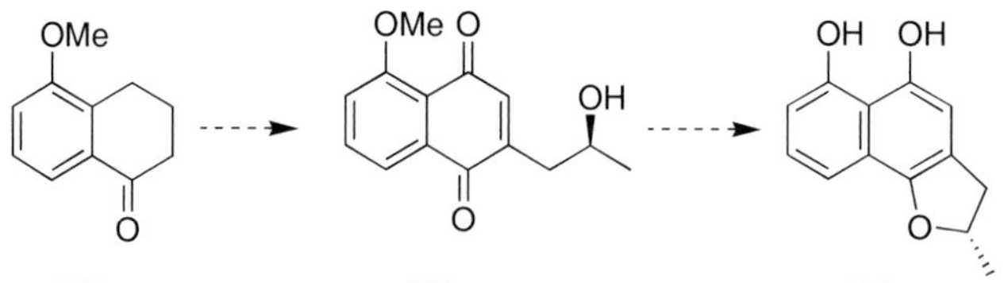

7.6

7.7

7.1

Scheme 7.2

Lithium enolate derived from **7.6** was reacted with (*S*)-propylene oxide **7.5** in the presence of Sc(OTf)₃ and the resulting alcohol **7.8** was protected as the 2,2,2-trichloroethoxycarbonate (Troc) derivative. Compound **7.9** was then oxidized to the corresponding quinone **7.10** with SeO₂. Reduction of quinone **7.10** concomitantly eliminated the Troc-protecting group giving the easily oxidized hydroquinone **7.11**. Ring closure of **7.11** to give tricyclic naphthol **7.12** was accomplished with inversion of the configuration under the standard Mitsunobu conditions in poor yield (28%). The oxidation of **7.12** to nocardione B **7.2** was effected with benzeneseleninic anhydride (PhSeO)₂O achieving the preparation of this product. Finally, demethylation of **7.2** was accomplished in the presence of AlCl₃ in DCM. In this way, nocardione A **7.1** was obtained in quantitative yield but as a racemic mixture. All the attempts to effect the demethylation without racemization were unsuccessful (Scheme 7.3). Evidently, the conditions to remove the methyl group were too harsh to be compatible with the stereochemical integrity of the stereocenter present in **7.2**.

The obvious solution to this problem is to use a more readily removable group to protect the phenolic hydroxy group. The group of choice was the benzyl group and the full synthetic scheme was repeated from compound **7.13**. After obtaining benzyl-protected *o*-quinone **7.14** the benzyl group was removed by hydrogenolysis leading to (–)-**7.1** (Scheme 7.3).

Evaluation
The racemization during the removal of a protecting group in the last step of the synthesis of nocardione A, **7.1**, does not result in a strategic change but in a major drawback. In fact, the choice of the MeO-protecting group instead of a benzyl group to access both nocardiones A and B, results in the necessity of repeating the full synthetic scheme to prepare nocardione A, since its removal resulted in complete racemization.

Scheme 7.3

7.2   (±)-Breynolide [4]

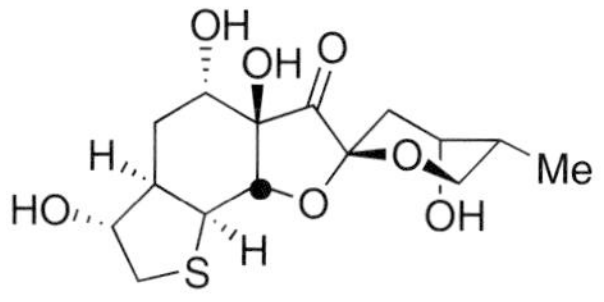

7.15 (±)-breynolide

Target relevance

Aglycon nucleus of the orally active hypocholesterolemic agent Breynin A (**7.16**). Breynins A (**7.16**) and B (**7.17**) are isolated from the Taiwanese woody shrub *Breynia officinalis* Hemsl [5].

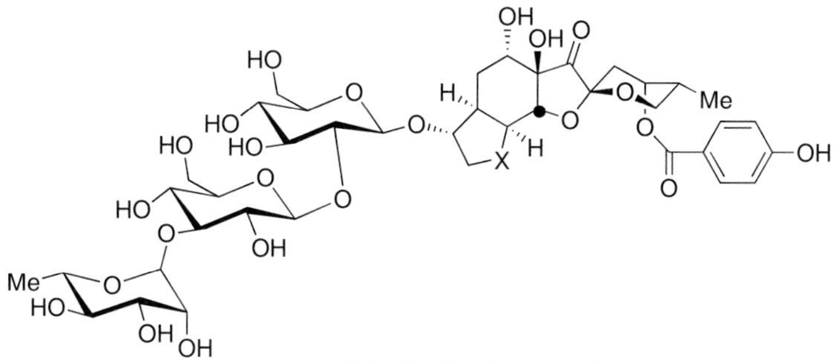

7.16 X = S,   **breynin A**
7.17 X = SO, **breynin B**

Concept

Stereochemically linear synthetic approach. A stereochemically linear approach employs a series of substrate-controlled operations to derive the relative configuration of the remaining stereocenters, from the chirality of a racemic or enantiomerically pure starting material. However, compared with a convergent synthetic strategy the stereochemically linear approach may entail additional steps. But, the overall efficiency may be increased by the use of a single chiral substrate. The key point of this kind of strategy is that a stereochemically linear synthesis of a racemate circumvents the formation of unwanted diastereomers. This is an operation that frequently accompanies the coupling of racemic fragments in a convergent synthesis.

Synthetic plan

Based on previous work, by the same authors, during the synthesis of phyllanthocin [6]. The key point of the synthetic planning would be the spiroketalization-equilibration protocol to access 7.19.

The implementation of this process requires:
- Introduction of the methyl group at carbon C12.
- Chemo- and stereoselective reduction of the C11 carbonyl group to the axial alcohol.
- Installation of the C6-C7 *trans*-vicinal diol.

Spiroketal **7.19** would derive from the coupling of lithiated dihydropryranone **7.21** and the perhydrobenzothiofene aldehyde **7.22**. Preparation of compound **7.22** would involve the thioannulation of enone **7.23**.

Predictable problems

- The introduction of the sulfur atom *anti* to the adjacent hydrogen in **7.23**.
- Introduction of the C6-C7 *trans*-vicinal diol.
- Incorporation of C12 methyl group in the ketone **7.19**.

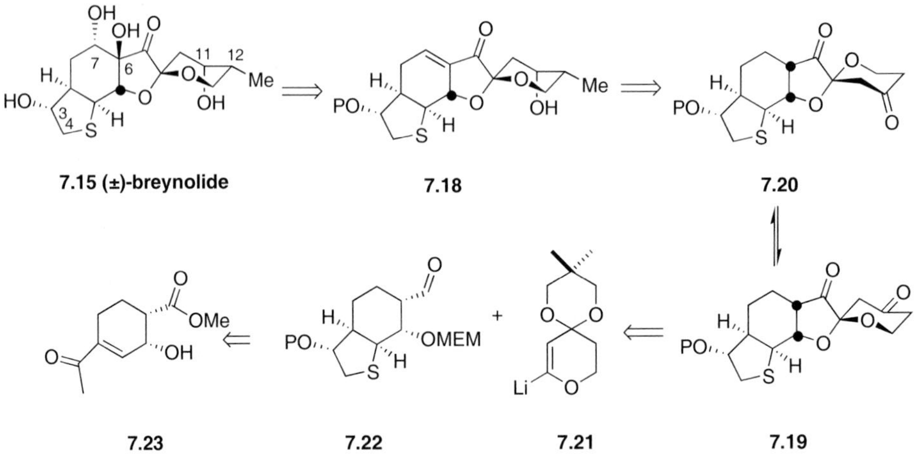

| 7.15 (±)-breynolide | 7.18 | 7.20 |

| 7.23 | 7.22 | 7.21 | 7.19 |

Scheme 7.4

Synthesis

Step 1. Thioannulation (Scheme 7.5):

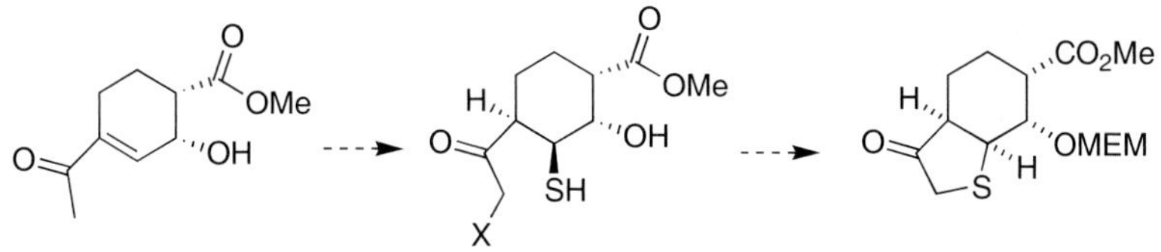

Scheme 7.5

Thioanulation of enone **7.23** involved the following steps:

- Introduction of the sulfur atom by functionalization of the enone *via* 1,4-addition with a suitable sulfur nucleophile.
- Installation of the leaving group at the Me group α to the ketone.
- Liberation of the thiol function and cyclization.

Enone **7.23** was reacted with thiolacetic acid to form a mixture (3:1) of epimers **7.26** and **7.27**. Both epimers embodied the required C17 configuration, which should arise from the preferred axial attack *anti* to the alkoxy and methoxy groups. However, the subsequent and required introduction of the halogen leaving group in the methyl ketone of **7.26** and **7.27** proved to be unfeasible. Thus, the sequence of events was reversed. Next the bromine group present in **7.28** was introduced via the silyl enol ether **7.29** obtained by deprotonation of **7.23** with LDA and further treatment with TESCl. Exposure of **7.29** to NBS effected the desired α-bromination to yield enone **7.28**. Standard treatment of **7.28** with H_2S rendered a mixture of diastereomeric disulfides **7.30**. Other agents were useless to effect this apparently simple transformation (Scheme 7.6). Therefore, access to the planned intermediate **7.31** was not possible by this route.

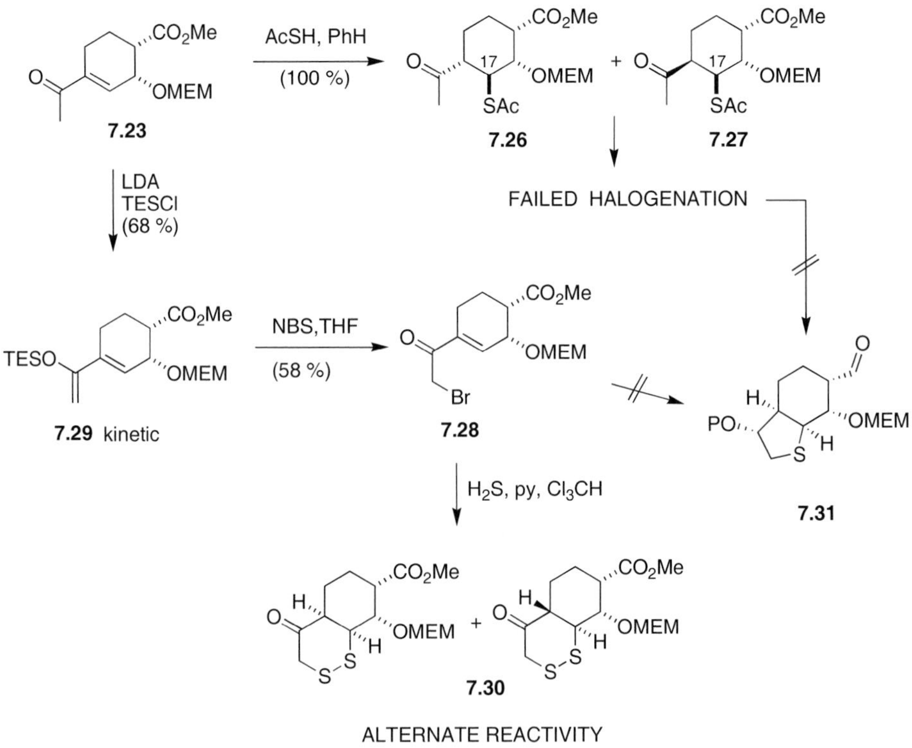

Scheme 7.6

Clearly, not only the C=C bond but also the highly electrophilic α-bromo ketone moiety of **7.14** have reacted with H$_2$S, inducing an alternate reaction. Since the reaction with the halogen atom was undesirable, the less reactive chlorine group was used instead of the bromide. Treatment of the dianion derived from **7.23** with NCS formed the α-chloro ketone **7.32** as the major product (59%) in a mixture of mono-, di- and trichloro ketones. The chlorination of the MEM-derivative was even less efficient. Next, thiolacetic acid was added as expected to **7.32**, giving the mixture of epimers **7.33**, which was cyclized and equilibrated to the desired *cis*-fused product **7.34** by base treatment (the basic medium hydrolyzed the thioester, liberating the sulfur nucleophile). Finally, the secondary alcohol was protected as its MEM-derivative (Scheme 7.7).

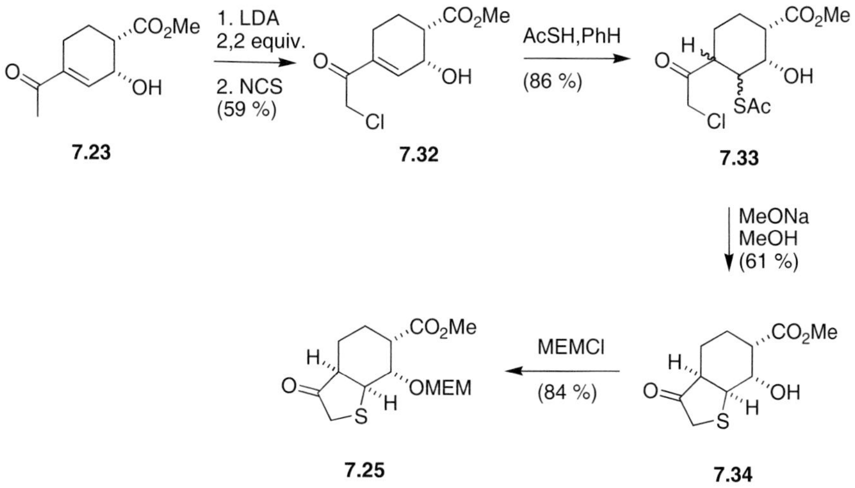

Scheme 7.7

The introduction of the sulfur group caused unexpected problems. In fact, a compromised, inefficient step (the chlorination of **7.23**) was unavoidable. Nevertheless, no deviations in the designed synthetic plan were derived from this tactical change. The strategy compiled in Scheme 7.5 was accomplished.

Step 2. Installation of the C3 α–hydroxyl group (Scheme 7.8):

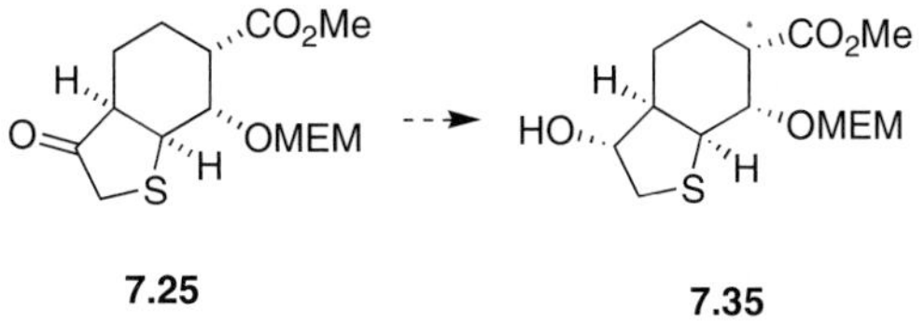

Scheme 7.8

At this point of the synthetic sequence, intermediate **7.25** embodies four of the five contiguous stereocenters of the advanced **7.35**. The remaining stereocenter will be derived from the C3 carbonyl group to form the desired α-isomer **7.35**, on paper an apparently simple task. The convex character of bicyclo **7.25** should force the production of the undesired β-isomer **7.36** in the hydride reduction of the ketone. However, it was assumed, based on the solid-state conformational analysis of **7.25** that the α-hydroxyl would be pseudoequatorial and, therefore, accessible by protocols directed to afford the more stable epimer. Nevertheless, consistent experimental results showed that this was not the case. The best ratio of epimers was obtained with DIBALH, which furnished a 2.5:1 mixture favoring the β-isomer **7.35**. Besides, all attempts to invert the configuration of the alcohol using standard Mitsunobu methodology or any nucleophilic attack in the mesylate **7.37** met with no success. Those reactions formed a major product **7.38** arising from the intramolecular alkylation of sulfur, followed by fragmentation of the resulting sulfonium ion (Scheme 7.9).

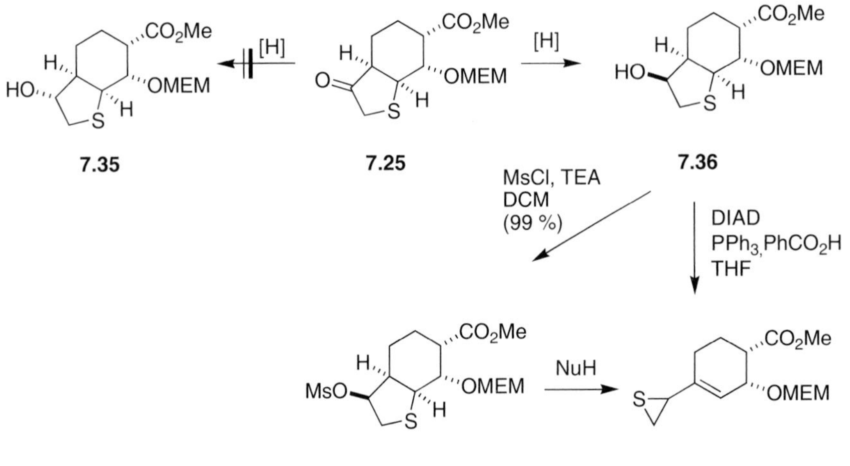

Scheme 7.9

Reversible oxidation of the sulfur moiety seemed the obvious solution and it was in the end, although not in the way predicted. In fact, the oxidation of alcohol **7.36** or its mesylate **7.37** with Davis phenyl oxaziridine [7] formed **7.39** and **7.40** both as mixtures of diatereomeric sulfoxides. Attempted inversion of the alcohol group at C3 in **7.39** or **7.40** gave the vinyl sulfoxide **7.41**. The elimination of the alcohol moiety occurred even in non-basic conditions. This unexpected result was the indication to continue with the synthesis. Reduction of **7.41** to the vinyl sulfide **7.42** followed by the sequence hydroboration-oxidation, exploiting the convex topography of the substrate, this time gave the desired α-alcohol **7.35** (Scheme 7.10). Compound **7.35** finally had the required C3 α-stereochemistry. Silylation with TBDPSOTf, followed by reduction with DIBALH afforded the advanced intermediate **7.43**.

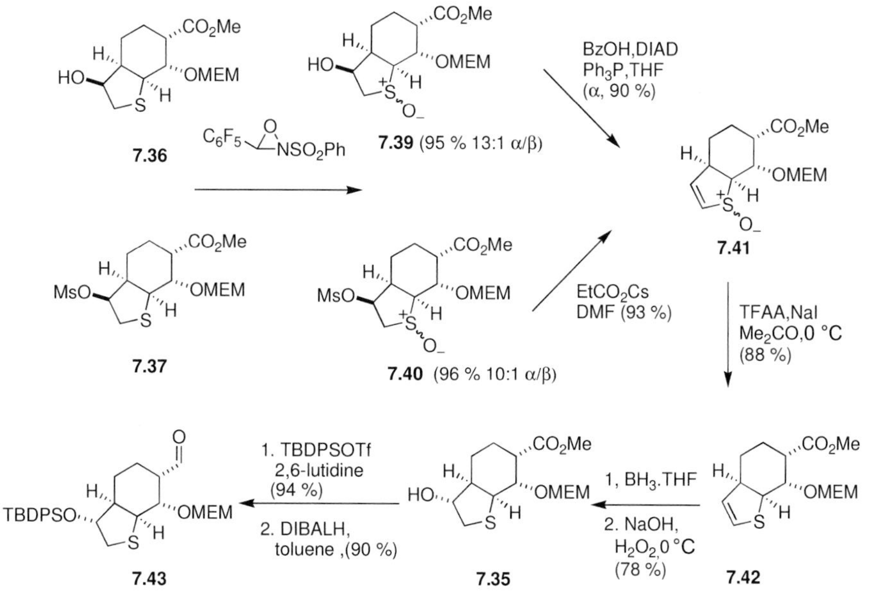

Scheme 7.10

This detour does not result in significant deviations from the synthetic scheme planned in Scheme 7.4 but requires the incorporation of five additional synthetic steps (Scheme 7.11). What is the origin of the unpredictable inability to effect the inversion of the alcohol at C3? Could it be predicted?

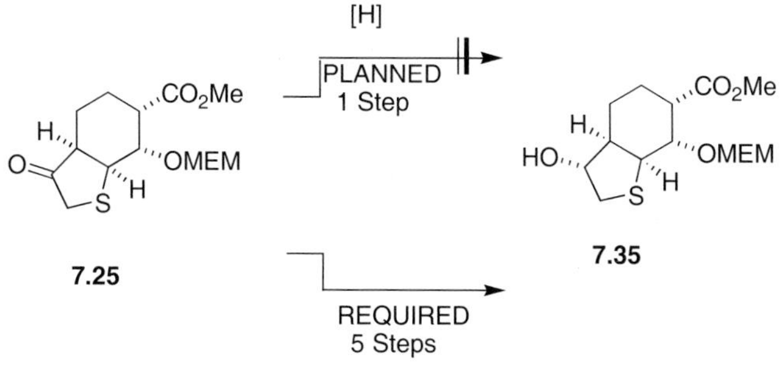

Scheme 7.11

Step 3. Spiroketal formation (Scheme 7.12):

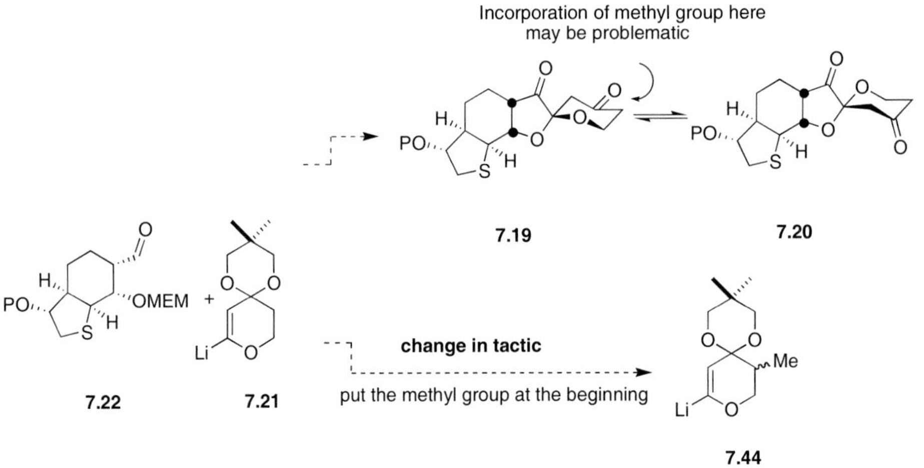

Scheme 7.12

The next key transformation in the synthesis of (±)-breynolide is the coupling of aldehyde **7.43** with the lithiated dihydropyrane **7.21**. As commented under the section on predictable problems, the chemo- and regioselective installation of the C12-Me group onto an advanced intermediate, like **7.19**, may be problematic. Instead, the incorporation of the potentially conflicting methyl group could be effected prior to spiroketalization, using a lithium derivative **7.44** (Scheme 7.12). Therefore, lithium derivative **7.44** was reacted with aldehyde **7.43** leading to **7.45**. The ketal group was removed and the secondary alcohol oxidized to ketone **7.46** using the Swern conditions [8]. Ketone **7.46** was the substrate in which the key spiroketalization step would be effected and so the stereocenter on C12 should now be established concurrently. Treatment of **7.46** with $ZnBr_2$ removed the MEM group and the free alcohols were exposed to TsOH. The product was the desired spiroketal **7.47** showing that the intended equilibration has occurred (Scheme 7.13).

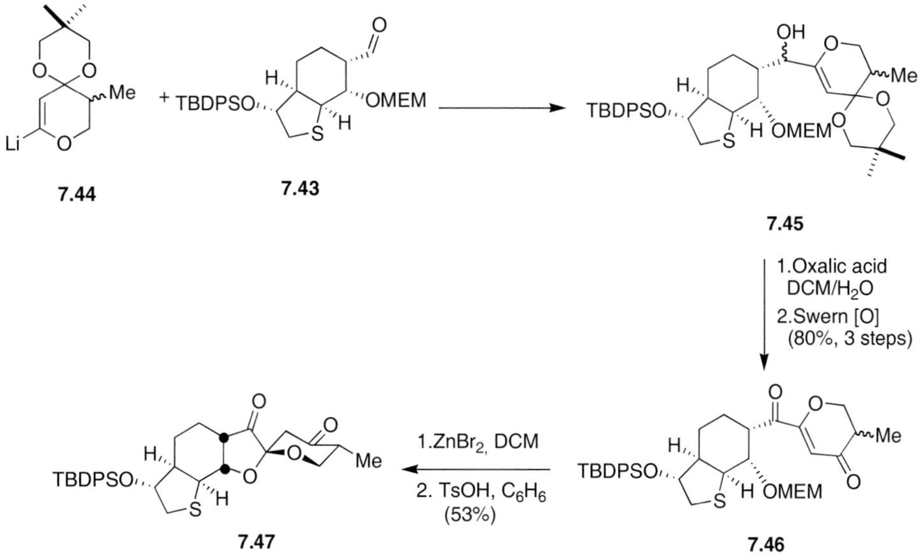

Scheme 7.13

Step 4. Incorporation of the trans-vicinal diol moiety and completion of the synthesis (Scheme 7.14):

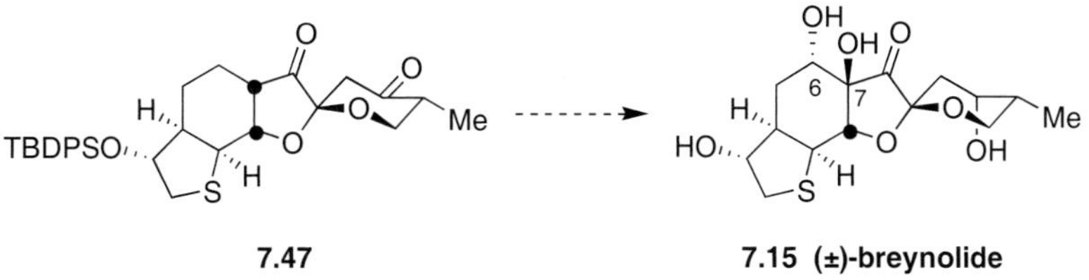

Scheme 7.14

To complete the task of accessing (±)-breynolide **7.15** it is necessary to selectively reduce the pyranone carbonyl group to the axial alcohol and to install the C6-C7 *trans*-diol moiety. The first transformation could be achieved chemoselectively by using the bulky L-Selectride to reduce **7.47**. The chemoselectivity of this transformation yielding exclusively alcohol **7.48**, derived from the higher accessibility of the pyranone ketone of **7.47**, compared with the more hindered five-membered ring ketone. Again the topology of the molecule determines the result of the reaction, but this time in our favor. Sylilation of **7.48** formed **7.49**, which was elaborated to the C6,C7-unsaturated enone **7.50** by treatment of the enolate with PhSe(O)Cl [9] (the use of this reagent avoids the oxidation of the sulfur) (Scheme 7.15).

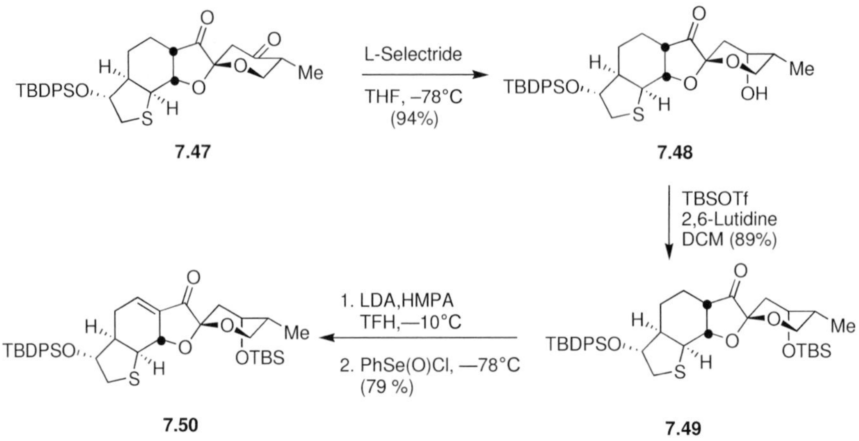

Scheme 7.15

The incorporation of the vicinal diol moiety was effected on the C6,C7-unsaturated enone **7.50**. Enolization of **7.50**, followed by α-hydroxylation of the extended enolate with (+)-camphorsulfonyl oxaziridine [7], leads to alcohol **7.51**. Reduction of the carbonyl group was the single result obtained in all the attempts to hydroborate-oxidize this alcohol **7.51**. Therefore, it was necessary to attempt dihydroxylation of enone **7.50** with stoichiometric amounts of OsO$_4$, reactions which produced extensive oxidation of the sulfur atom, and a side reaction which substantially decreased the reaction yields of diol **7.52**. Desilylation with acidic methanol then afforded 6-epibreynolide **7.53**. Again, any attempt to invert the C6 β-hydroxyl group were futile (Scheme 7.16).

Finally, the serendipitous discovery that enone **7.50** isomerizes to the β,γ-unsaturated isomer **7.55** upon exposure to aqueous K$_2$CO$_3$, opened up a new way of solving the problem. This isomerization is a manifestation of the highly strained architecture of the conjugated enone **7.50**. Due to this strain, the Michael addition of alcohols to this enone also occurred easily. A variety of alcohols in conjunction with several base catalysts were then examined. The best results were obtained with cesium carbonate as catalyst. Treatment of **7.50** with benzyl alcohol furnished **7.56**. The completion of the *syn*-diol fragment would entail hydroxylation at C7 followed by deprotection of the benzyl group at C6. The hydroxylation of the enolate derived from **7.56** placed the needed hydroxyl group at C(7). At this point, deprotection of the benzyl group seems the single remaining task to finally obtaining (±)-breynolide. However, none of the methods used to remove the benzyl group succeeded (Scheme 7.17).

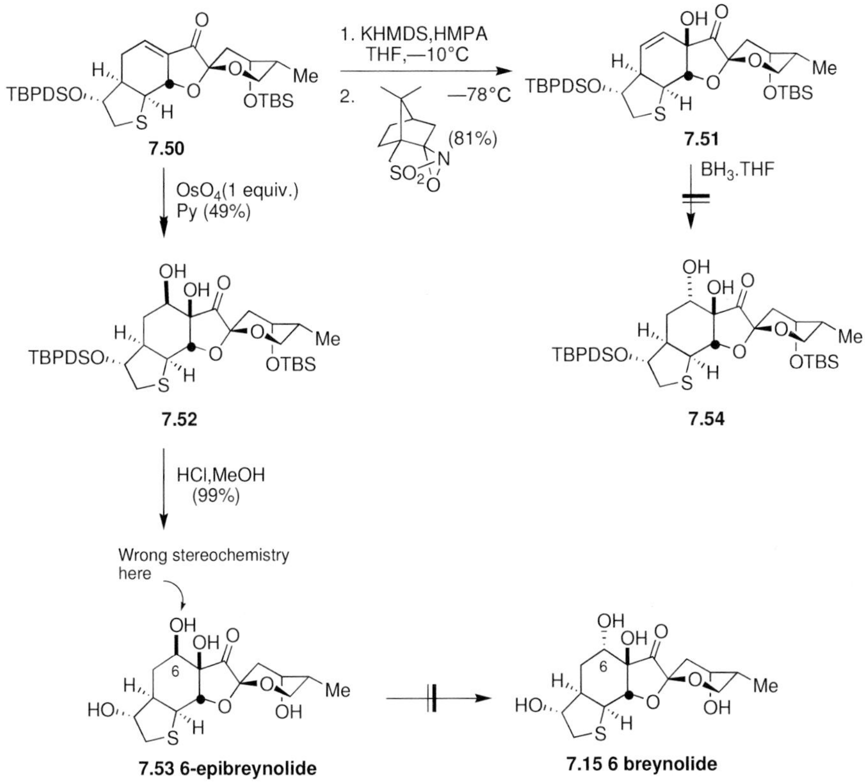

Scheme 7.16

Alternatively, the sequence can be repeated with allyl alcohol in the presence of CsCO$_3$ to obtain a 3:1 mixture of α- and β-adducts **7.58**. Hydroxylation of the α-adduct at C7 formed the allyl protected diol **7.59**. Deallylation was effected employing [(Ph$_3$P)$_3$RhCl] to isomerize the allyl ether to vinyl ether [10] followed by hydrolysis. This treatment simultaneously eliminated the silyl protecting groups, finally affording synthetic (±)-breynolide **7.15** (Scheme 7.18).

This method of accomplishing the last steps of the synthesis, represents a minor detour from the original synthetic plan that cannot even be considered as a tactical change. However, the enormous effort devoted in tuning up the introduction of the diol moiety is not truly reflected in the final work, even if the problem was predicted from the beginning. Further, the impossibility of eliminating the benzyl group in **7.57** the fully elaborated precursor of (±)-breynolide, has to be considered a major drawback. Evidently, changing the protecting groups would solve the problem. However, it means repeating a series of reactions on a very advanced intermediate, with the logistical problems of availability of advanced materials associated with the last steps of a multi-step synthesis.

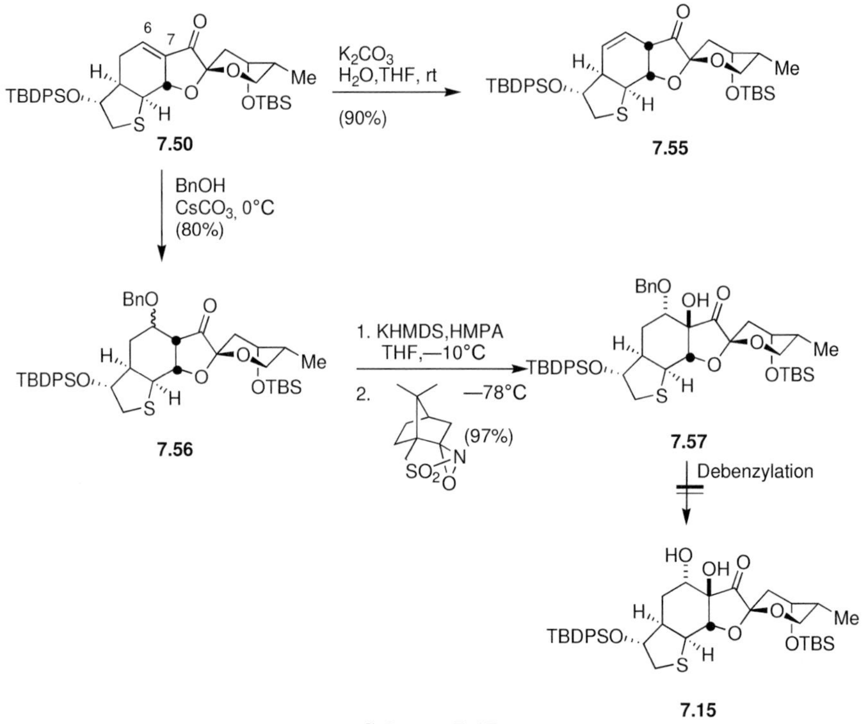

Scheme 7.17

Evaluation

- No strategic changes have been made during the synthesis of (±)-breynolide. The synthesis plan depicted on Scheme 7.4 has been fully implemented.
- Major tactical changes have been made during the synthesis, especially to install the C(3)-α-OH.
- The planned synthesis is complicated by the concave topology of many of the synthetic intermediates.

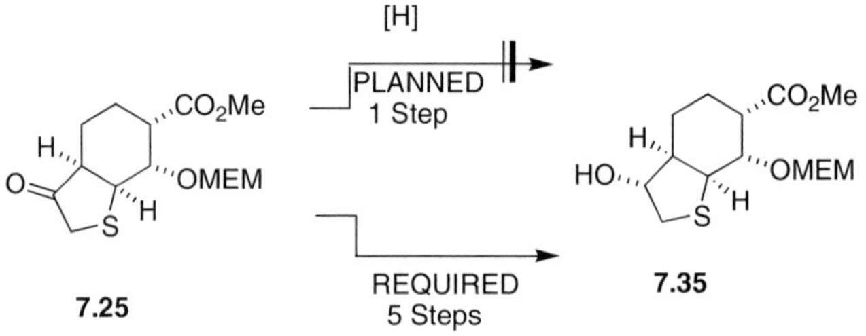

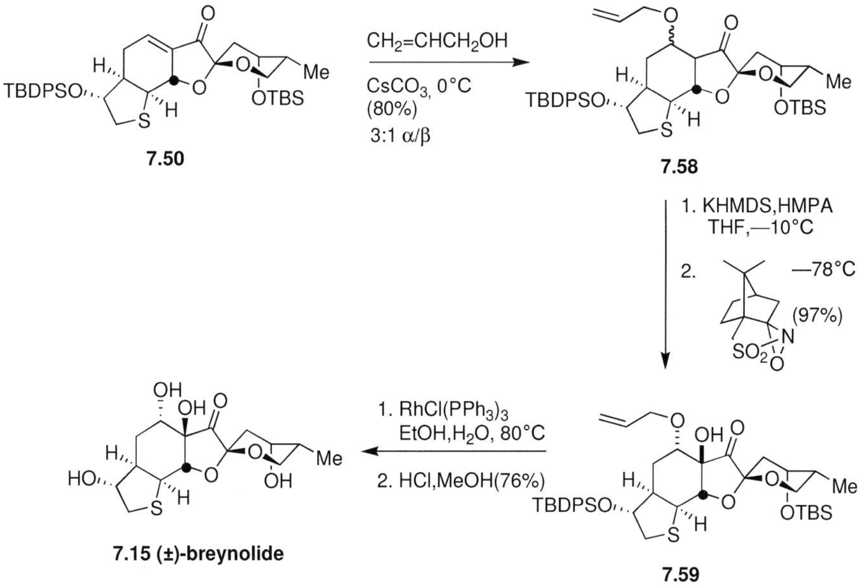

Scheme 7.18

Especially interesting reagents

PhSe(O)Cl [9]: Obtaining of α,β-enones from ketones via their enolates. The use of benzeneseleninyl chloride allows the presence of functional groups susceptible to oxidation.

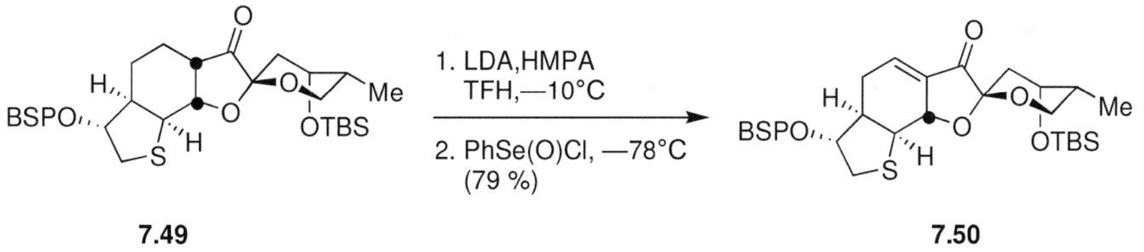

Key synthetic reactions

Davis oxidation: α-oxidation of an enolate by a 2-sulfonyloxaziridine. Using a chiral 2-sulfonyloxaziridine produces good enantiomeric excesses [7].

7.3   Hamigerans A and B [11]

7.60 hamigeran A   **7.61 hamigeran B**

Target relevance

Hamigerans were isolated from the poecilosclerid sponge *Hamigera tarangaensis* Bergquist & Fromont (family Anchinoidae, syn. Phorbasidae) from the Hen and Chicken Islands off the coast of New Zealand [12]. These compounds show a moderate cytotoxicity against P-388 leukemia cells and a strong antiviral activity. In spite of the small size of the molecules, the complexity of their structures offers a challenge for synthesis as well as the possibility of obtaining analogs to improve their biological activities.

Synthetic plan

The cornerstone of the synthetic plan to access to hamigerans was the photo-enolization and intramolecular trapping of hydroxy-*o*-quinodimethanes **7.63** generated from the corresponding *o*-tolualdehydes **7.62** having a tethered dienophile [13] (Scheme 7.19).

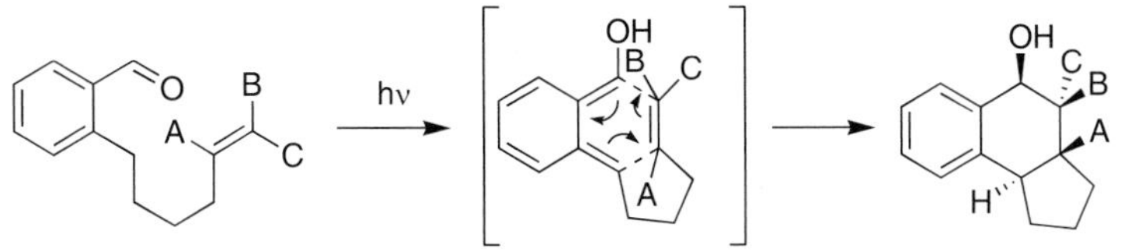

Scheme 7.19

Based on this general construction of the tricyclic skeleton of hamigerans, the approach to hamigeran A **7.60** was devised from the benzaldehyde **7.65**. In fact, the photocyclization of **7.65** will form tricycle **7.66** having the full carbon framework of hamigeran A **7.60**. This compound will be derived from **7.66** by adjusting the oxidation levels and placing the bromine group in the aromatic ring of an intermediate derived from **7.67**. Moreover, the oxidative cleavage of the hydroxy acid derived from hydroxy ester **7.67** would lead, after protecting group manipulations, to hamigeran B **7.61** (Scheme 7.20). Clearly, and according to this synthetic plan, the epimerization of the C5 center has to be effected during the synthesis (**7.66 → 7.77**).

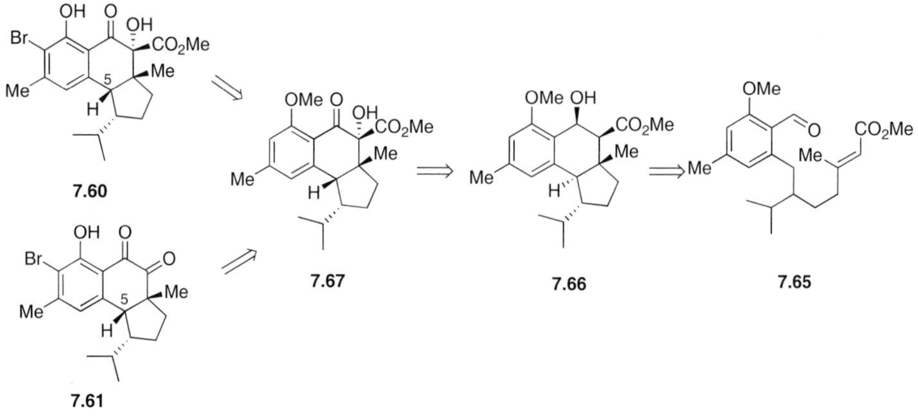

Scheme 7.20

Predictable problems

- The establishment of the relative stereochemistry at the four contiguous stereocenters (C5, C6, C9 and C10) would be predictably difficult.
- The suitability of the required inversion of the stereochemistry at C5, at any stage of the synthesis, is not clear.

Step 1. Synthesis of hydroxy ester 7.67: Following the synthetic plan depicted in Scheme 7.23, the irradiation of aldehyde **7.65** produced in an exceptional 91% yield, the expected hydroxy ester **7.66**. The relative configuration of the substituents at the C5, C6, C9 and C10 carbons was fully controlled by the single stereocenter at C6. Compound **7.66** was formed as a mixture of 3:1 epimers at C10 as a consequence of 3:1 *E/Z*-mixture used as the starting material **7.65**. This is irrelevant since the configurations of C10 and C11 will be further removed. Most important is that the stereochemistry of C5 in compound **7.66** was the wrong one, causing a necessary correction to a further step of the synthesis. Acid treatment of **7.66** efficiently formed **7.68**, which was dihydroxylated with OsO$_4$/NMO in the presence of pyridine to form the dihydroxy derivative **7.69**. Dihydroxylation occurred mainly by the α-face (12:1) and the benzylic alcohol of **7.69** was oxidized to ketone **7.70** with SO$_3$ · py in DMSO. Any attempts to epimerize the C5 stereocenter on **7.70** or its precursors **7.66** or **7.69** were unsuccessful. Therefore, access to hamigeran A by this route was thwarted (Scheme 7.21).

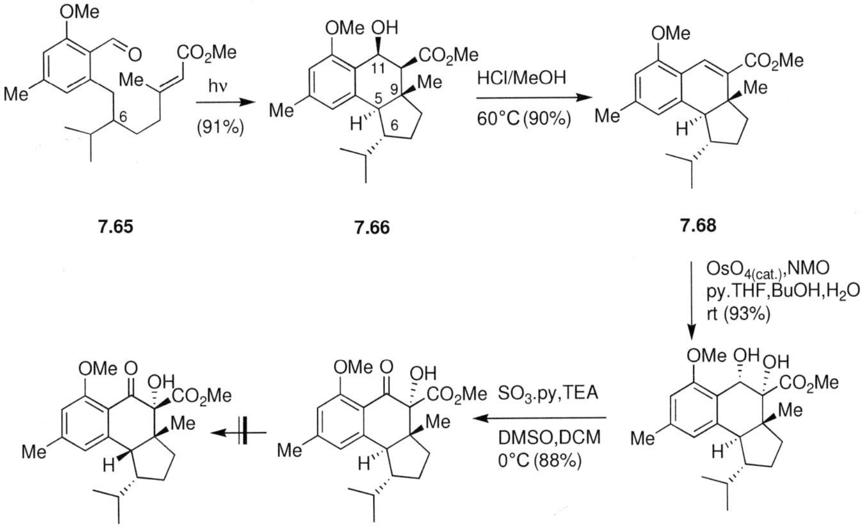

Scheme 7.21

Step 2. New entry to hamigerans by placing a hydroxyl group on C6 to facilitate C5-epimerization: Since the problem with the first synthetic design was its inability to effect the necessary epimerization of C5, it was thought that the introduction of one adjacent ketone moiety to C5 would facilitate this epimerization while allowing for the further introduction of the isopropyl group. In this regard, compounds **7.71** and **7.72** will be used as key synthetic intermediates (Scheme 7.22).

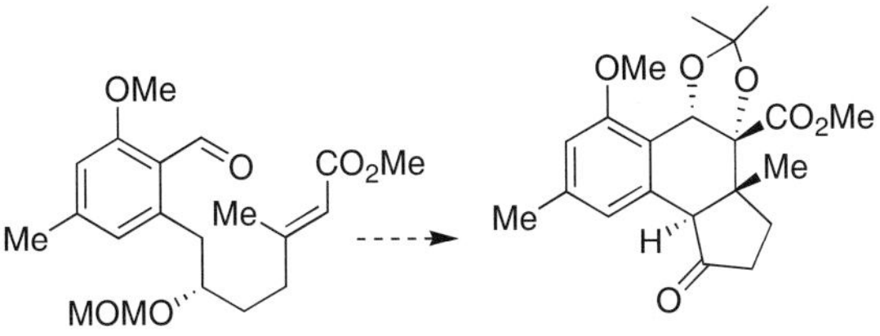

Scheme 7.22

Photo-irradiation of **7.71** led to the hydroxy ester **7.73**, which was transformed to diol **7.74** following an analogous sequence to that depicted in Scheme 7.23. Diol **7.74** was protected as the acetonide **7.75**, the free alcohol oxidized and, as predicted, with the carbonyl group in place, the base-induced isomerization at C5 required only a short exposure to DBU at 0°C to give the *cis*-fusion compound **7.76**. The isopropyl moiety was attached to the main skeleton using the cerium-mediated Grignard addition, giving the product derived from the attack by the *exo* face **7.77**. Reductive removal of the tertiary alcohol proved to be unfeasible. Ether **7.78** was formed, probably in all the conditions tested, by intramolecular trapping of the tertiary carbonium ion being developed α- to the pendant methoxycarbonyl group by the proximal C5 hydroxy group.

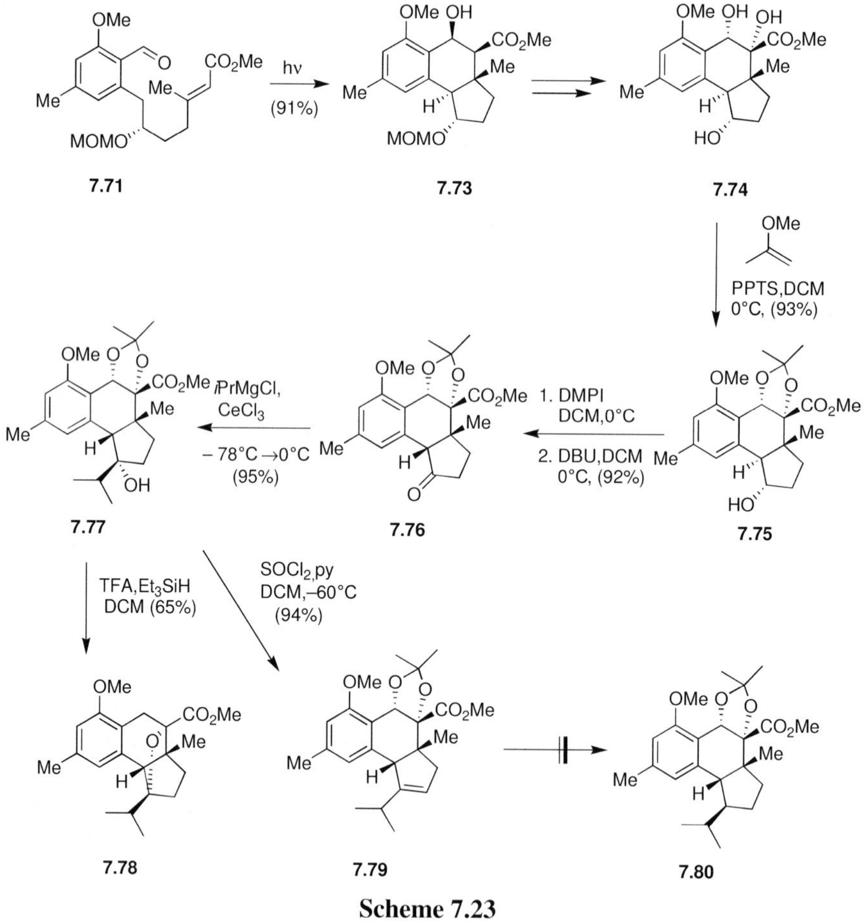

Scheme 7.23

Alternatively, the elimination of the C5 alcohol formed olefin **7.79** together with its conjugated and exocyclic double bond isomers (10:2:1 ratio). No reduction methods used to reinstall the correct stereochemistry of the isopropyl group were satisfactory since they led to complex mixtures of products together with the desired **7.80**. Therefore, this approach to hamigerans was also discarded (Scheme 7.23).

Step 3. Installation of the correct C6 stereochemistry on **7.79** *and final entry to hamigerans:* The failure to establish the right C6 stereochemistry discussed above forced authors to search for a detour from the original sequence leading to hamigerans. The problem increased during the attempts to manipulate the five-membered ring double bond. This was attributable to the facial selectivity of **7.79**. Now the intrinsic facial selectivity of **7.79**, together with the regioselectivity of the hydroboration of double bonds, could be used to access hamigerans (Scheme 7.24).

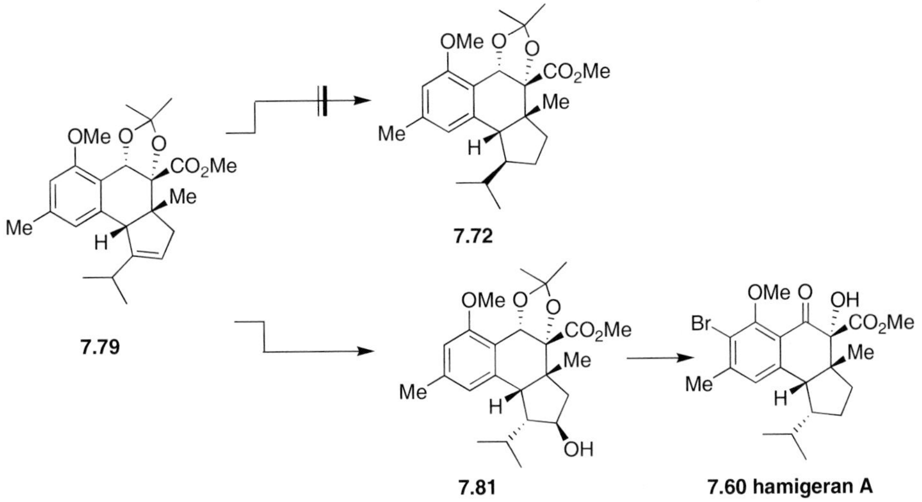

Scheme 7.24

Therefore, **7.79** was hydroborated with BH$_3$.SMe$_2$ under sonication to form, after the standard oxidative work-up, the needed alcohol **7.81** (44% yield) together with its α-estereoisomer **7.82** (24% yield). Alcohol **7.81** has the stereochemistry of hamigerans, therefore it was separated and submitted to the radical (TBTH,AIBN) deoxygenation [14] of its phenylthiocarbonate, and to removal the ketal moiety to yield **7.83**. The diol **7.83** was oxidized to ketone **7.67** with PDC. Debromohamigeran **7.84** was obtained by BBr$_3$ induced cleavage, and transformed to hamigeran A **7.61** by NBS bromination. Oxidative decarboxilation of **7.61** formed the second target, hamigeran B **7.62** (Scheme 7.25). Therefore, both synthetic targets were finally achieved.

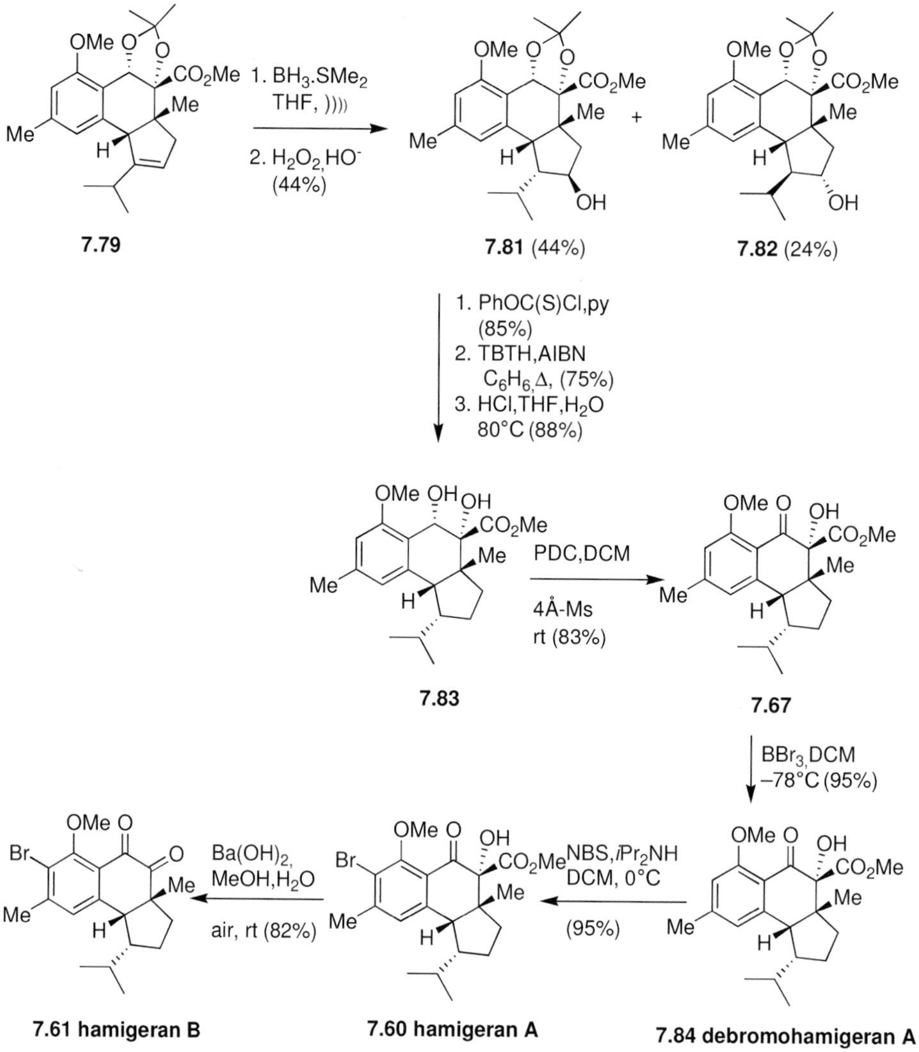

Scheme 7.25

Evaluation

- The synthesis of hamigerans A and B represents a concatenation of problems. This is due to the difficulties encountered in the installation of the desired stereochemistry at different points of the synthetic plan. The predictable problem of installing the relative stereochemistry at the four contiguous stereocenters (C5, C6, C9 and C10) was cleverly solved by the intramolecular *o*-quinodimethane cycloaddition. However, the task of epimerizing the C5 center on **7.70** was not accomplished, which forces a strategic change by placing a group at C6, which would facilitate such epimerization. This new approach matches perfectly the task for which it was designed. Nevertheless, the installation of the isopropyl pendant group with the right stereochemistry could not be done. This new failure resulted in a significant tactical detour, involving construction of two stereocenters in compound **7.79** with low selectivity *via* hydroboration-oxidation. When the isomer has the right stereochemistry, the synthesis of hamigerans is finally achieved.

7.70   **7.67**

7.79   **7.72**

- It is interesting to note that a very powerful synthetic transformation like the intramolecular *o*-quinodimethane cycloaddition happens exactly as planned. However, the synthesis becomes really complicated by stereochemical problems caused by the intrinsic facial selectivity of different intermediates (for example **7.79**) or the inertia of, *a priori*, reactive positions (the benzylic position of **7.68** to **7.70**).

A clear situation that may cause unresolvable problems during a synthesis is the intrinsic selectivity during the coupling of two enantiomerically pure intermediates. This is a key step in many convergent syntheses, and with regard to controlling the outcome of the process, causes serious problems, which result in significant deviations from or drawbacks to the original planning.

7.4   Calphostin A [15]

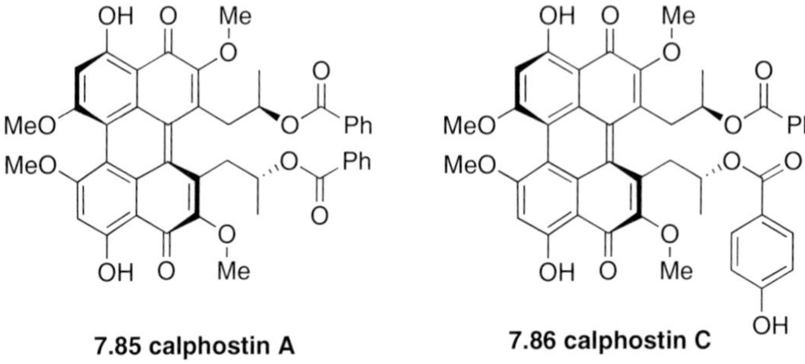

7.85 calphostin A   **7.86 calphostin C**

Target relevance

The calphostins are isolated from the phytoparasitic mold *Cladosporium cladosporioides* [16]. Calphostins A **7.85** and C **7.86** are potent and selective inhibitors of protein kinase C, a regulatory enzyme important for cellular differentiation and proliferation, playing a key role in the *trans*-activation event in HIV infected T-lymphocytes.

Synthetic plan

The plan to access calphostin A **7.85** relies on the atropdiastereoselective Cu(I) promoted coupling [17] of two units of a chiral naphthalene moiety like **7.87**. The stereogenic side chains will be introduced in the naphthalene moiety by addition of a chiral lithium reagent **7.89** to a naphthalene aldehyde **7.88** (Scheme 7.26).

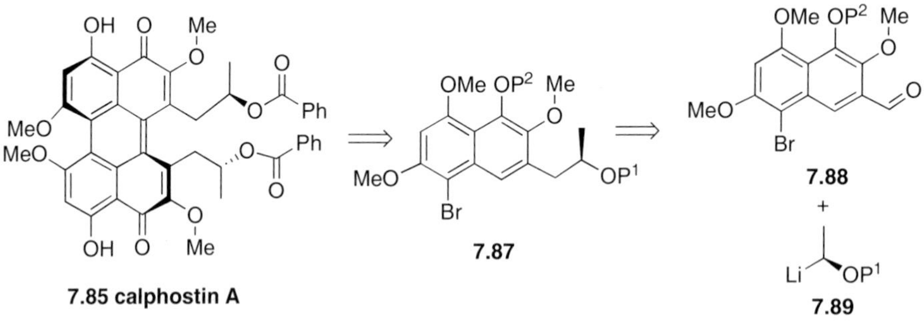

7.85 calphostin A   **7.87**   **7.88**
                                                            +
                                                     7.89

Scheme 7.26

Predictable problems

The compromising step of the synthetic plan is the sense of the atropdiastereoselection during the joining of the two chiral fragments derived from **7.87** to build the perylenequinone skeleton of calphostin A. This process should give the right stereochemistry, assuming that it is based on the previous work by Broka [18] who reported the closely related dimerization of the lithium derivatives **7.90** promoted by FeCl$_3$ to yield **7.91** (Scheme 7.27).

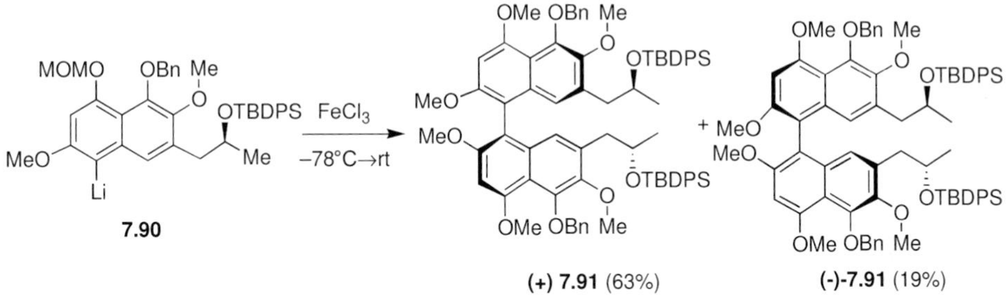

Scheme 7.27

*Step 1. Synthesis of naphthalene **7.92** and its dimerization reaction:* Compound **7.92** was prepared by the addition of the configurationally stable lithium derivative **7.93**, prepared by treatment of stannane **7.94** with BuLi, to aldehyde **7.95**. Alcohol **7.96** was obtained as a 1:1 mixture of diastereomers. The hydroxyl group of **7.96** was then removed by using the Barton–McCombie [14] procedure to provide naphthalene **7.97**. The regioselective bromination of this compound was achieved by treatment with NBS affording enantiomerically pure naphthalene **7.92**.

The halogen–metal exchange was realized on **7.92** by treatment with BuLi, and the dimerization was accomplished in the presence of CuCN/TMDA followed by oxygen. In this way, an astonishing 8:1 atropselectivity was obtained with compound **7.98** as the main isomer. Unfortunately, this compound had the wrong *R*-configuration around the axial chiral axis. Therefore, the choice of the configuration of the chiral center in the side chains was adequate because it is the one present in calphostin A. However, the complete absence of stereocontrol in the formation of the new chiral element (the chiral axis) makes the stereochemical outcome a question of chance that, in this case, acted against the authors (Scheme 7.28).

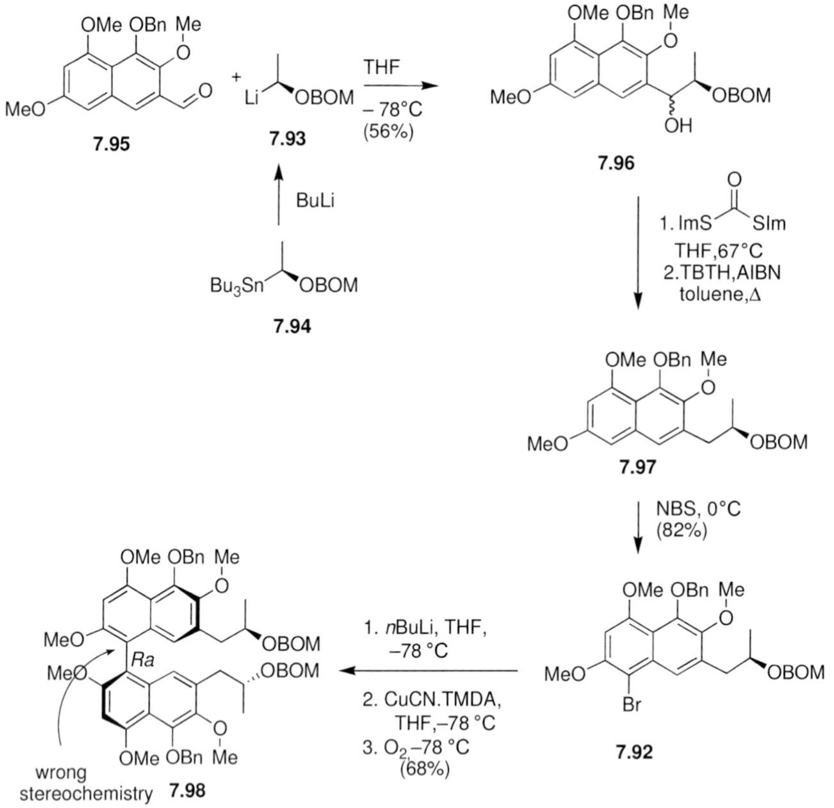

Scheme 7.28

Step 2. Changing the stereochemistry of the chiral side chain and accessing to calphostin A: Obviously, the enantiomer of compound **7.92**, naphthalene ***ent*-7.92** should afford the desired axial stereochemistry. However, the side-chains of binaphtyl **7.99** would now bear the wrong configuration at both stereocenters. Therefore, additional steps to invert this stereochemistry have to be included to reach calphostin A **7.85**. For this reason, naphthalene ***ent*-7.92** was prepared following a reaction sequence identical to that depicted in Scheme 7.31 but now using ***ent*-7.93**. Binapthyl **7.99** was prepared in this way, and the correction of the stereochemistry of the side chains was effected by a double Mitsunobu inversion under carefully controlled conditions, which concomitantly introduced the required benzoate groups, affording **7.100**. Removal of the benzyl-protecting groups, followed by *p*-phenoxy radical cyclization and oxidation, gave the perylenequinone **7.101**. Finally, selective cleavage of the methyl ethers formed calphostin A **7.85** (Scheme 7.29).

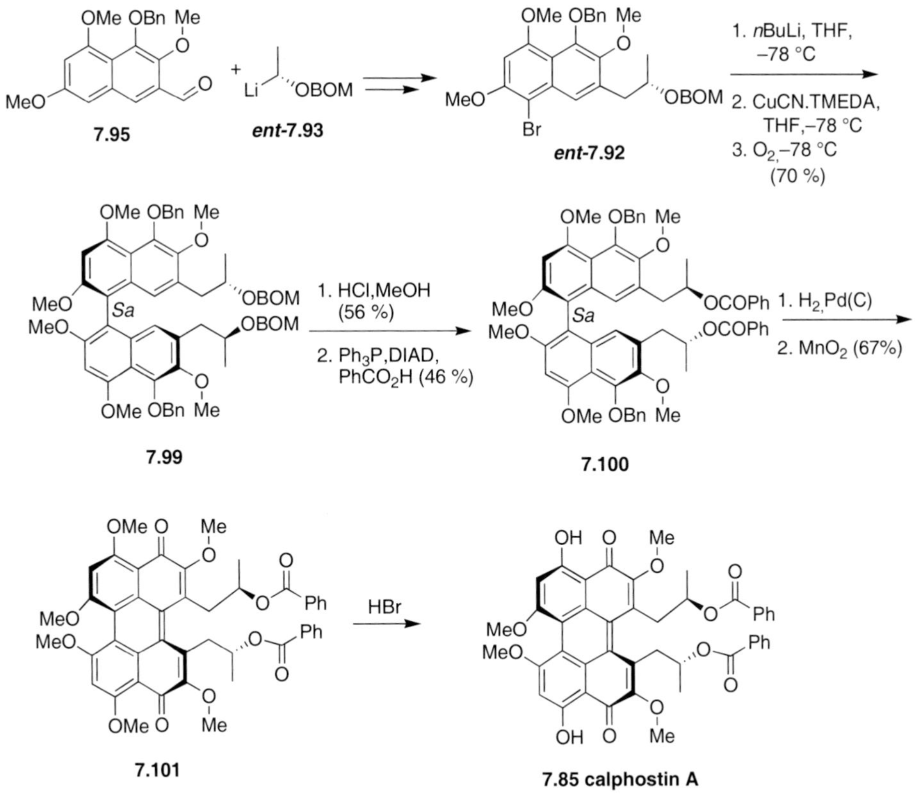

Scheme 7.29

Evaluation

Efforts to control the troublesome intrinsic selectivity of the oxidative coupling of naphthalene **7.92** were not made. The route to the final product was chemically efficient but it involved repeating the sequence and introducing an additional inversion step to achieve the desired target. No ways of exercising this kind of control efficiently are yet known [19].

Key synthetic reactions

The Barton–McCombie radical desoxygenation: Desoxygenation of secondary alcohols to methylene fragments. The process is sequential and involves the preparation of a xanthate (many times isolable) by reaction with Im$_2$C=S/base (usually NaH), followed by radical reduction by reaction with TBTH/AIBN in boiling toluene [14] (Scheme 7.33).

The use of enantiomerically pure reagents to overwhelm the presumed modest diastereo-facial preferences of the chiral substrate and to control the stereochemical outcome is now standard synthetic methodology. In these cases, especially for aldol-type and related condensations the configuration of the products is highly predictable [20]. Nevertheless, surprises are still encountered.

7.5 Fragment C3-C10 of Cytovaricin [21]

7.102 cytovaricin

7.103

Target relevance
The macrolide cytovaricin **7.102** is produced by *Streptomyces diastatochromogenes* [22] and is highly active *in vitro* against Yoshida sarcoma cells. The successful synthesis of a molecule having the extreme complexity of cytovaricin represents one landmark among the many synthetic successes in the late 80's of the 20th century. The example that follows will be devoted to the preparation of the fragment C3-C10 **7.103** to exemplify the problem of well-established techniques for stereocontrol failing, in a key trans-formation. This situation requires the undertaking of risky detours to achieve the final goal.

Synthetic plan
To prepare the fragment **7.103**, the asymmetric aldol reaction is the obvious method. This would lead to compounds **7.104** and **7.105** as precursors, both derivable from easily available products (Scheme 7.30).

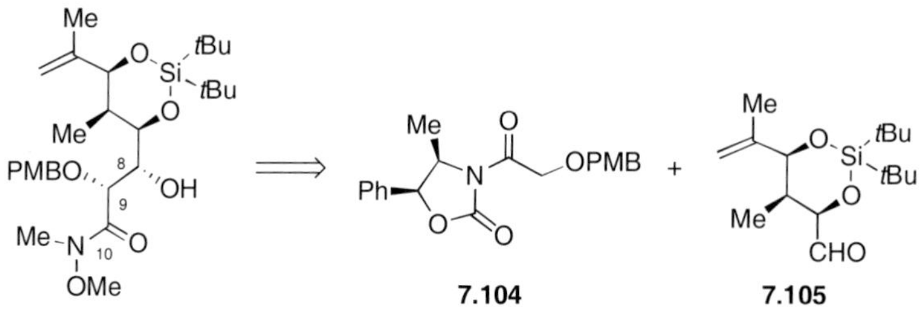

7.103

7.104   **7.105**

Scheme 7.30

Predictable problems
The extensive body of knowledge for this kind of reaction predicts complete control of the stereochemical outcome [20][23], by the facial bias of the chiral enolate derived from **7.104**, its geometry, and the cyclic transition states for the aldol condensation, regardless of the configuration of the aldehyde **7.105**. Therefore, no problems are foreseeable.

Step 1. Aldol condensation and synthesis of fragment 7.103: The boron enolate derived from **7.104** was condensed with aldehyde **7.105**. Surprisingly, the *anti* aldol adduct **7.106** was obtained as a single diastereomer instead of the expected *syn* aldol adduct **7.107**. This reversal of the stereochemistry was unprecedented in this condensation using this chiral enolate reagent. Apparently the configuration of C9 is defined by the chiral enolate, while the inherent bias (predicted by the Felkin–Anh model [24]) of the chiral aldehyde, determines the formation of the C8 stereocenter. Other organometallic reagents such as MeLi and MeMgBr add to the same carbonyl diastereoface with selectivity greater than 20:1. All efforts to override the diastereofacial preference of aldehyde **7.105** were fruitless. In the words of the authors: *Ironically, all three of the undesired aldol diastereomers could be obtained as major products by variation of reaction conditions. A variety of metal enolates (B, Li, Ti, Zr, Mg) were surveyed. Stoichiometry, solvent, and auxiliary were also varied to no avail.*

This unexpected result led to the difficult problem of how to invert the stereochemistry of the C8 center in a very sensitive molecule. This task was undertaken on transaminated intermediate **7.108** formed by the reaction of **7.106** with AlMe₃/MeONHMe.HCl (formation of the Weinreb amide). The key was using the diastereofacial bias of a trigonal carbon of **7.109** at C8. Oxidation of the alcohol at C8 followed by immediate reduction of the β-ketoamide **7.109** with L-Selectride gave a single diastereomer of the desired *syn* alcohol **7.103**. In spite of the fact that β-ketoamide **7.109** is the intermediate in the transformation of **7.108** to **7.103**, no epimerization at C9 was observed (an explanation of this fact is that allylic 1,3-strain forces the molecule into a conformation that sufficiently reduces the acidity of C9 methyne hydrogen, to avoid the epimerization at this center) (Scheme 7.31).

Evaluation

Was the formation of *anti*-aldol **7.103** instead of its *syn*-diastereomer **7.104** predictable? Probably, after exhaustively studying the facial selectivity of aldehyde **7.102** towards the nucleophilic addition and aldol condensation reactions, the answer would be "yes". At the beginning, when designing an extremely difficult task such as preparing synthetic cytovaricin, previous experience of the stereochemical outcome of the aldol reactions of enolates derived from oxadolidinones like **7.101**, is overwhelming and therefore makes the experimental result surprising.

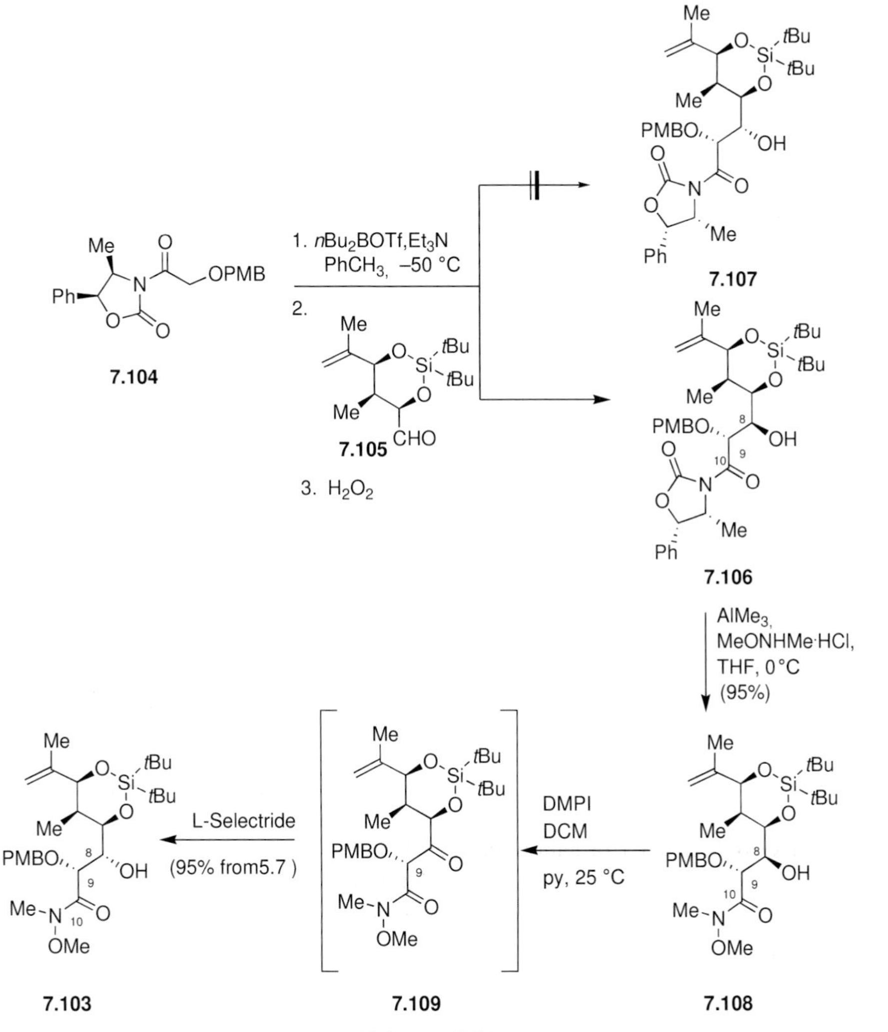

Scheme 7.31

The inability to obtain the right stereochemistry in a synthetic sequence may even totally thwart the approach to the synthetic target, leading to a completely new synthetic design. The examples above represent four different situations in which the stereochemical outcome of one synthetic step is wrong, but can be corrected by introducing additional synthetic steps. Some times this is not the case and the planned route has to be abandoned due to the impossibility of obtaining the stereoisomer needed. This situation may be intrinsic to the methodology used.

7.6   Goniocin [25]

7.110 goniocin

7.111 cyclogoniodenin T

Target relevance

Goniocin **7.110** and cyclogoniodenin T **7.111** have been isolated from *Goniothalamus giganteus* [26]. They show remarkable cytotoxic, antimalarial, immunosuppressive, pesticidal and antifeedant properties, like other Annonaceous acetogenins. Compounds **7.110** and **7.111** are structurally similar but stereochemically very different, yet they share the same biological activity. Therefore, the fact of encountering goniocin and cyclogoniodenin T in the same plant may lead to fascinating biogenetic conclusions. Since the absolute configuration of **7.110** and **7.111** was unknown, prior to discussing biogenetic implications [27], the stereochemistry of both products has to be confirmed by synthesis.

Synthetic plan

Goniocin **7.110** will be accessed by the Wittig joining of two fragments **7.112** and lactone **7.113**. The fragment **7.112** will be prepared by the oxidative cascade cyclization of alcohol **7.114**. This is the key transformation and the point is to transfer the chirality of the single estereocenter of alcohol **7.114**, derived by the Sharpless asymmetric epoxidation of alcohol **7.115** to the *all-trans* tricyclic tetrahydrofuran rings of **7.110** (Scheme 7.32).

Predictable problems

The key point of the synthetic design represented in Scheme 7.32 is the chirality transfer from the single alcohol stereocenter in **7.115** to the *tris*-tetrahydrofuran system. This has to be accomplished during the oxidative tandem ring closure in an impressive "tour de force", which creates seven stereogenic centers in a single synthetic step. Although previous work by these authors had shown the possibility of carrying out two consecutive cyclizations [28], to translate these findings to the scenario depicted in Scheme 7.36 is risky at the very least.

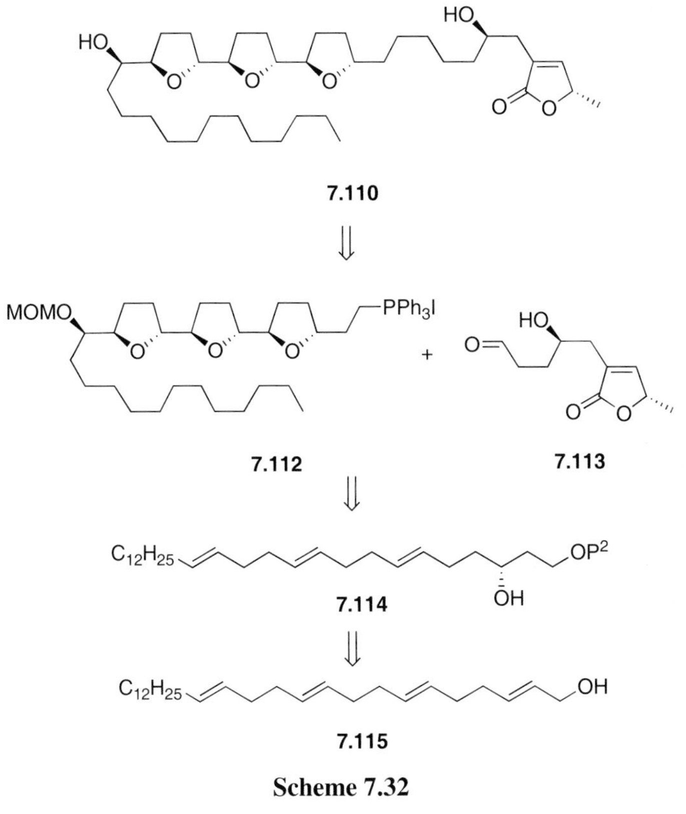

7.110

7.112   **7.113**

7.114   ŌH

7.115

Scheme 7.32

*Step 1. Building of the alcohol **7.114** and its oxidative cyclization:* The asymmetric epo-xidation of alcohol **7.115** using Ti(O*i*Pr)$_4$ and DIPT produces epoxy alcohol **7.116**. The reductive cleavage of the epoxy ring of **7.116** using Red-Al affords 1,3-diol **7.117**. This diol was selectively protected at the primary position to give **7.114**. The secondary alco-hol **7.114** was submitted to oxidative cyclization [29] using CF$_3$CO$_2$ReO$_3$/(CF$_3$CO$_2$)O. These reaction conditions produced a single *tris*-THF product **7.118**. However, this product has a *trans-threo-cis-threo-cis-threo* stereochemistry, not the desired *trans-threo-trans-threo-trans-threo* as in **7.119**. Therefore, this entry to goniocin was thwarted (Scheme 7.33).

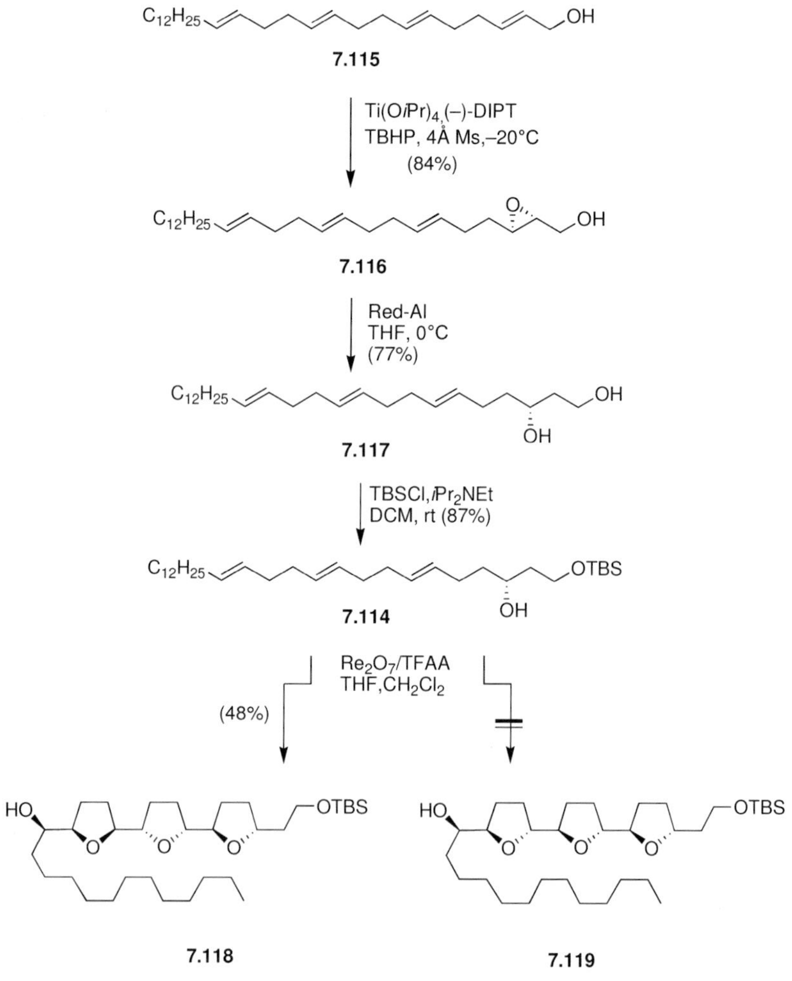

Scheme 7.33

Step 2. Second route to goniocin, deferring the cyclization step to a latter step on the synthesis: Since the tandem oxidative cyclization reaction induced by Re(VII) seemed to be fully compatible with many functional groups, it was though that delaying this cyclization to a much more advanced substrate, **7.120**, would led to a different and much more satisfactory stereochemical result. This assumption was made on biomimetic grounds. Therefore, alcohol **7.114** was elaborated to phosphonium salt **7.121** by protecting the secondary alcohol as its MOM-derivative, removing the silyl protecting group and final reaction with I_2/PPh_3. Wittig coupling with aldehyde **7.122** formed the cyclization precursor **7.120** having the complete carbon skeleton of goniocin **7.110**. The substrate **7.120** was submitted to $Re_2O_7/TFAA$ cyclization and again the tris-THF compound **7.123** having the stereochemistry of compound **7.118** was obtained. Therefore, access to goniocin was not possible by this route (Scheme 7.34).

Step 3. Re-evaluation of the synthetic planning. Designing a stepwise oxidative route to goniocin **7.110**: It is clear that the tandem oxidative THF-cyclization produced the trans-threo-cis-threo-cis-threo stereochemistry, not the needed trans-threo-trans-threo-trans-threo stereochemistry, which is present in goniocin. This unfruitful approach led to a re-evaluation of the synthetic strategy [25b]. Now, the problematic tris-THF moiety of the key phosphine **7.112** will be step-wise constructed. The first THF-ring will be closed in diene-ol **7.125** through oxidative ring closure, and subsequently the second and third THF-rings will arise from the bis-mesylate **7.126**. This new entry to goniocin is drastically different form the first failed approach (Scheme 7.35).

Step 4. First THF-ring closure and synthesis of mesylate 7.126: The oxidative cyclization of alcohol **7.125** in the presence of $CF_3CO_2ReO_3$ and lutidine produced the *trans*-THF derivative, which, by protection of the free hydroxyl group as TBS-derivative, formed **7.127**. Asymmetric dihydroxylation using AD-mix-α and double mesylation of **7.127** yielded the key mesylate **7.126** (Scheme 7.36).

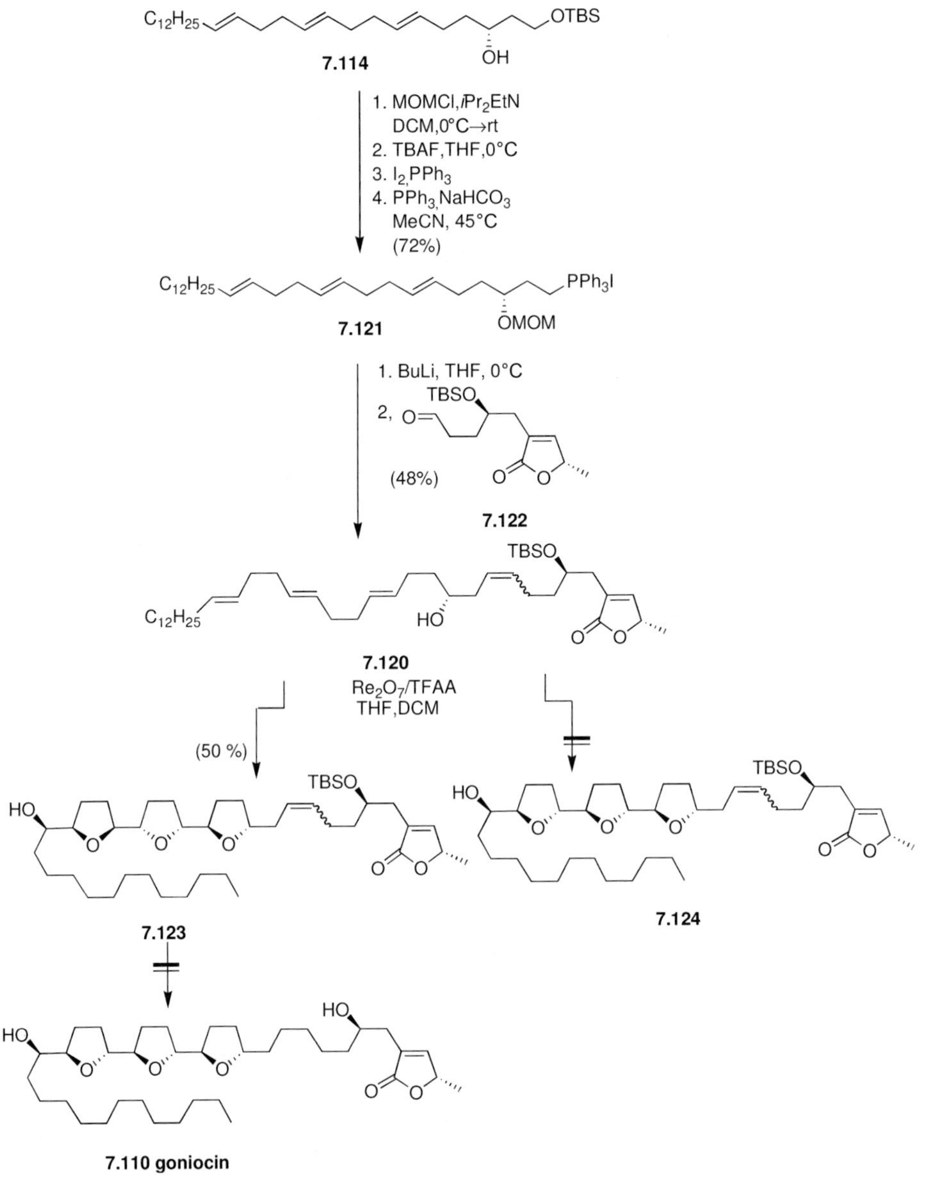
C12H25
OTBS
OH
7.114
1. MOMCl,iPr2EtN
DCM,0°C→rt
2. TBAF,THF,0°C
3. I2,PPh3
4. PPh3,NaHCO3
MeCN, 45°C
(72%)
C12H25
PPh3I
OMOM
7.121
1. BuLi, THF, 0°C
TBSO
2, O
(48%)
O
O
7.122
TBSO
C12H25
HO
O
O
7.120
Re2O7/TFAA
THF,DCM
(50 %)
TBSO
HO
O
O
O
O
O
7.123
HO
TBSO
O
O
O
O
O
7.124
HO
HO
O
O
O
O
O
7.110 goniocin

Scheme 7.34

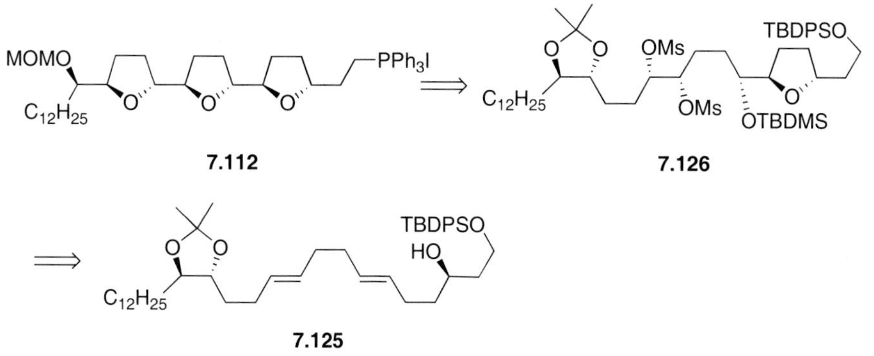

Scheme 7.35

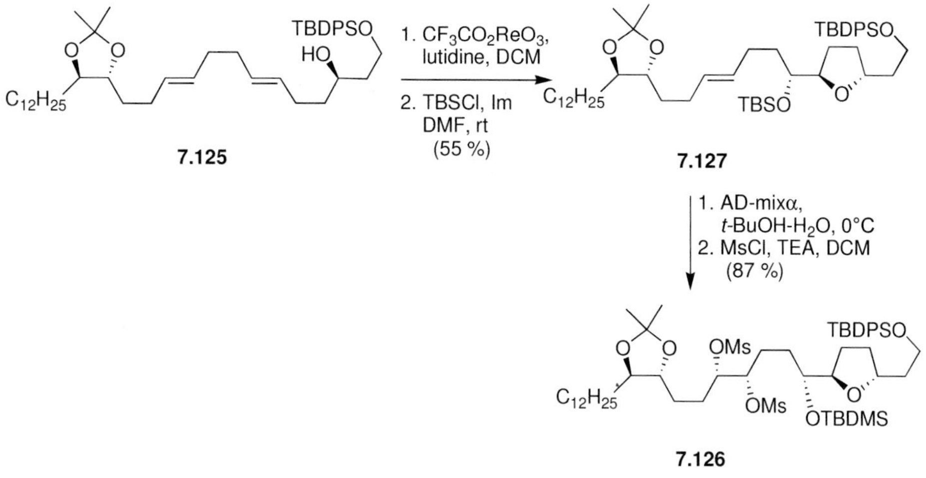

Scheme 7.36

Step 5. Building the two remaining THF rings and preparation of phosphonium iodide **7.112**. The synthesis of the two remaining THF rings requires simply the hydrolysis of the acetonide moiety as well as of the silyl ethers. This was achieved by heating **7.126** in the presence of TsOH and resulted in the formation of the *all-trans*-tris-THF **7.128**. To prepare **7.112** it was necessary to protect the primary alcohol of **7.128** as TBS-derivative, followed by the protection of the secondary alcohol as MOM-derivative. Hydrolysis of the silyl-protecting group was followed by iodination of the primary alcohol (I_2/PPh$_3$) and formation of **7.112** by treatment with PPh$_3$ (Scheme 7.37).

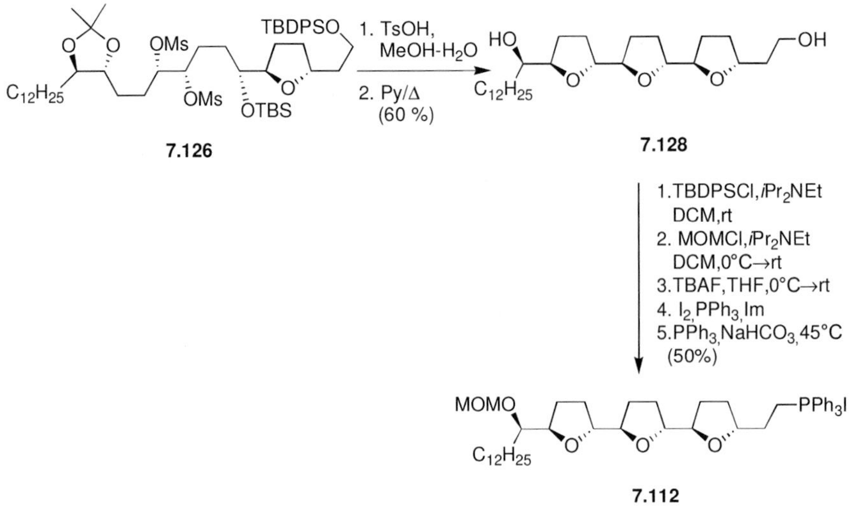

Scheme 7.37

Step 6. Wittig reaction and completion of the synthesis of goniocin **7.110**: Phosphonium salt **7.112** has the full tris-THF skeleton of goniocin with the correct stereochemistry in all its stereocenters. The Wittig reaction of the ylide derived from **7.112** with aldehyde **7.122** leads to the full carbon skeleton of the target. Homogeneous hydrogenation followed by removal of the protecting groups lead finally to goniocin **7.110** (Scheme 7.38).

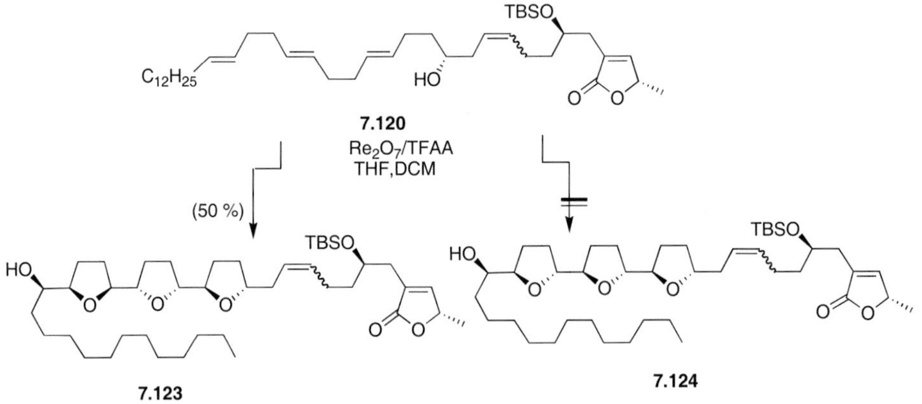

Evaluation

The change in strategy derived from the inability to achieve the desired stereochemistry in the key cascade oxidative formation of the *tris*-THF moiety of goniocin, resulted in a conceptually different approach to this compound. It is impressive to see the major troubles that a failed key transformation causes in a synthetic plan. In this case, it simply hampers access to the desired molecule.

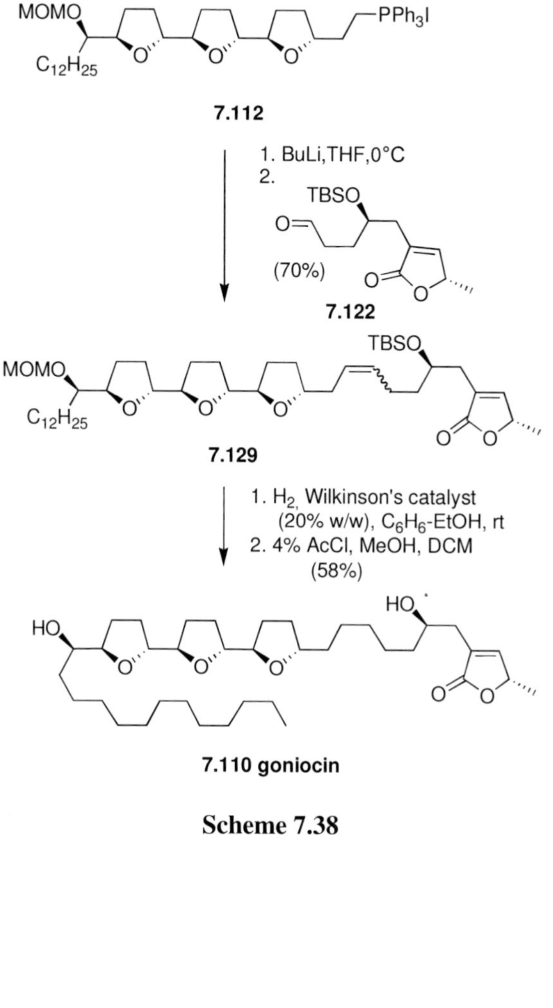
MOMO
$C_{12}H_{25}$
PPh_3I
7.112
1. BuLi,THF,0°C
2.
TBSO
(70%)
7.122
TBSO
MOMO
$C_{12}H_{25}$
7.129
1. H_2, Wilkinson's catalyst (20% w/w), C_6H_6-EtOH, rt
2. 4% AcCl, MeOH, DCM
(58%)
HO
HO
7.110 goniocin

Scheme 7.38

References

1. a) Among the plethora of books dedicated to asymmetric synthesis the five volumes of the series edited by Morrison are still fully illustrative, see *Asymmetric Synthesis, Vol. 1–5* (Ed. J. D. Morrison), Academic Press, Orlando-Florida, **1983– 1985**; b) see also the thematic issues of *Chemical Reviews* dedicated to enantioselective synthesis: *Chem. Rev,***1992**, *92*, 741–1140, and enantioselective catalysis: *Chem. Rev.* **2003**, *103*, 2761–3400.
2. Tanada, Y.; Mori, K. *Eur. J. Org. Chem.* **2001**, 4313.
3. Otani, T.; Sugimoto, Y.; Aoyagi, Y.; Igarashi, Y.; Furumai, T.; Saito, N.; Yamada, Y.; Asao, T.; Oki, T. *J. Antibiot.* **2000**, *53*, 337.
4. Smith, A. B.; Empfield, J. R.; Rivero, R. A.; Vaccaro, H. A.; Duan, J. J.-W.; Sulikowski, M. M. *J. Am. Chem. Soc.* **1992**, *114*, 9419.
5. Sasaki, K.; Hirata, Y.; *Tetrahedron Lett.* **1973**, 2439.
6. Smith, A. B.; Fukui, M.; Vaccaro, H. A.; Empfield, J. R. *J. Am. Chem. Soc.* **1991**, *113*, 2071.
7. a) Davis, F. A.; Jenkins, R. H.; Yocklovich, S. G. *Tetrahedron Lett.* **1978**, 5171. b) Davis, F. A.; Townson, J. C.; Vashi, D. B.; ThimmaReddy, R.; McCauley, J. P.; Harakal, M. E.; Gosciniak, D. J. *J. Org. Chem.* **1990**, *55*, 1254. c) Davis, F. A.; Chen, B. C. *Chem. Rev.* **1992,** *92*, 919.
8. a) Tidwell, T. T. *Synthesis* **1990**, 857. b) Tidwell, T. T. *Org. React.* **1990**, *39*, 297. c) Mancuso, A. J.; Huang, S.-L.; Swern, D. *J. Org. Chem.* **1979**, *44*, 4148.
9. a) Ayrey, G.; Barnard, D.; Woodbridge, D. T. *J. Chem. Soc.* **1962**, 2089; b) Reich, H. J.; Renga, J. M.; Reich, I. L. *J. Am. Chem. Soc.* **1975**, *97*, 5434.
10. a) Gent, P. A.; Gigg, R. *J. Chem. Soc., Chem. Commun.* **1974**, 277.; b) Corey, E. J.; Suggs, J. W. *J. Org. Chem.* **1973**, *38*, 3224.
11. Nicolaou, K. C.; Gray, D.; Tae, J. *Angew. Chem. Int. Ed.* **2001**, *40*, 3675.
12. Wellington, K. D.; Cambie, R. C.; Rutledge, P. S.; Bergquist, P. R. *J. Nat. Prod.* **2000**, *63*, 79.
13. a) Quinkert, G.; Stark, H. *Angew. Chem. Int. Ed. Engl.* **1983**, *22*, 637; b) Kraus, G. A.; Wu, Y. *J. Org. Chem.* **1992**, *57*, 2922; c) Hasimoto, K.; Horikawa, M.; Shirahama, H. *Tetrahedron Lett.* **1990**, *31*, 7074; d) Oppolzer, W.; Keller, K. *Angew. Chem. Int. Ed. Engl.* **1972**, *11*, 728; e) Kraus, G. A.; Chen, L. *Synth. Commun.* **1993**, *14*, 2041; f) Nicolaou, K. C.; Gray, D. *Angew. Chem. Int. Ed.* **2001**, *40*, 761; g) Kraus, G. A.; Zhao, G. *J. Org. Chem.* **1996**, *61*, 2770.
14. a) Barton, D. H. R.; Jaszberenyi, J. Cs. *Tetrahedron Lett.* **1989**, *30*, 2619; b) Barton, D. H. R.; McCombie, S. W. *J. Chem. Soc., Perkin Trans. I* **1975**, 1574.
15. Coleman, R. S.; Grant, E. B. *J. Am. Chem. Soc.* **1994**, *116*, 8795.
16. a) Iida, T., Kobayashi, E.; Yishida, M.; Sano, H. *J. Antibiot.* **1989**, *42*, 1475; b) Kobayashi, E.; Ando, K.; Nakano, H.; Iida, T.; Ohno, H.; Morimoto, M.; Tamaoki, T. *J. Antibiot.* **1989**, *42*, 1470; c) Kobayashi, E.; Ando, K.; Nakano, H.; Tamaoki, T. *J. Antibiot.* **1989**, *42*, 153; d) Arnone, A.; Assante, G.; Merlini, L.; Nasini, G. *Gazz. Chim. Ital.* **1989**, 557.

17. a) Coleman, R. S.; Grant, E. B. *J. Org. Chem.* **1991**, *56*, 1357; b) Coleman, R. S.; Grant, E. B. *Tetrahedron Lett.* **1993**, *34*, 2225.

18. Broka, C. A. *Tetrahedron Lett.* **1991**, *32*, 859.

19. Duplantier, A. J.; Nantz, M. H.; Roberts, J. C.; Short, R. P.; Somfai, P.; Masamune, S. *Tetrahedron Letts.* **1989**, *30*, 7357.

20. Reviews of the aldol reaction: a) Cowden, C. J.; Paterson, I. *Org. React.* **1997**, *51*, 1; b) Franklin, A. S.; Paterson, I. *Contemp. Org. Synth.* **1994**, *1*, 317; c) Heathcock, C. H. in *Comprehensive Organic Synthesis*; Heathcock, C. H., Ed.; Pergamon Press: New York, 1991; Vol. 2, p 181; d) Kim, B. M.; Williams, S. F.; Masamune, S. in *Comprehensive Organic Synthesis*; Heathcock, C. H., Ed.; Pergamon Press: New York, 1991; Vol. 2, p 239; e) Paterson, I. in *Comprehensive Organic Synthesis*; Heathcock, C. H., Ed.; Pergamon Press: New York, 1991; Vol. 2, p 301; f) Braun, M. *Angew. Chem., Int. Ed. Engl.* **1987**, *26*, 24; g) Heathcock, C. H. *Asymmetric Synth.* **1984**, *3*, 111; h) Evans, D. A.; Nelson, J. V.; Taber, T. R. *Top. Stereochem.* **1982**, *13*, 1.

21. Evans, D. A.; Kaldor, S. W.; Jones, T. K.; Clardy, J.; Stout, T. J. *J. Am. Chem. Soc.* **1990**, *112*, 7001.

22. Kihara, T.; Kusukabe, H.; Nakamura, G.; Sakurai, T.; Isono, K. *J. Antibiot.* **1981**, *34*, 1073.

23. Masamune, S.; Choy, W.; Petersen, J. S.; Sita, L. R. *Angew.Chem., Int. Ed. Engl.* **1985**, *24*, 1.

24. a) Ahn, N. T. *Top. Curr. Chem.* **1980**, *88*, 146; b) Paddon-Row *J. Chem. Soc., Chem. Commun.* **1990**, 456.

25. a) Sinha, S. C.; Sinha, A.; Sinha, S. C.; Keinan, E. *J. Am. Chem. Soc.* **1997**, *119*, 12014; b) Sinha, S. C.; Sinha, A.; Sinha, S. C.; Keinan, E. *J. Am. Chem. Soc.* **1998**, *120*, 4017.

26. Gu, Z.-M.; Fang, X.-P.; Zeng, L.; McLauhlin, J. L. *Tetrahedron Lett.* **1994**, *35*, 5367.

27. de la Torre, M. C.; Sierra, M. A. *Angew. Chem. Int. Ed.* **2004**, *43*, 160.

28. Sinha, S. C.; Sinha-Bagchi, A.; Keinan, E. *J. Am. Chem. Soc.* **1995**, *117*, 1447.

29. a) Edwards, P.; Wilkinson, G. *J. Chem. Soc., Dalton Trans.* **1984**, 2695; b) Hermann, W. A.; Thiel, W. R.; Kühn, F. E.; Fisher, R. W.; Kleine, M.; Herdtweck, E.; Scherer, W.; Mink, J. *Inorg. Chem.* **1993**, *32*, 5188.

Chapter 8
Reluctant Ring Closures

One of the most exciting topics in total synthesis is the formation of rings. Many synthetic routes devise a ring closure reaction at some stage, since many synthetic targets have at least one cycle in their core structure. Not surprisingly, the development of methodologies for ring formation has been (and still is) a priority in chemistry. The following cases exemplify that cyclization processes are not totally under our control. On the other hand, it is also worth noting that failures in closing the desired ring may cause greater changes to the planned synthesis than any of the problems discussed previously. In other words, the changes in strategy produced by the reluctant closure of one ring will lead, in many cases, to a complete redesign of the synthetic planning. This is especially serious since the key ring closure is commonly delayed until the final stages of the synthesis.

8.1   Discorhabdin C [1]

8.1 discorhabdin C

Dead Ends and Detours: Direct Ways to Successful Total Synthesis
Miguel A. Sierra and María C. de la Torre
Copyright © 2004 WILEY-VCH Verlag GmbH & Co. KGaA, Weinheim
ISBN: 3-527-30644-7

Target relevance
The discorhabdin alkaloids were isolated from the sponge of *Latrunculia* du Bocage in New Zealand [2]. Discorhabdin C is the first of the discorhabdin alkaloids reported and it exhibits extreme toxicity toward tumor cells (P388 and L1210 leukemia).

Synthetic plan
Access to discorhabdin C was planned by intramolecular imine formation between the tryptamine nitrogen and the quinone carbonyl group on intermediate **8.2**. This intermediate would be prepared from tricyclic dibromophenol **8.3**. This key intermediate will be synthesized from quinone **8.4**, which is accessible from 2-hydroxy-4-methoxy-benzaldehyde by using conventional methodologies (Scheme 8.1).

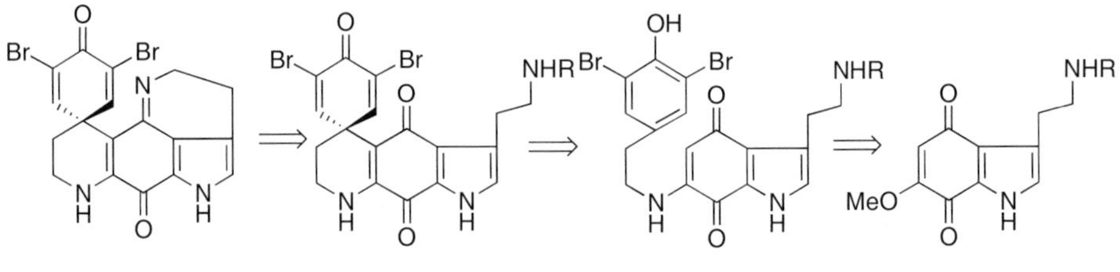

8.1 discorhabdin C   **8.2**   **8.3**   **8.4**

Scheme 8.1

Predictable problems
The construction of highly fused ring system of discorhabdin C may be problematic. The formation of the acid-sensitive indoloquinone imine moiety present in discorhabdin C will be difficult to handle.

Synthesis

Step 1. Preparation of indoloquinone amine 8.2 (Scheme 8.2):

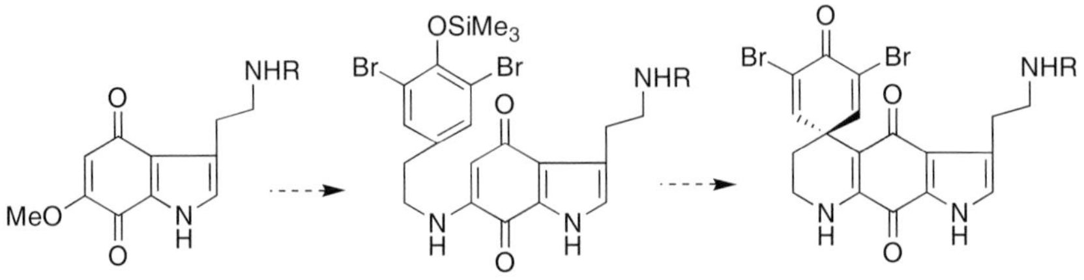

8.4   **8.6**   **8.2**

Scheme 8.2

The starting materials for the synthesis of compounds **8.2** were indoloquinone amines **8.4a** and **8.4b**. Both compounds differ exclusively on the primary amine protecting group, trifluoracetyl for **8.4a** and [(trimethylsilyl)ethoxy]carbonyl (TEOC) group for **8.4b**. Both compounds were treated with 3,5-dibromotyramine hydrobromide **8.5** in the presence of TEA to form phenol derivatives **8.3a,b**. Sylilation of the phenol group of compounds **8.3** was effected with MeCH=C(OMe)($OSiMe_3$) to yield **8.6**, which was oxidized with PhI($OCOCF_3$)$_2$ (PIFA) [3], effecting the oxidative phenol coupling while unmasking the quinone moiety on the phenol moiety. Compounds **8.2a** and **8.2b** were obtained in this way (Scheme 8.3).

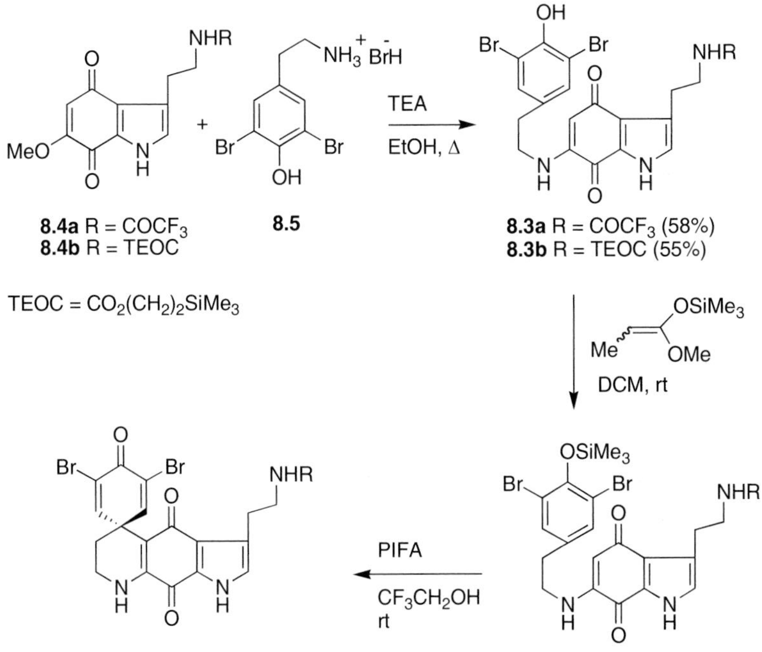

8.4a R = $COCF_3$
8.4b R = TEOC

8.5

8.3a R = $COCF_3$ (58%)
8.3b R = TEOC (55%)

8.2a R = $COCF_3$ (58%)
8.2b R = TEOC (62%)

8.6a R = $COCF_3$
8.6b R = TEOC

Scheme 8.3

Step 2. Failed intramolecular imine formation leading to a major strategic change (Scheme 8.4):

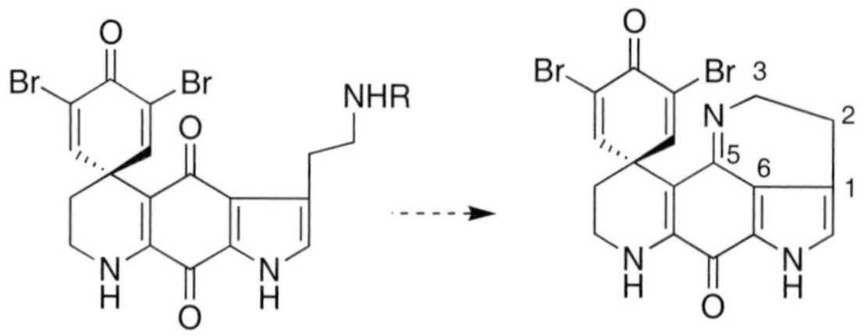

Scheme 8.4

To complete the synthesis of discorhabdin C **8.1** the apparently simple task of closing the last ring, remained. This final step was to be effected by formation of an imine between the tryptamine nitrogen atom and the quinone carbonyl group of **8.2**. Neither **8.2a** nor **8.2b** formed discorhabdin C in the attempted conditions. The failure of the imine formation from compounds **8.2** may be due to the weakly electrophilic nature of the carbonyl group of the quinone nucleus. Furthermore, compound **8.2b** may be considered a vinilogous urea and it is, additionally, very sterically hindered. Activation of the carbonyl group by decreasing the electron-donating ability of the vinilogous nitrogens was attempted by protecting these nitrogens with strong electron-withdrawing groups. This tactic was also fruitless due to the instability of compounds **8.2** under the basic conditions (LDA or NaH) used to effect this transformation (Scheme 8.5).

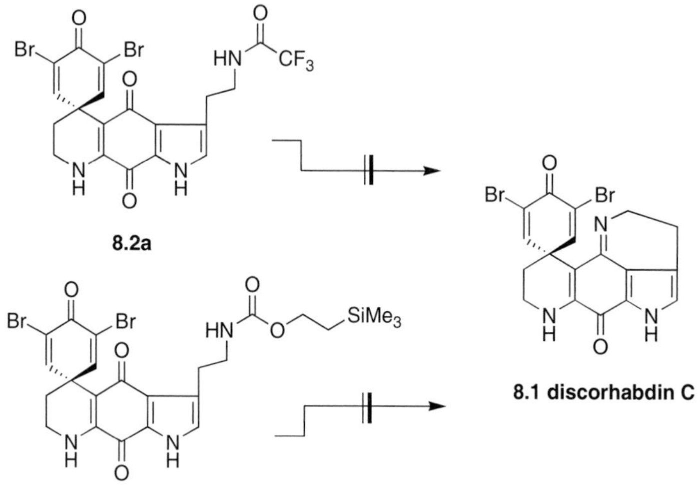

Scheme 8.5

Step 3. Re-evaluation of the strategy: An alternate route to discorhabdin C (Scheme 8.6):

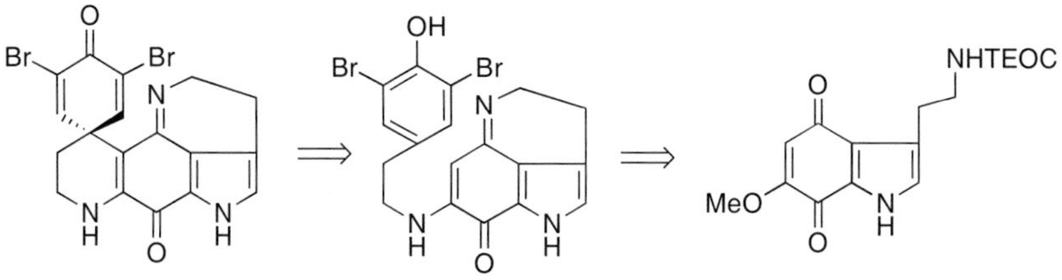

Scheme 8.6

The sensitivity of the quinoneimine moiety of discorhabdin C was a predictable problem. However, it was not finally responsible for the failure of the synthesis of discorhabdin C **8.1**. The failure to close the remaining ring to access compound **8.1** from **8.2** was attributable to the low electrophilia of the quinone carbonyl group of the last compound. The solution to this problem would be, in principle, to effect the imine formation in an earlier intermediate, such as **8.7**, having the carbonyl group activated by the presence of the methoxy group conjugated with the quinone-carbonyl. Therefore, the new approach to discorhabdin C carried the sensitive quinone-imine group from a very early intermediate **8.9** (Scheme 8.6).

Step 4. Formation of the tricyclic quinone-imine 8.8 (Scheme 8.7):

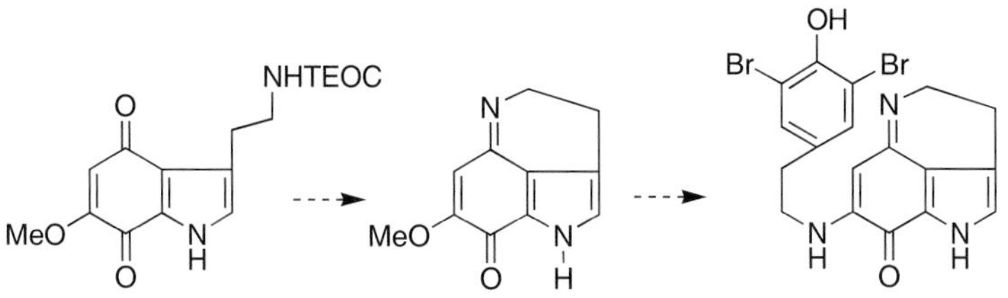

Scheme 8.7

Direct formation of imine **8.9** by deprotection of the TEOC group of **8.7**, followed by condensation under diverse acidic or basic dehydrative conditions, failed. Probably, the intermediate amine **8.10** is still very slightly electrophilic due to the vinilogous indole nitrogen. The donating ability of the nitrogen group was diminished by tosylation of **8.7** and the TEOC group of the tosylated derivative **8.11** was deprotected with TsOH/NaHCO$_3$/3Å-Ms to yield the highly unstable trycyclic imine **8.9**. This imine was not isolated but reacted with 3,5-dibromotyramine hydrobromide **8.5**. Compound **8.8** was obtained in this way by substitution of the C6-MeO group of **8.9** and subsequent detosylation (Scheme 8.8).

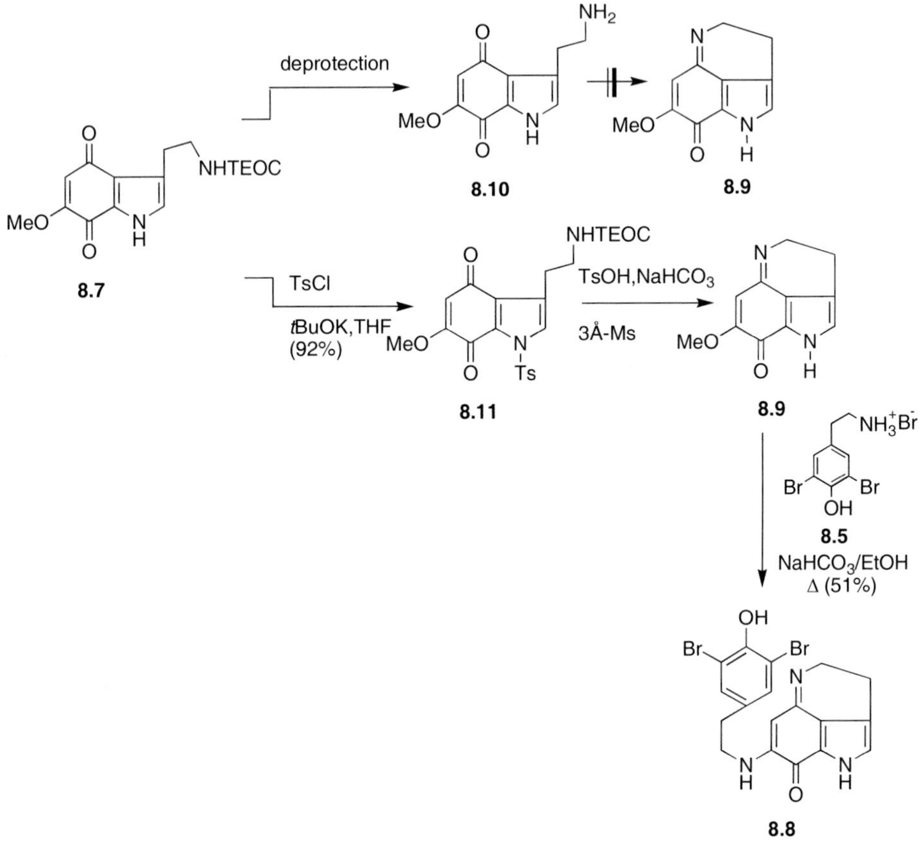

Scheme 8.8

*Step 5. Intramolecular oxidative coupling of **8.8** and completion of the synthesis of discorhabdin C (Scheme 8.9):*

8.8   **8.1 discorhabdin C**

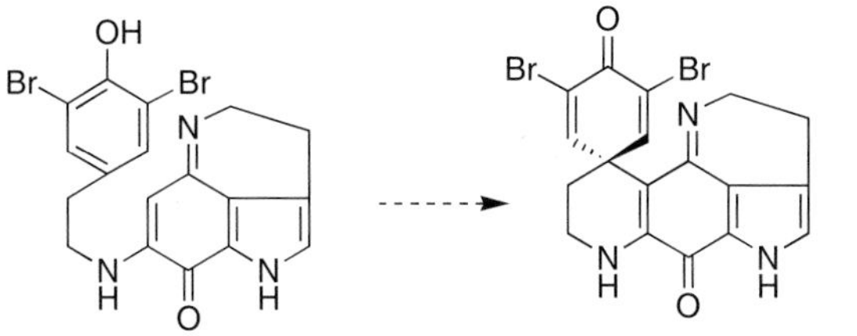

Scheme 8.9

Completion of the synthesis of discorhabdin C was achieved by sylilation of the phenol group of aminoindoloquinone imine **8.8** to yield protected derivative **8.12** and its further oxidative coupling using PIFA. Discorbhadin C **8.1** was obtained in this way in a 42% yield (Scheme 8.10).

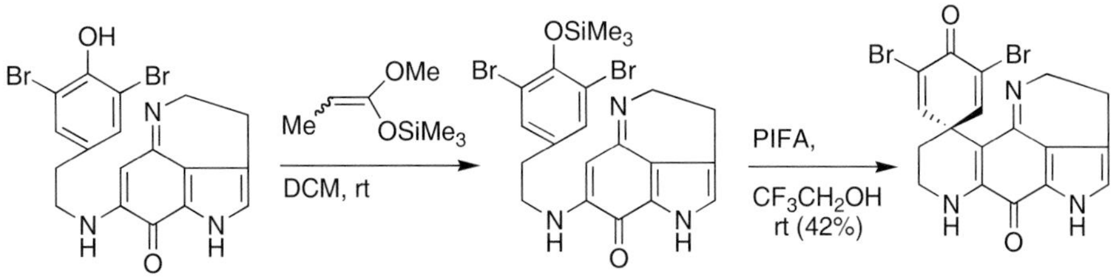

8.8   **8.12**   **8.1 discorhabdin C**

Scheme 8.10

Evaluation

The problem presented by the construction of the ring system of discorbhadin C was not the major challenge of its synthesis. The oxidative coupling methodology resolved this problem. In fact, the main drawback found on the synthesis of discorbhadin C was the failure to close the last ring by forming a quinone-imine. This situation was, in principle, predictable due to the vinilogous urea nature of **8.2**, the substrate in which the cyclization was planned. In fact, in the successful approach to discorbhadin C the formation of the quinone-imine feature onto a quinone carbonyl group (**8.11**) required not only one activating methoxy group, but the decrease in the donating ability of the vinilogous indole nitrogen by tosylation. Only in these conditions could the quinone imine **8.9** be formed.

8.2

8.1 discorhabdin C

8.2   Discorhabdin A [4]

8.13 discorhabdin A

The discussion of some of the steps of the synthesis of discorhabdin A **8.13** is of relevance at this point to illustrate how closely related cyclization processes may lead either to success or failure. Discorhabdin A **8.13** synthesis was planned by using the methodology for effecting the oxidative ring closure developed for the preparation of discorhabdin C **8.1**. In this regard, tritilated aminoester **8.14** was reacted with pyrroloquinone **8.9** used as an advanced intermediate in the synthesis of discorhabdin C (see above) to form **8.15**. Interestingly, the PIFA oxidative coupling used efficiently in discorhabdin C synthesis to prepare discorhabdin **8.1**, gave only complex reaction mixtures with **8.15** (Scheme 8.11). The presence of the pendant ester group seemed to be incompatible with the PIFA 1e⁻ oxidation. Therefore, the amino ester **8.14** was reduced with DIBALH to the corresponding alcohol and protected as TBS-derivative. The aminoalcohol **8.16** was attached to the tosilated pyrroloquinone **8.17** to produce **8.18** which, by submission to PIFA treatment in the presence of MK10, yielded the desired pyrroloiminoquinone **8.19** (Scheme 8.11).

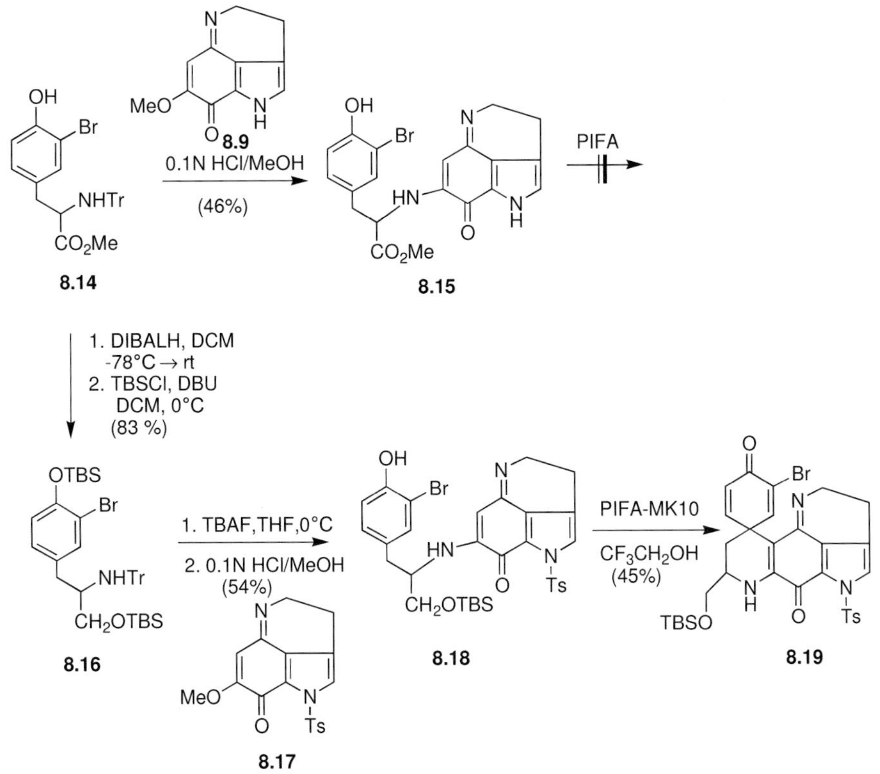

Scheme 8.11

This example clearly shows another failure of a well-tested reaction (in this case PIFA oxidative cyclization) to effect a ring closure in very similar substrates. In this case, it is probably the ester group of **8.15** that is responsible for the failure of the cyclization.

8.3   Frondosin B [5]

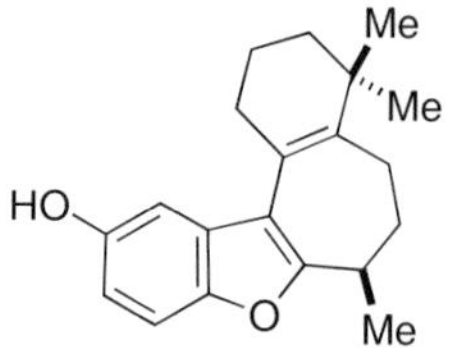

8.20 frondosin B

Target relevance

Frondosins A–D are isolated from the sponge *Dysidea frondosa* [6]. Each of them inhibits the binding of interleukin-8 (IL-8), a chemo-attractant for neutrophiles that is produced by macrophages and endothelial cells, at the low-µM range. Since IL-8 has been implicated in different acute inflammatory disorders, including psoriasis and rheumatoid arthritis, an IL-8 receptor antagonist such as frondosin is a good target for the development of novel drugs against autoimmune illness.

Synthetic plan

Synthetic planning to access to frondosin B has as the paramount feature, the installation of the skeleton of the natural product by the Claisen rearrangement of compound **8.21**, obtained by joining two fragments **8.22** and **8.23**. The closure of the seven-membered ring present in the frondosin B skeleton would be effected on the Claisen product **8.24**, thereafter leading to frondosin B **8.20** (Scheme 8.12).

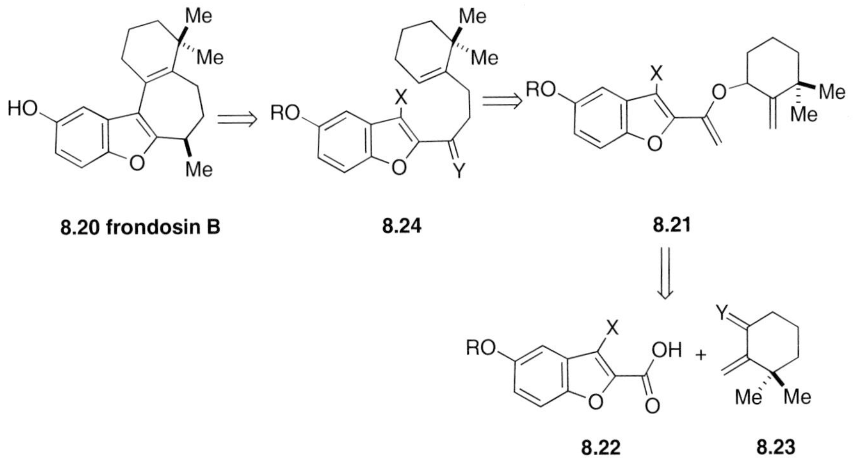

8.20 frondosin B   **8.24**   **8.21**

8.22   **8.23**

Scheme 8.12

Step 1. Coupling of fragments 8.22 and 8.23 and Claisen rearrangement (Scheme 8.13):

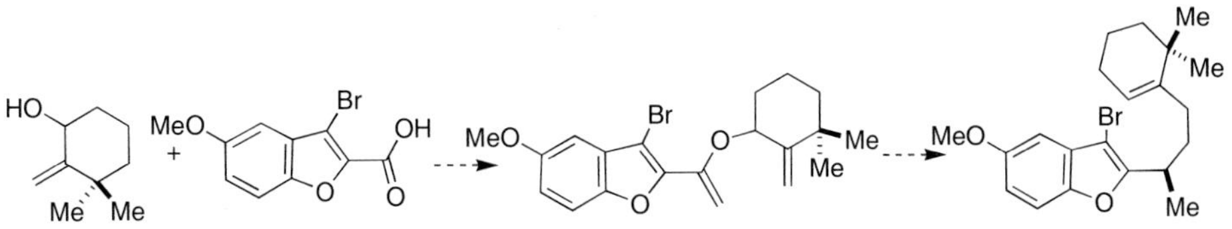

8.25   **8.26**   **8.27**   **8.28**

Scheme 8.13

Standard esterification of cyclohexanol **8.25** and benzofuran carboxylic acid **8.26**, formed ester **8.29** in 78% yield. Ester **8.29** was transformed to the precursor of the Claisen rearrangement **8.27** by reaction with Tebbe's reagent [7]. Heating (toluene, 80°C) of **8.27** smoothly formed the product derived from Claisen rearrangement **8.30** in excellent yields. The ketone **8.30** was converted to the required C19 methyl derivative **8.28** by successive addition of BrMgMe and reduction of the resultant carbinol with NaCNBH$_3$ in the presence of anhydrous ZnI$_2$ (Scheme 8.14).

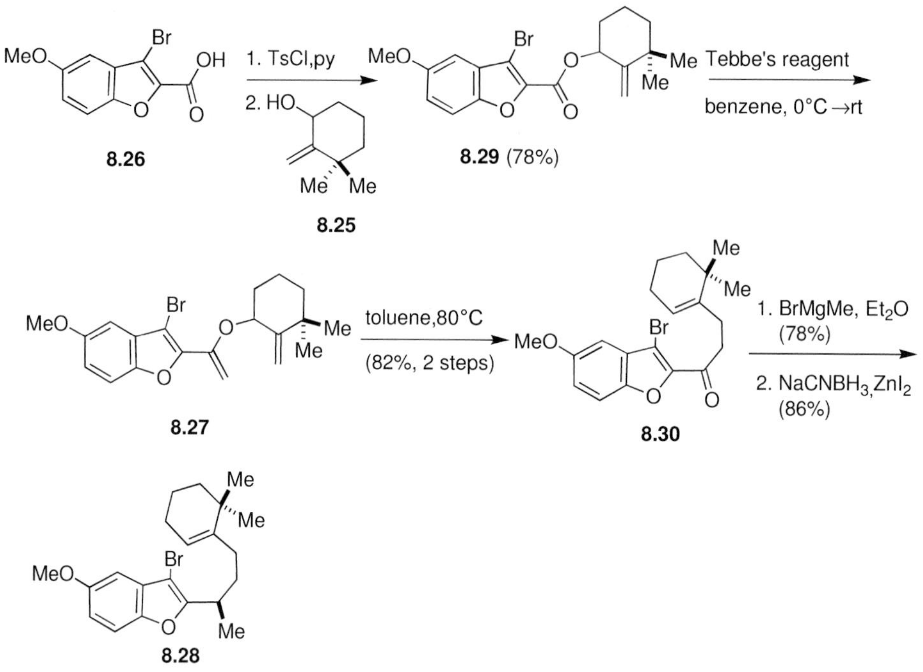

8.26 1. TsCl,py **8.29** (78%) Tebbe's reagent
2. HO... benzene, 0°C →rt

8.25

8.27 toluene,80°C **8.30**
(82%, 2 steps)

1. BrMgMe, Et$_2$O
(78%)

2. NaCNBH$_3$,ZnI$_2$
(86%)

8.28

Scheme 8.14

Step 2. Failed ring closure of 8.28 and re–evaluation of the synthetic strategy (Scheme 8.15):

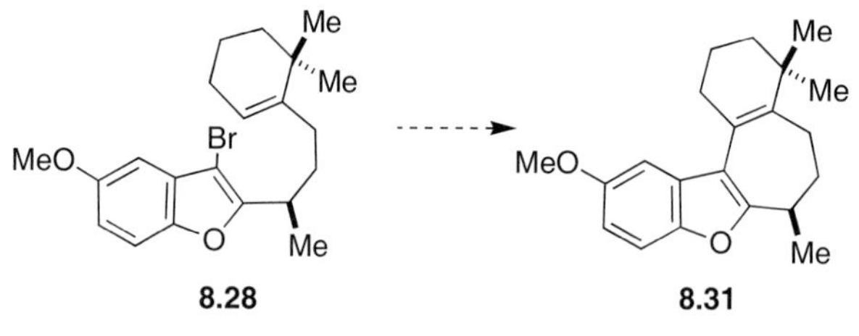

8.28   **8.31**

Scheme 8.15

The structure of benzofurane **8.28** is nicely suited to close the fourth ring of the frondosin B skeleton through a Heck reaction. Treatment of vinyl bromide **8.28** under different conditions and Pd-catalysts was, however, unsuccessful in effecting the ring closure to yield **8.31**. Stille-type reactions were devised as the alternative. To meet this objective, benzofurane **8.29** was transformed into vinyl triflates **8.33** and **8.34** through ketone **8.32**. This ketone was accessed from **8.29** via hydroboration-oxidation, followed by Jones oxidation of the alcohol thus formed. Enol triflation of ketone **8.32** with triflic anhydride (Tf$_2$O) in the presence of 2,6-di-*tert*-butylpyridine as the base gave a 2:3

mixture of the needed triflate **8.33** and its isomer **8.34**. The mixture of triflates was inseparable and it was submitted to different Pd-catalyzed Stille-type reactions. Each of the experiments were successful (Scheme 8.16). This failure to close the remaining ring of the skeleton of Frondosin B requires the design of new strategy to reach the target molecule.

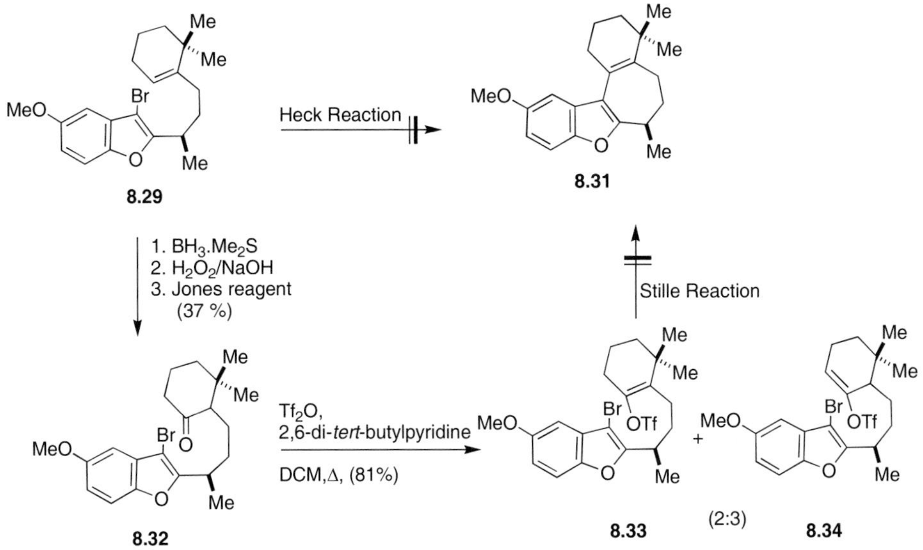

Scheme 8.16

*Step 3. A new strategy: formation of the troublesome C10–C11 bond by intermolecular coupling of **8.35** and **8.36** followed by Claisen rearrangement (Scheme 8.17):*

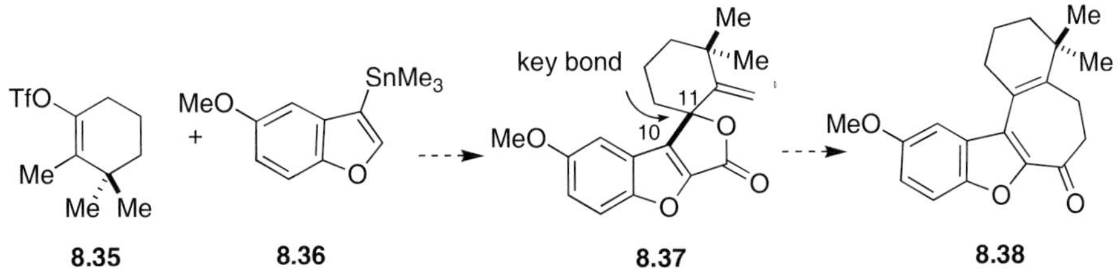

Scheme 8.17

Since the previous Claisen rearrangement works very efficiently in the above reactions, it was decided to alter the sequence of events: first forming the C10–C11 linkage by intermolecular coupling of fragments **8.35** and **8.36** of lactone **8.37**. Introduction of a methylene group in the lactone carbonyl, followed by Claisen rearrangement should lead to ketone **8.38** having the tetracyclic skeleton of frondosin B.

Compounds **8.35** and **8.36** were coupled under ligandless conditions and the adduct **8.39** was successfully isolated, albeit in low yields. Installation of a carboxylic acid at C9 on the bezofuran ring was achieved by Vilsmeier formylation [8] followed by $AgNO_3$ oxidation yielding acid **8.40**. The building of the spirocyclic lactone moiety of the desired tetracyclic compound **8.37** was attempted by activation of the tetrasubstituted alkene of **8.40**, with the idea of effecting the halolactonization reaction. Unfortunately, this lactonization step did not progress. This disappointing result reflects the difficulties of attacking with a viable iodonium equivalent at C5, even in the presence of a proximal, and presumably participating, carboxylate function (Scheme 8.18). It is difficult to express this situation in words other than those used by the authors: *This negative outcome provided further testimony to the high risks in attempting to conduct chemistry proximal to the gem-dimethyl quaternary center.*

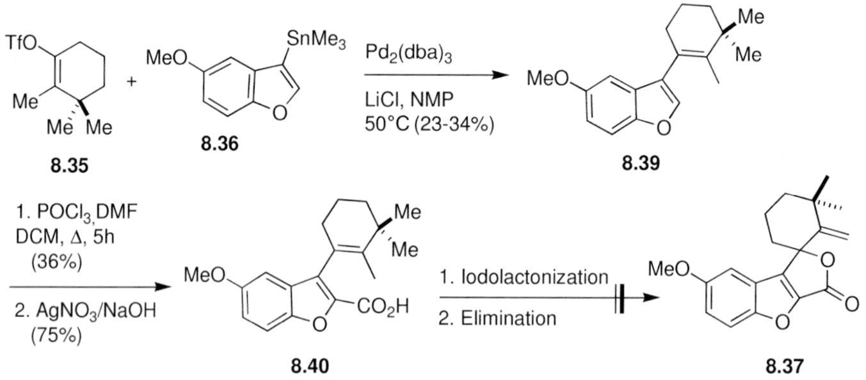

Scheme 8.18

Step 4. A totally revised approach to the problem, following a major change in strategy (Scheme 8.19):

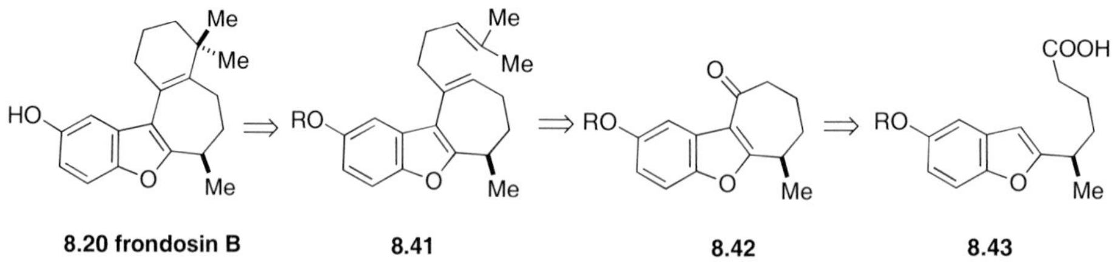

Scheme 8.19

The revised approach to frondosin B **8.20** involved incorporation of the cyclohexene ring, having the gem-dimethyl group onto a prebuilt skeleton like **8.41**, which has the troublesome seven-membered ring already in place. Now, a homoprenyl group will be used to construct the cyclohexene ring and it will be added to a tricyclic ketone like **8.42** derived from benzofuran **8.43**. The remaining cycle will be closed on **8.41** by acid catalyzed cyclization, exploiting the activated benzofuran ring to promote and guide the sense of the electrophilic attack on the cycloheptenyl double bond. One premise for the success of this new and radically different approach is that the tetrasubstituted olefin between C5 and C11 would be markedly more stable than trisubstituted isomers and could therefore be generated under equilibrating conditions.

In this regard, the benzofuran **8.44**, having the side chain from which the seven-membered ring will derive, was treated with oxalyl chloride and subsequently with $SnCl_4$ to effect a Friedel–Crafts ring closure. Ketone **8.45** was obtained in good yield. The cerium reagent derived from the treatment of 4-methyl-3-pentene magnesium bromide **8.46** with $CeCl_3$ was added to ketone **8.45**, and the tertiary alcohol formed was dehydrated by dissolving the compound in chloroform to form the diene **8.47** as a 5:1 mixture with its exo-isomer. Various acid combinations ($BF_3.OEt_2$, HCOOH, H_3PO_4) systematically resulted in the formation of the six-membered ring in an approximately 2.5:1 mixture favoring the desired isomer **8.31**. This ratio reflects the thermodynamic stability order of rapidly equilibrating isomers. In any case, frondosin B **8.20** was prepared, together with its olefinic isomers **8.49** in 87 % yield. An analogous result was achieved starting from either separated olefin **8.31** or **8.48**. This fact indicates that, in order to control the synthesis, the cyclohexene ring has to be formed strictly excluding acidic conditions (Scheme 8.20).

*Step 5. Finally, a totally regiocontrolled entry to frondosin B **8.20** (Scheme 8.21):*

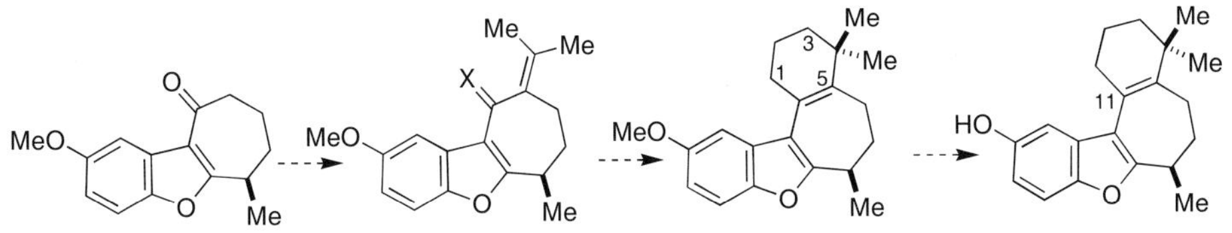

Scheme 8.21

The key to installing the C5–C11 double bond of frondosin B without resorting to acid reagents that cause the formation of isomeric mixtures, was to use a Diels–Alder reaction. The cyclohexene ring would be formed in this way concomitantly with the C1–C2 and C3–C4 bonds. To implement this new strategy, a diene **8.50** would be built onto ketone **8.45**. The Diels–Alder adduct **8.31** derived from the diene **8.50** will meet all the requirements to lead to frondosin B, provided that acid treatment is avoided.

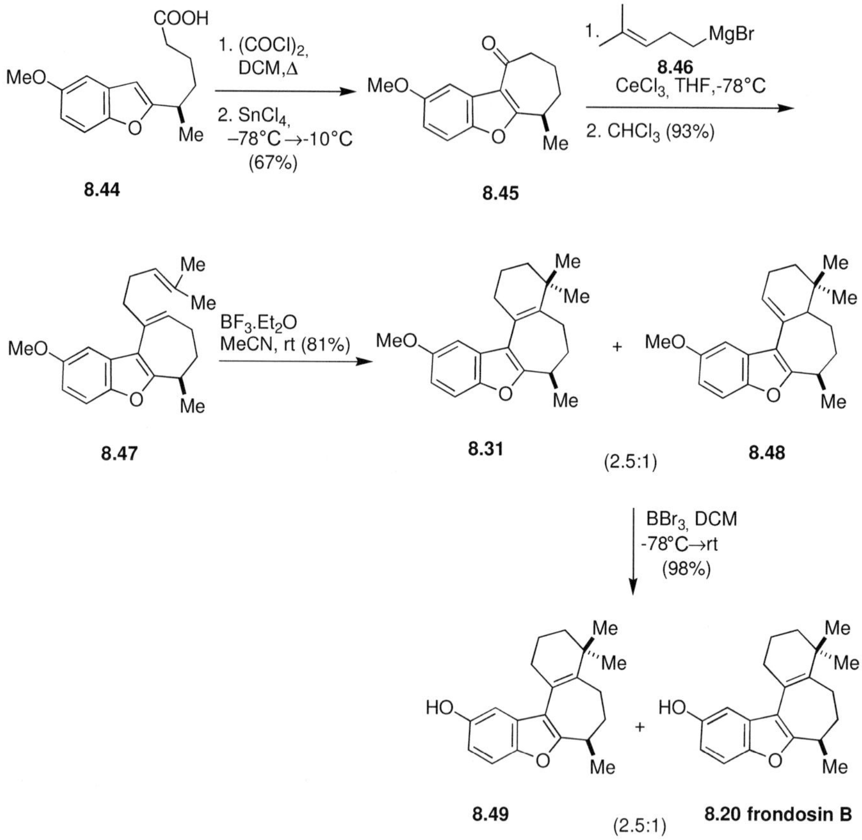

Scheme 8.20

The Zn-enolate derived from tricyclic ketone **8.45** was reacted with acetone to yield the sensitive aldol adduct **8.51**. Compound **8.51** was prone to retro-aldol fragmentation, especially in acidic media. The best conditions to effect its dehydratation were treatment with mesyl chloride/TEA followed by isomerization of the 1:1 mixture of olefins thus obtained with NaOMe. The base treatment effected isomerization to provide the ketone **8.52**. The diene **8.53** was formed by reaction of **8.52** with Tebbe's reagent buffered with pyridine. Reaction of this diene with nitroethylene in the presence of di-*tert*-butyl pyridine at 80°C formed adduct **8.54** in excellent yields. Radical reduction of the nitro group of **8.54** was achieved with TBTH/AIBN to form **8.31** from which frondosin B **8.20** was obtained by MeO-removal with NaSEt. Therefore, this route accomplished the total synthesis of frondosin B without the concomitant formation of additional isomers (Scheme 8.22). An asymmetric synthesis of (+)-frondosin could also be achieved using this synthetic scheme, using chiral (−)-**8.45**.

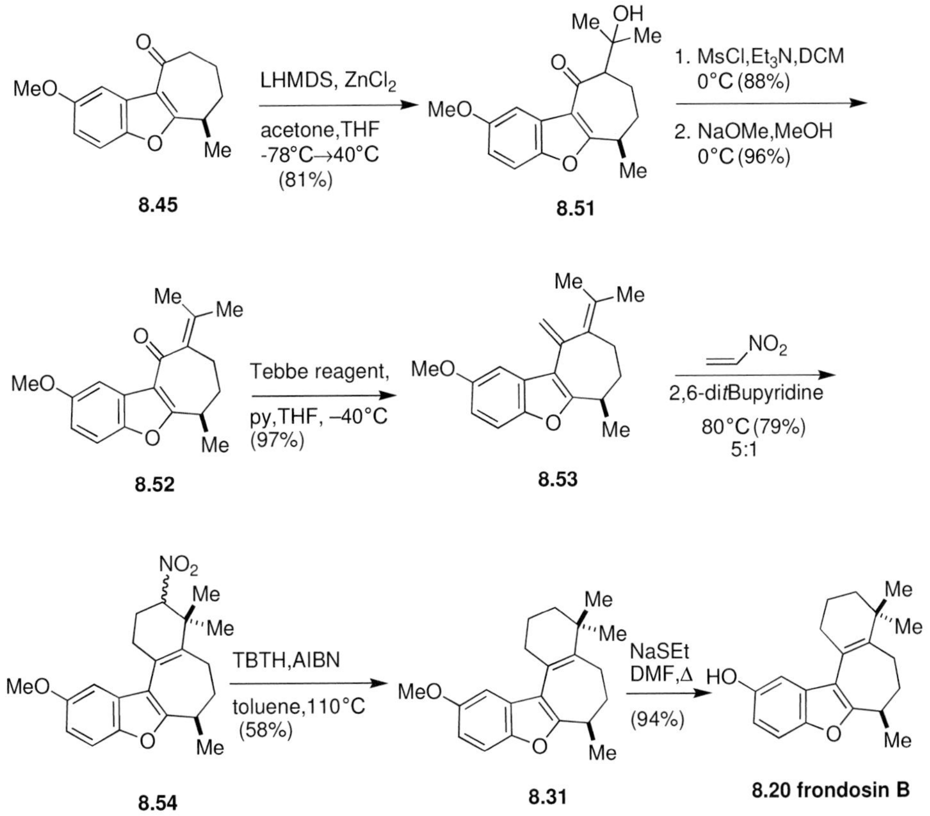

Scheme 8.22

Evaluation

- Two major strategic changes were made during frondosin B **8.20** synthesis. Both changes were due the difficulty in effecting a ring closure by bond formation proximal to the dimethyl group of the six-membered ring of frondosin B.

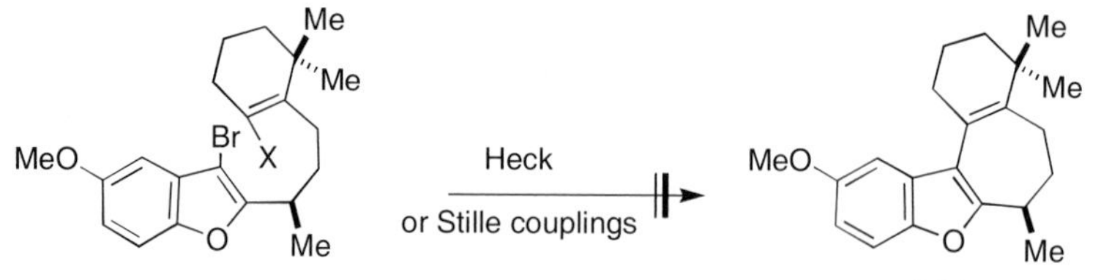

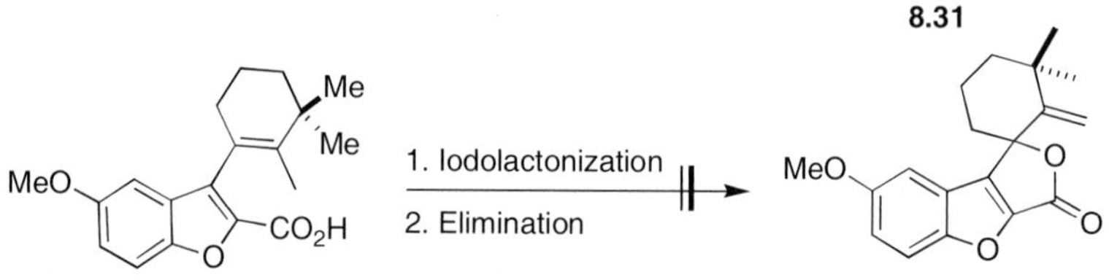

8.40 → **8.37**

- Frondosin B six-membered ring building through a cationic ring closure on a substrate, having the preformed seven-membered ring, occurred efficiently. However, the acidic conditions needed to effect the Friedel–Crafts ring closure produced a regioisomeric mixture of olefins. Only one of the two regioisomers led to frondosin B.

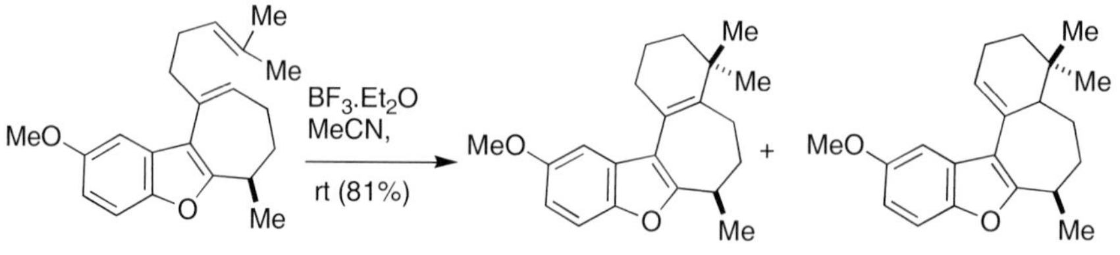

8.47   **8.31**   (2.5:1)   **8.48**

- The final and most successful approach to frondosin B involved the construction of the six-membered ring through a Diels–Alder process. This approach unambiguously placed the double bond on the pertinent position of the seven-membered ring.

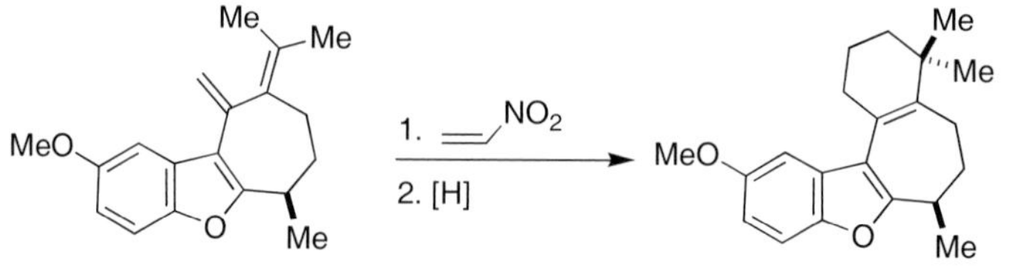

8.53   **8.31**

The synthesis of laulimalide, discussed below, exemplifies the problems encountered during a macrocyclization process, one step that is delayed in many cases to the last stages of the synthesis. Failure in this step results, in general, in serious synthetic problems.

Especially interesting reagents

Tebbe's reagent: The reagent "Cp$_2$Ti=" generated *in situ* by reacting Cp$_2$TiCl$_2$ with MeAl$_3$. This reaction forms dimetallacyclobutane **8.55** which is able to methylenate ketones, esters or even amides with very good efficiencies [7]. The Petasis' methylenation reagent [9] **8.56** is an alternative to Tebbe's. In this case, the reagent is a solution of Cp$_2$TiMe$_2$ generated from MeLi and Cp$_2$TiCl$_2$ in an ethereal solvent, usually THF. Heating this solution in the presence of a carbonyl group effects its methylenation.

$$Cp_2TiCl_2 + Me_3Al \longrightarrow Cp_2Ti\overset{\frown}{\underset{Cl}{}}AlMe_2$$

8.55 Tebbe reagent

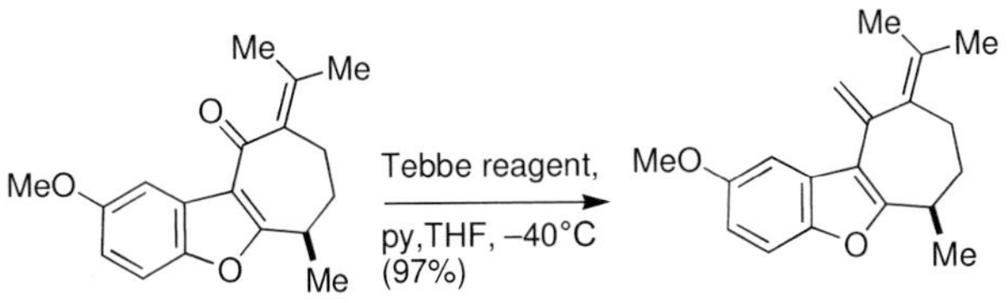

8.52 → Tebbe reagent, py, THF, –40°C (97%) → **8.53**

$$Cp_2TiCl_2 + MeLi \longrightarrow Cp_2TiMe_2$$

8.56 Petasis reagent

8.4   (–)-Laulimalide [10]

8.57 (–)-laulimalide

Target relevance

Laulimalide **8.57** (*aka* as figianolide B) was isolated from the marine sponge *Cacospongia mycofijiensis* [11]. Laulimalide exhibits very potent antitumor activity against numerous NCI cell lines and it has maintained a high level of potency against the multidrug-resistant cell line SKVLB-1. The mechanism of action of compound **8.57** is analogous to that encountered on taxol, making laulimalide one of the few non-taxane natural products that share the ability to stabilize microtubules.

Synthetic planning

Laulimalide **8.57** was to derive from the metathesis ring closure of the open-chain fully functionalized advanced intermediate **8.58**. This metathesis would set the structure of laulimalide with the unravelling of the epoxide moiety remaining, to accomplish the synthesis. The intermediate **8.58** will be prepared by Julia olefination [12] joining moieties **8.59** and **8.60** (Scheme 8.23).

*Step 1. Synthesis of the metathesis precursor **8.61** and its attempted cyclization:* Fragments **8.62** and **8.63** were prepared through multi-step sequences and submitted to Julia olefination to generate the C16–C17 double bond of laulimalide. Thus, sulfone **8.62** was first lithiated with 2.1 equiv of BuLi in THF at –78°C and the dianion reacted with aldehyde **8.63** to yield the mixture of α-hydroxy sulfone derivatives.

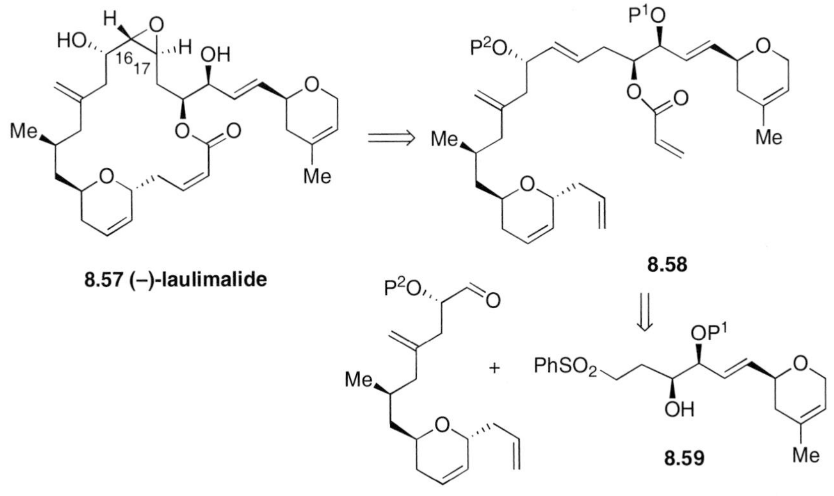

8.57 (−)-laulimalide

8.58

8.60

Scheme 8.23

Acylation of this mixture with Ac₂O/TEA in the presence of a catalytic amount of DMAP, followed by exposure of the diacetate to Na(Hg) in methanol, furnished an *E/Z* mixture of the C16-C17 olefin **8.64** (unreported yield). To incorporate the acryloyl side chain needed to effect the ring-closing metathesis [13], the free alcohol group of compound **8.64** was protected as its TIPS-derivative, the PMB group was removed by DDQ treatment and the free alcohol of **8.65** was acylated with acryloyl chloride in the presence of base to proportionate the metathesis substrate **8.61**. All the conditions designed to effect the ring-closing metathesis, led exclusively to decomposition (Scheme 8.24). No traces of the desired metathesis product were detected. Therefore, this approach was abandoned and the strategy used to construct laulimalide was re-evaluated.

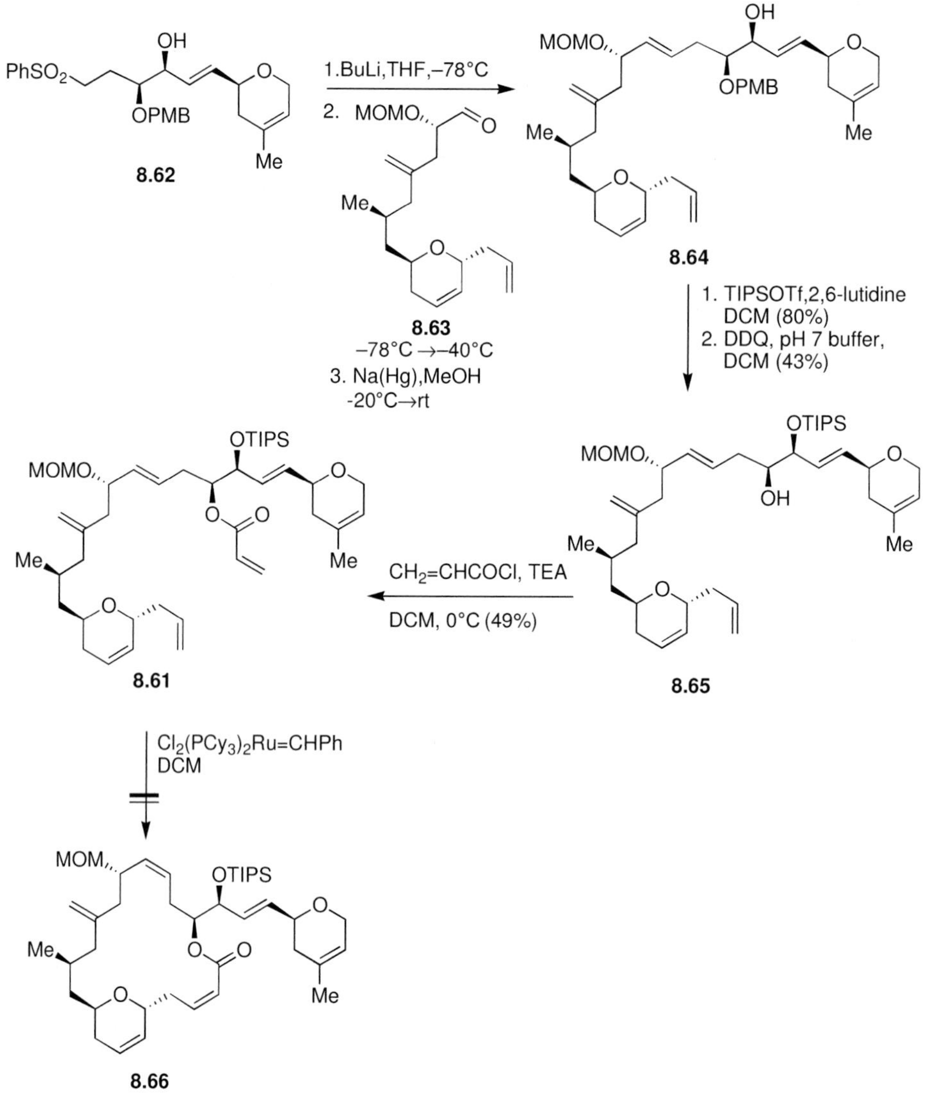
8.62
1.BuLi,THF,–78°C
2. 8.63
–78°C→–40°C
3. Na(Hg),MeOH
-20°C→rt
8.64
1. TIPSOTf,2,6-lutidine DCM (80%)
2. DDQ, pH 7 buffer, DCM (43%)
8.65
CH2=CHCOCl, TEA
DCM, 0°C (49%)
8.61
Cl2(PCy3)2Ru=CHPh
DCM
8.66

Scheme 8.24

Step 2. First re-evaluation of the synthetic approach to laulamide: an intramolecular Horner–Emmons reaction (Scheme 8.25): Since the failure of the metathesis route, depicted in Scheme 8.24, to produce the macrocyclic ring of laulimalide **8.66**, the ring–closure methodology was re-evaluated. Now, a Horner–Emmons ring closure on phosphonate **8.67** should produce the compound **8.68** having the macrocyclic structure of laulamide. At first glance the approach represented in Scheme 8.25 looks very similar to the failed metathetical route, even more so, considering that the phosphonate **8.67** derives from the Julia coupling of fragments **8.69** and **8.70**. However, the preparation of aldehyde **8.70** required again a multi-step synthesis, which was quite different from that used to prepare aldehyde **8.63**.

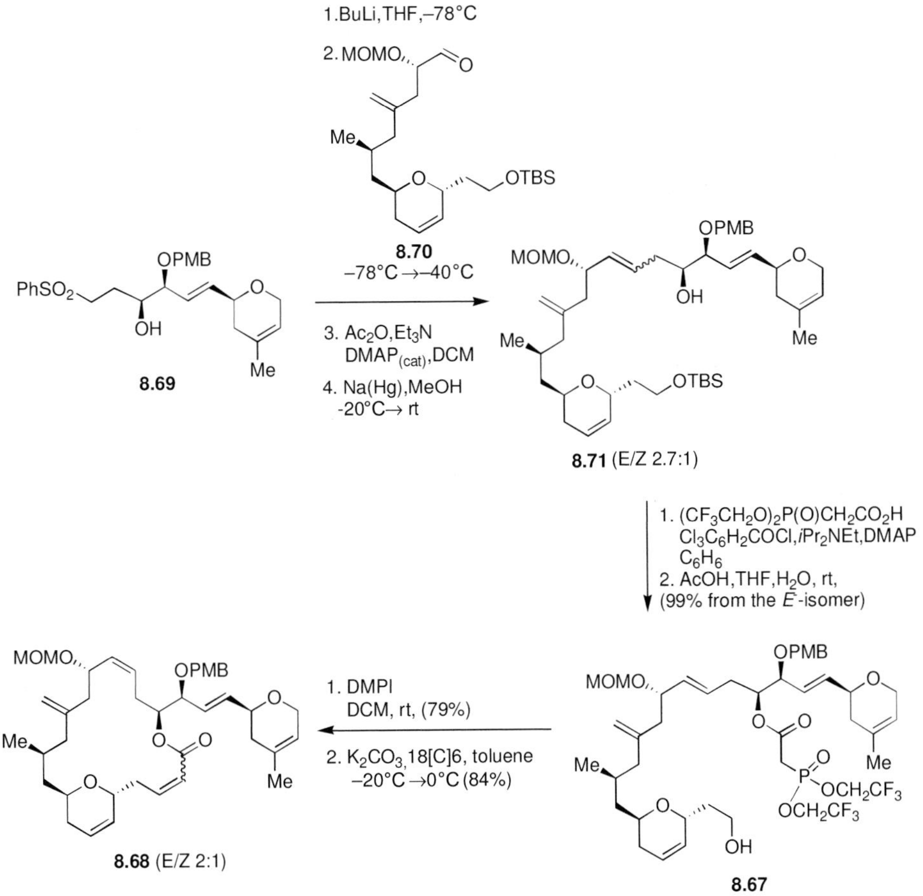

Scheme 8.25

The Horner–Emmons ring closure of the phosphonate **8.67** occurred very efficiently. However, compound **8.68** was obtained as a 2:1 isomer mixture. Even worse, the major compound was the undesired *trans*-isomer. This chemically acceptable, but stereochemically poor, result forced the authors to redesign the ring closure. Before designing a different way to close the ring, many attempts were made to improve the selectivity, which would favor the formation of the minor *cis*-isomer. The best results were obtained when the PMB-protecting group was substituted by a MOM-protecting group (1.7:1). Nevertheless, the selectivity of this reaction is far from the optimum.

Step 3. Second re-evaluation of the synthetic approach to laulimalide, cyclization using the Yamaguchi protocol (Scheme 8.26): Now, the sequence of events that should lead to the macrocycle precursor of laulimalide was inverted. Instead of manipulating the north hemisphere, the side chain of the hydropyrane ring was selected for modification. In fact, the alcohol **8.72** was protected as the THP derivative and the primary alcohol in the side chain of the hydropyrane ring deprotected to yield alcohol **8.73**. Alcohol **8.73** was oxidized to the corresponding aldehyde and reacted with $(PhO)_2P(O)CH_2CO_2CH_2CH_2SiMe_3$ under Horner–Emmons conditions to yield the Z-olefin **8.74**. Removing the PMB-protecting group and further treatment with TBAF resulted in the formation of acid **8.75** that was submitted to $Cl_3PhCOCl$, iPr_2NEt (Yamaguchi conditions) to yield again an *E/Z* mixture of macrocycles **8.76** favoring (2:1) the undesired *E*-isomer (Scheme 8.26). This new drawback, which derived from the isomerization of the double bond in the dihydropyrane side chain, was not an unexpected result, taking into account analogous observations during Roush's synthesis of verrucarin B [14].

In fact, Roush had proposed that the reason for this olefin isomerization was probably the reversible Michael addition of the acylating catalyst (DMAP) to the active acylating agent. Both, the Horner–Emmons approach (Scheme 8.25) and the Yamaguchi macrolactonization (Scheme 8.26) of the *cis*-α,β-unsaturated acid **8.75** succeeded in assembling the macrocyclic structure of laulimalide. However, both were unable to produce a single isomer of the macrocyle, or at least the desired *cis*-isomer, as the main reaction product. These drawbacks resulted in a tactical change based on the successful application of the Yamaguchi macrocyclization to the preparation of the macrolactone ring of laulamide using an alkyne, and in the end the selective *cis*-hydrogenation of the triple bond resulted in the successful entry to laulimalide.

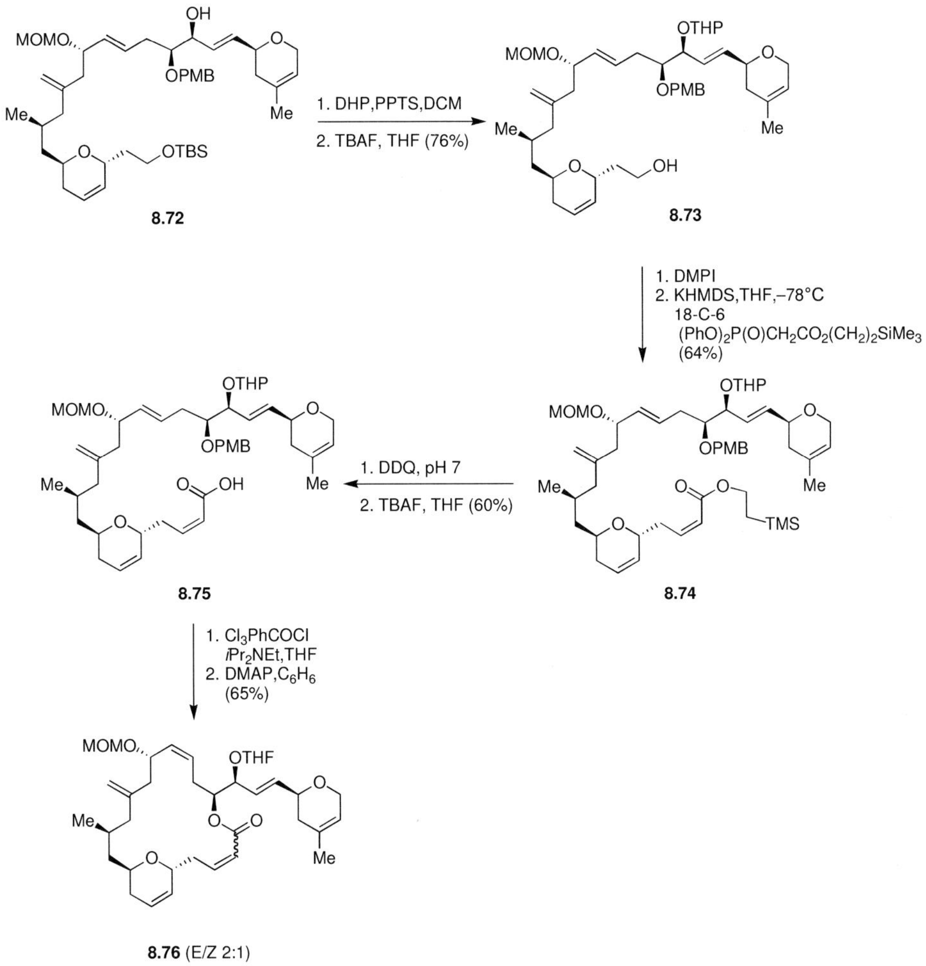

MOMO,,, — OH / OPMB ... Me ... O ... Me / OTBS

8.72

1. DHP,PPTS,DCM
2. TBAF, THF (76%)

MOMO,,, — OTHP / OPMB ... Me ... O ... Me / OH

8.73

1. DMPI
2. KHMDS,THF,–78°C
   18-C-6
   (PhO)$_2$P(O)CH$_2$CO$_2$(CH$_2$)$_2$SiMe$_3$
   (64%)

MOMO,,, — OTHP / OPMB ... Me ... O ... Me / O O—O—TMS

8.74

1. DDQ, pH 7
2. TBAF, THF (60%)

MOMO,,, — OTHP / OPMB ... Me ... O ... Me / O OH

8.75

1. Cl$_3$PhCOCl
   *i*Pr$_2$NEt,THF
2. DMAP,C$_6$H$_6$
   (65%)

MOMO,,, — OTHF ... Me ... O ... Me / O O

8.76 (E/Z 2:1)

Scheme 8.26

Step 4. Yamaguchi macrolactonization of a hydroxy alkynoic acid and the successful entry to laulimalide (Scheme 8.27): The starting material for this approach was again alcohol **8.71**, which was protected as its THP-derivative and the TBS-group were removed by reaction with Bu$_4$NF to yield the primary alcohol **8.77**. Oxidation of the primary alcohol to the corresponding aldehyde, followed by Corey–Fuchs homologation [15] afforded dibromo olefin **8.78**. Compound **8.78** was converted to alkynyl ester by lithiation to form the corresponding alkynyl anion, which was trapped with methyl chloroformate at –78°C. Removal of the THP group was effected by CSA treatment in MeOH and was followed by ester hydrolysis by treatment with aqueous LiOH. Hydroxy acid **8.79** was obtained and submitted again to Yamaguchi macrolactonization conditions (2,4,6-trichlorobenzoyl chloride, *i*Pr$_2$EtN, DMAP) affording macrolactone **8.80** in 68% yield.

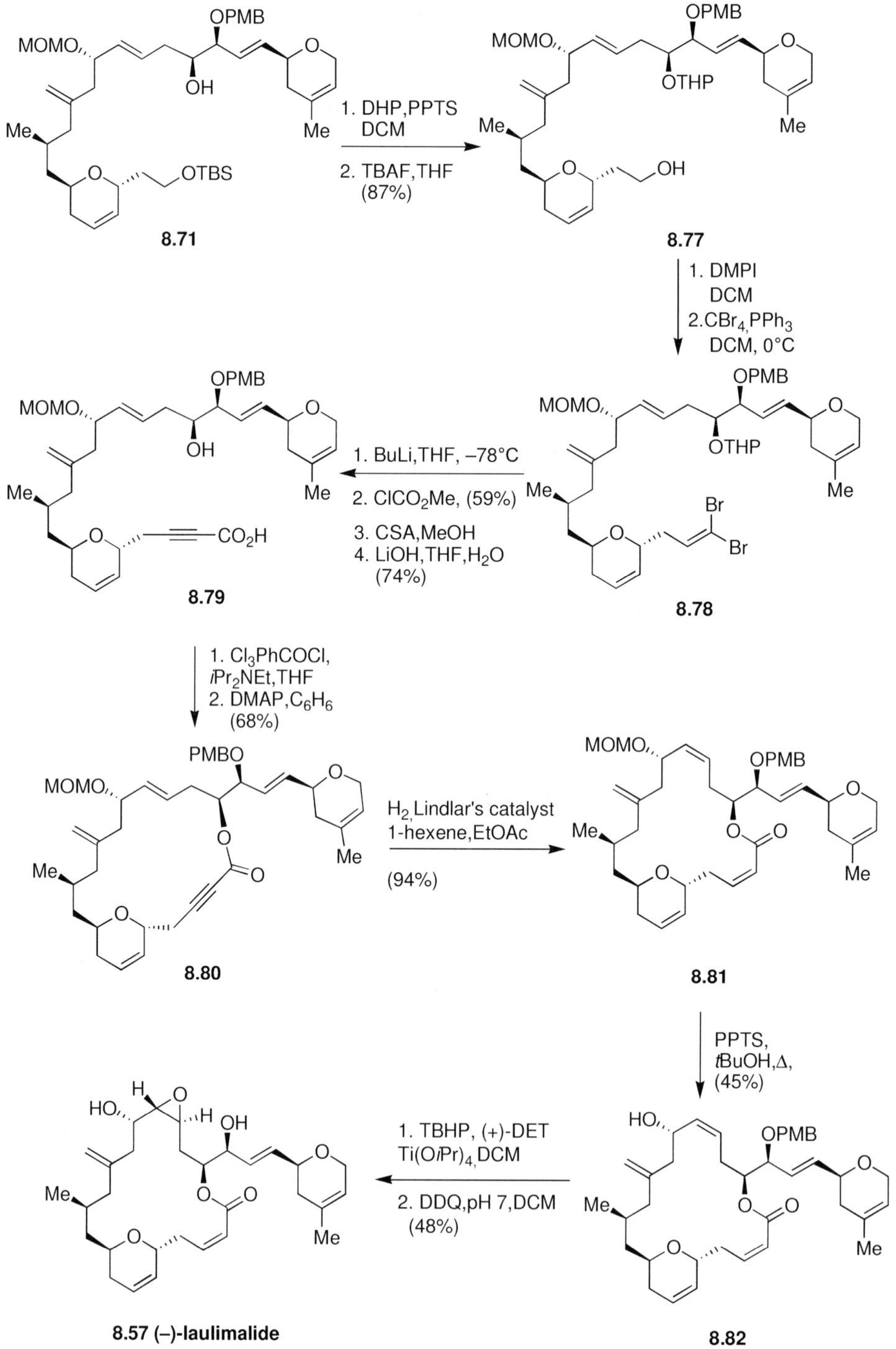
OPMB
MOMO
OH
Me
OTBS
8.71
1. DHP,PPTS DCM
2. TBAF,THF (87%)
OTHP
OH
8.77
1. DMPI DCM
2.CBr4,PPh3 DCM, 0°C
Br
Br
8.78
1. BuLi,THF, –78°C
2. ClCO2Me, (59%)
3. CSA,MeOH
4. LiOH,THF,H2O (74%)
CO2H
8.79
1. Cl3PhCOCl, iPr2NEt,THF
2. DMAP,C6H6 (68%)
PMBO
8.80
H2,Lindlar's catalyst
1-hexene,EtOAc
(94%)
8.81
PPTS, tBuOH,Δ, (45%)
HO
8.82
1. TBHP, (+)-DET Ti(OiPr)4,DCM
2. DDQ,pH 7,DCM (48%)
H
8.57 (–)-laulimalide

Scheme 8.27

Now, compound **8.80** was hydrogenated over Lindlar's catalyst to ensure the adequate *cis*-stereochemistry, and compound **8.81** was obtained as a single isomer in 94% yield.

The completion of the total synthesis of laulimalide required the selective removal of the C15-MOM protecting group to direct the epoxidation of the allylic alcohol formed. This critical step was achieved by treatment of **8.81** with excess of PPTS in *t*BuOH followed by heating the resulting mixture under reflux for 8h. The desired alcohol **8.82** was obtained in 45% isolated yield. Sharpless epoxidation of **8.82** with (+)-DET furnished the corresponding epoxide as a single isomer. Removal of the PMB-ether (DDQ) yielded synthetic (–)-laulimalide **8.57** in 48% yield (Scheme 8.27).

Evaluation

- The progressive development of different approaches to building the macrocycle ring of laulimalide clearly represents the different problems encountered during macrocyclization steps.
- The first approach failed because of the inability of the metathesis reaction to effect the ring closure. In spite of the fact that this reaction has demonstrated its efficiency in macrocyclization reactions, on this occasion no results, other than decomposition, were obtained.

- The second approach, using a Horner–Emmons reaction, placing the phosphonate in the north hemisphere and the acceptor aldehyde over the dihydropyrane ring, succeeded in creating the macrocycle, but without control of the stereochemistry of the double bond.

- The control of the stereochemistry was attempted by using a fixed Z-configuration by attaching a double bond to the side chain of the dihydropyrane ring in the south hemisphere. Unfortunately, albeit that the ring was closed with excellent yield, *E/Z*-isomerization occurred. This type of isomerization has already been described [14].

1. Cl₃PhCOCl
   *i*Pr₂NEt, THF

2. DMAP, C₆H₆
   (65%)

8.75   **8.76** (E/Z 2:1)

- Finally, the successful synthesis of laulimalide recurred with the placement of the right stereochemistry, once the macrocycle was closed. The sequence required the introduction of a compromised cyclization step using an alkynoic acid. Lindlar's hydrogenation gave the right Z-stereochemistry.

8.5   *ent*-Rubifolide [16]

8.83 rubifolide   **8.84 *ent*-rubifolide**

Target relevance

The furanocembrane rubifolide **8.83** is a secondary metabolite of the soft coral, *Gersemia rubiformis*, a subtidal species that inhabits the cold temperate waters off the coast of British Columbia [17]. The synthesis of **8.83** or its enantiomer **8.84** will demonstrate the applicability of the methodology to construct 2-vinylfurans **8.85** by base treatment of 2-hexen-4-yn-1-ols **8.86** (Scheme 8.28). Additionally, the synthesis of **8.84** will establish the relative and absolute configuration of the natural product **8.83** and taxonomically related natural products. At the time of Marshall's synthesis the absolute configuration of no member of this family was known.

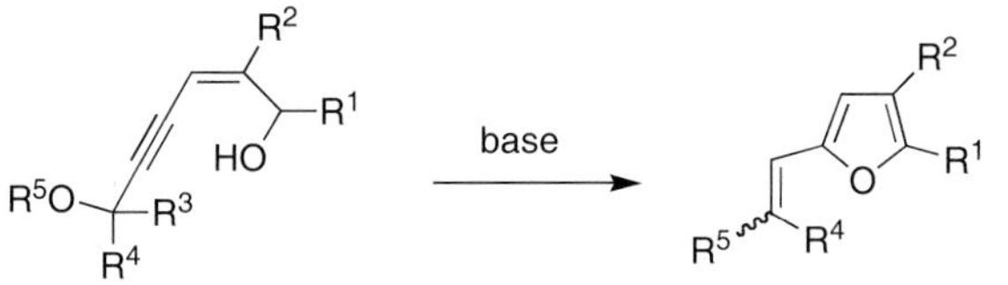

Scheme 8.28

Marshall's synthesis of *ent*-rubifolide **8.84** represents the reverse situation of the above-discussed synthesis of lauliamide **8.57**. Now, the drawbacks caused for a reluctant ring closure did not emerge during the closing of a macrocycle ring but in the process of constructing a furan ring onto an already-built macrocyclic system.

Synthetic plan

The synthetic planning to access to *ent*-rubifolide **8.84** was devised to effect the furan ring closure on the macrocyclic enynol **8.87** by base treatment to form **8.88**. The butenolide ring closure will be effected by manipulation of the enynol moiety of the south hemisphere of **8.88**. The assembling of precursor **8.87** would be done by the Horner–Emmons cyclization on open chain diynal **8.89**. This compound would be derived from (*S*)-(–)-perillyl **8.90**. The starting material was chosen arbitrarily since it was considerably cheaper than its enantiomer (*R*)-(+)-perillyl. The sequence connecting **8.90** with **8.89** involves ring fragmentation, Corey–Fuchs homologation to form **8.91** and the incorporation of the west part of the molecule from aldehyde **8.92** (Scheme 8.29).

   The proposal written in Scheme 8.29 was based on the exceptional yield obtained on the cyclization of enynol **8.92** that was reacted as a mixture of four diastereomers with KO*t*Bu to afford the bridged *Z*-vinylfuran **8.93** [18] (Scheme 8.30).

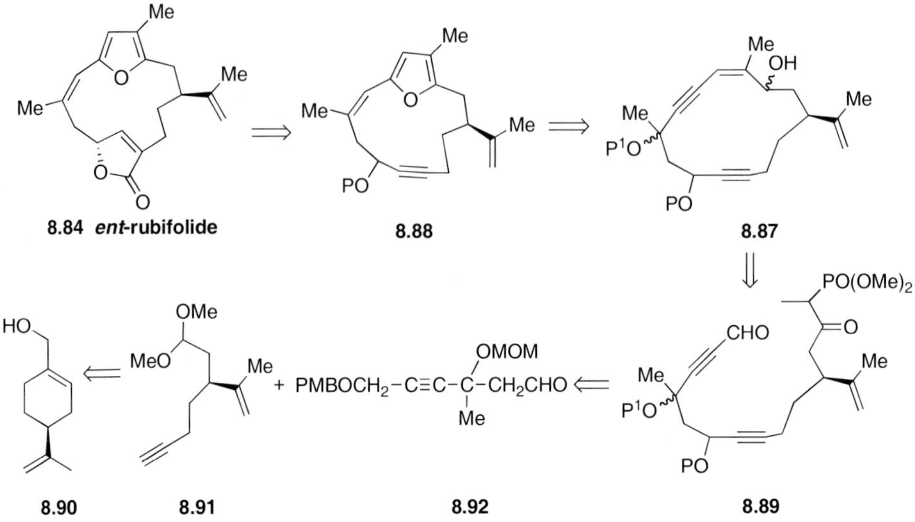

8.84 *ent*-rubifolide   **8.88**   **8.87**

8.90   **8.91**   +   PMBOCH$_2$–C≡C–C–CH$_2$CHO   **8.89**

8.92

Scheme 8.29

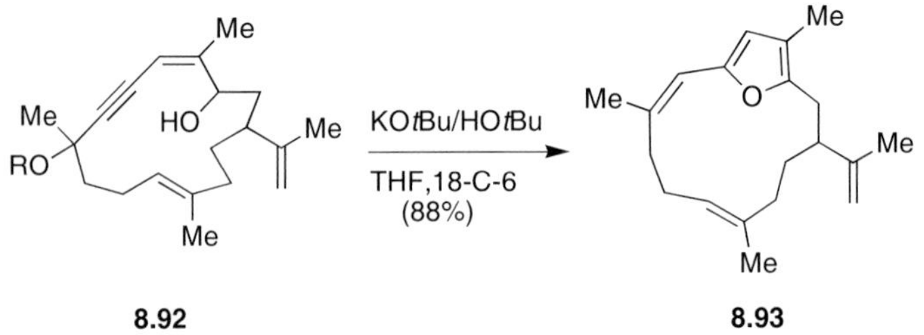

8.92   **8.93**

Scheme 8.30

*Step 1. Synthesis of phosphonate **8.94** from (S)-(–)-perillyl **8.90** (Scheme 8.31):*

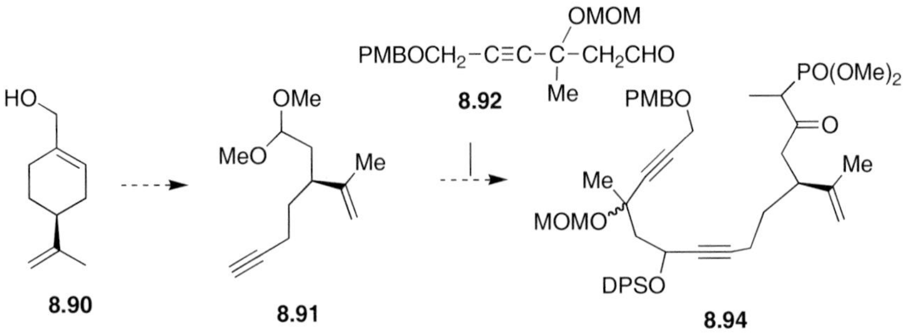

8.90   **8.91**   **8.94**

Scheme 8.31

Reaction of (*S*)-(–)-perillyl **8.90** with VO(acac)$_2$ effected a hydroxyl directed epoxidation that was followed by epoxide breakage with periodate and treatment with MeOH in the presence of TsOH to yield the ester-acetal **8.95** in 52% overall yield. DIBALH reduction of the ester group on **8.95** led to the corresponding aldehyde that was homologated following the Corey–Fuchs protocol. Then, treatment of the aldehyde derived from the reduction of **8.95** with CBr$_4$ in the presence of PPh$_3$ and finally BuLi led to the corresponding lithium acetylide **8.96**, which was reacted *in situ* with aldehyde **8.92** to yield adduct **8.97** as a mixture of the four possible diastereomers. Alcohol **8.97** was protected as the DPS-derivative, the acetal group was hydrolyzed and the corresponding aldehyde reacted with the lithium salt of MeCH$_2$PO(OEt)$_2$. Compound **8.94** was obtained in 70% yield (Scheme 8.32).

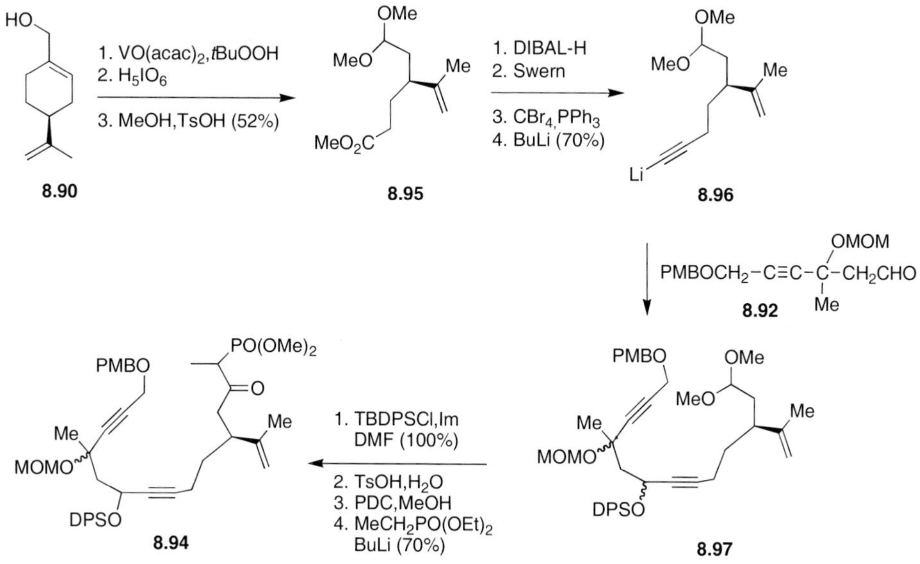

Scheme 8.32

Step 2. Macrocycle ring closure and initials attempts to close the furan ring (Scheme 8.33):

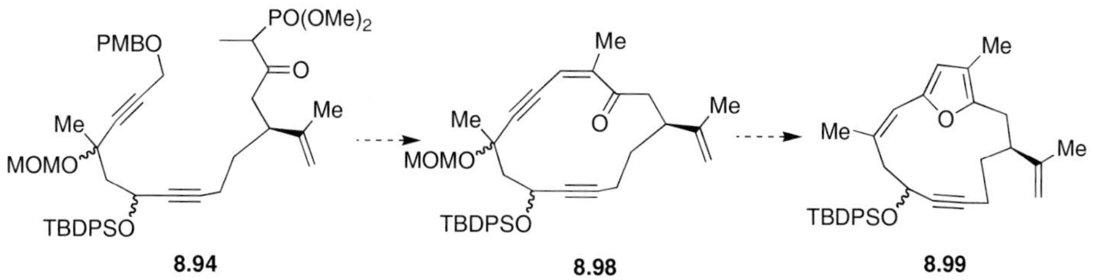

Scheme 8.33

The phosphonate **8.94** was designed to close the macrocycle ring using a Horner–Emmons reaction after removing the PMB protecting group and oxidizing the alcohol to aldehyde. This was achieved by treatment of **8.94** with DDQ and subsequent oxidation of the alcohol under Swern conditions to yield aldehyde **8.100**. Treatment of aldehyde **8.100** with DBU in the presence of LiCl resulted in the cyclization to compound **8.98**. Now, the enone **8.98** was reduced to the alcohol **8.101** providing a mixture of all the eight diastereomers. Treatment of **8.101** with the KO*t*Bu to form the furan ring did not lead to the expected bicycle **8.99** (Scheme 8.34).

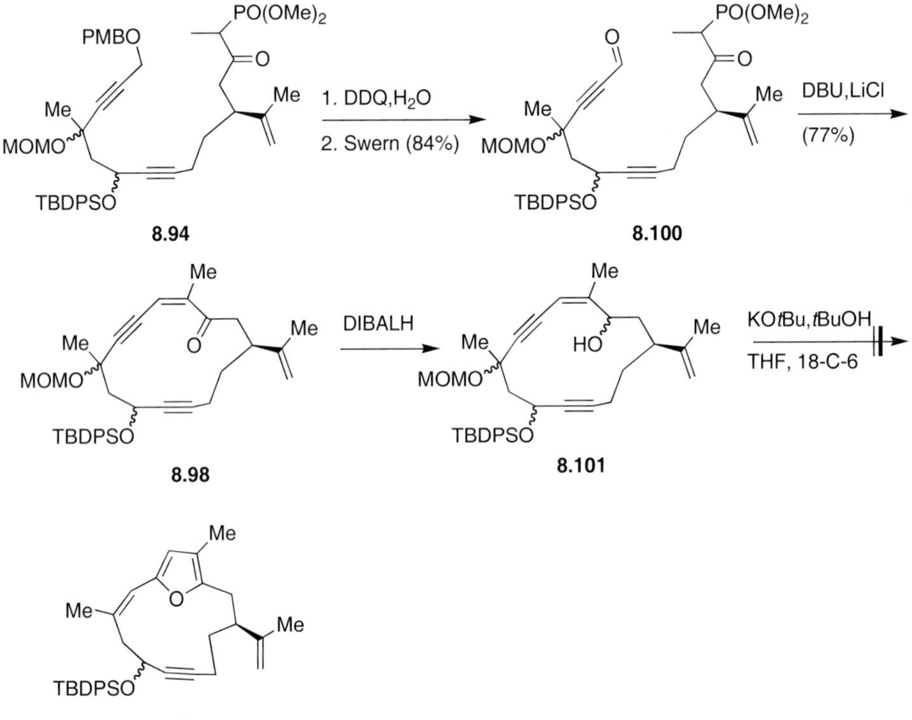

Scheme 8.34

These results are quite surprising since this strategy to construct the bicyclic furan was based on an, at least apparently, very closely-related model, namely the transformation of **8.92** to **8.93**. However, in the words of the authors ...*it soon became apparent that the similarity between 8.92 and 8.101 was superficial.* The obvious (*a posteriori*) difference between substrates **8.92** and **8.101** is clearly the triple bond present in **8.101**. This may cause severe steric constriction in **8.101** avoiding the cyclization, a restriction that is not present in **8.92**.

Step 3. Releasing the steric constrains of the macrocyclic endiyne ring by using an allene moiety and new attempts to built the furan ring: The severe steric constraints of macrocycle **8.101** are due to the presence of the triple bond in the south part of the molecule. The obvious solution for this problem would be to transform this triple bond into a, less sterically demanding but equivalent, functional group. The choice was the less constrained allenoate **8.102**. The allene moiety will, moreover be, used to build the butenolide moiety of rubifolide. Conversion of the alcohol derived from **8.98** into **8.102** was achieved by mesylation and Pd-catalyzed carbonylation of the mesylate in the presence of β-TMS-ethanol. The enone moiety of **8.102** was selectively reduced by LiBH₃NiPr₂ to form **8.103**. Hydroxy-ester **8.103** was reacted with AgNO₃ producing exclusively recovered starting material or, when the reaction conditions were forced, decomposition products. Furane **8.104** was not observed.

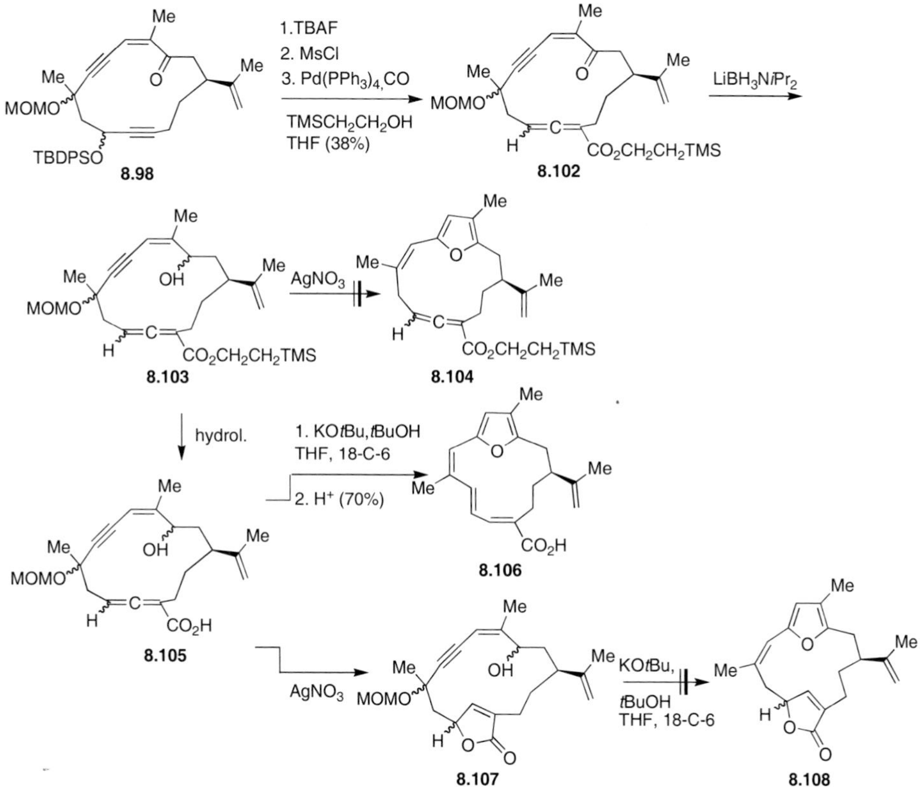

Scheme 8.35

The base treatment of **8.103** was not attempted since it was clear that it would cause isomerization of the allenoate to a conjugated dienoate. The corresponding acid **8.105** would be, however, less prone to this isomerization. Therefore, the β-ethylsilyl ether was hydrolyzed and the resulting acid **8.105** submitted to treatment with *t*BuOK leading to a furane derivative. Unfortunately, this compound **8.106** did not retain the allene moiety but it was the conjugated trienoate. To introduce an additional constraining element in order to avoid this isomerization, acid **8.105** was transformed into butenolide **8.107** which, upon treatment with *t*BuOK, led exclusively to decomposition products (Scheme 8.35). Furan **8.108** was not obtained. The base decomposition of the butenolide moiety is probably responsible for the lack of success of this transformation.

It was clear at this point that the enynol-furan cyclization methodology was not energetic enough to overcome the steric factors associated with the formation of highly strained furanocycles like those used above.

*Step 4. Re-evaluation of the synthetic approach. Construction of the furan ring by Ag-catalyzed ring closure on a macrocyclic allenone like **8.109** (Scheme 8.36):*

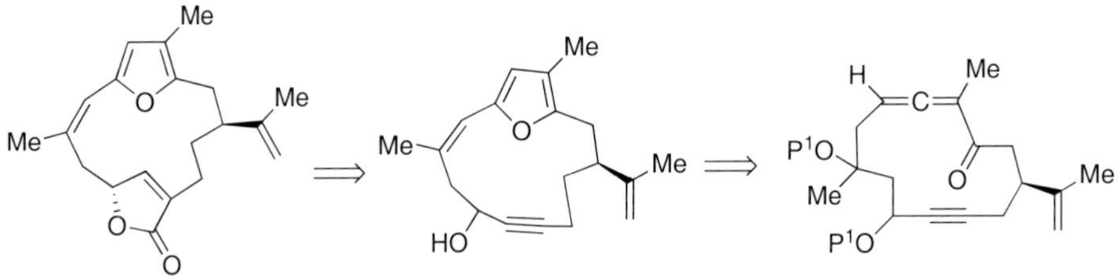

8.84 *ent*-rubifolide   **8.110**   **8.109**

Scheme 8.36

The above failed attempts to construct the furan ring, clearly show that a more efficient furan ring building methodology must substitute the enynol-furan cyclization. This could be the silver-catalyzed cyclization of allenones (like **8.109** → **8.110**), which has been applied to the building of 12- and 14-membered furanocycles under very mild conditions [19]. Furthermore, no strategic changes have to be introduced since this approach uses essentially the same starting materials as the failed approaches to *ent*-rubifolide **8.84**.

Now, macrocycle **8.111** was prepared starting again from (*S*)-(–)-perillyl alcohol as a mixture of eight diastereomers, and the intra-anular furan formation was effected in 84% yield by reaction with Ag⁺ on silica-gel. Furanocycle **8.112** was transformed to the 1:1 diastereomeric mixture of vinylic furans **8.113** and **8.114** by reaction with TsOH. Since only **8.113** has the right stereochemistry to effect the butenolide ring closure through a S_N2' reaction (which means an inversion of configuration at the alcohol center), **8.114** was epimerized through the ketone **8.115** by reduction with K-Selectride (the calculated lowest energy conformers of ketone **8.115** show shielding of the "top-face" of the carbonyl group by the vinylic Me-substitutent). Finally, alcohol **8.116** obtained by saponification of **8.113** was mesylated and treated with Pd(0)-catalysts to form the butanolyde ring and complete the synthesis of *ent*-rubifolide **8.84**. This compound was obtained in a 49% yield (Scheme 8.37).

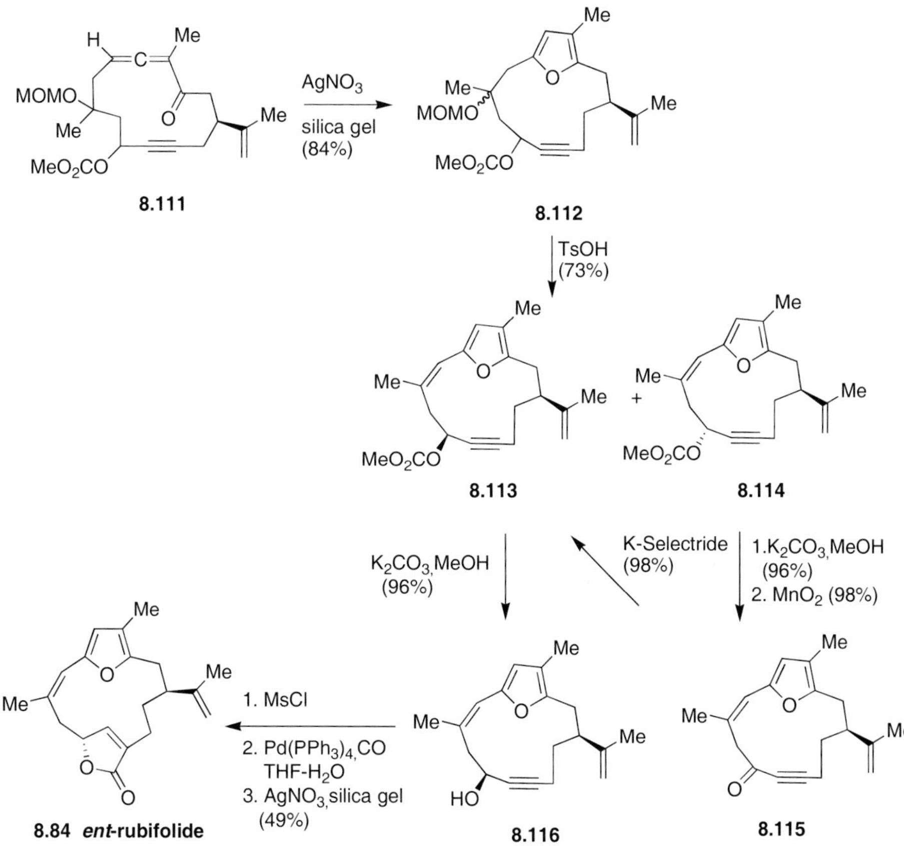

Scheme 8.37

Evaluation

The synthesis of rubifolide, discussed in this section, shows how troublesome it may be, to close a furanocycle embedded into a macrocycle system. Any of the solutions attempted, such as reducing the steric strain (moving from an alkyne to an allene, using a preformed butanolide, etc) were unsuccessful. A different procedure for closing the ring had to be found. In fact, the combination of different methodologies, which will effect the closure of five-membered rings, is the solution to the problem.

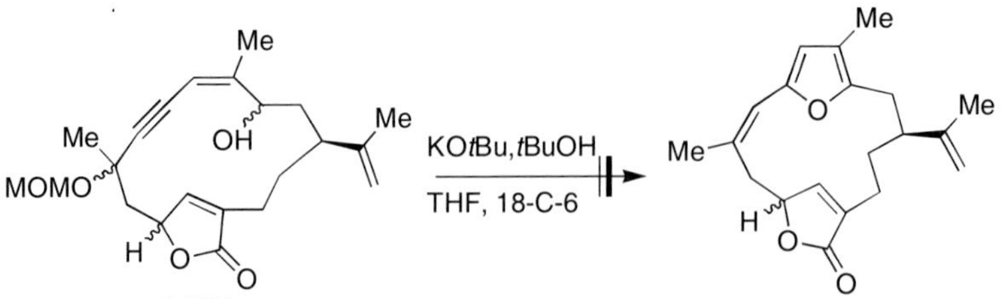

8.107   **8.108**

Key synthetic sequences

The Corey-Fusch protocol: The sequence aldehyde→ dibromo-olefin (+ 1C) → terminal alkyne effected by treatment of the aldehyde with CBr_4/PPh_3 (dibromo olefin formation), and further BuLi treatment. Since the product of this last step is an alkynyl anion (like **8.94**), it can be reacted with electrophiles [15].

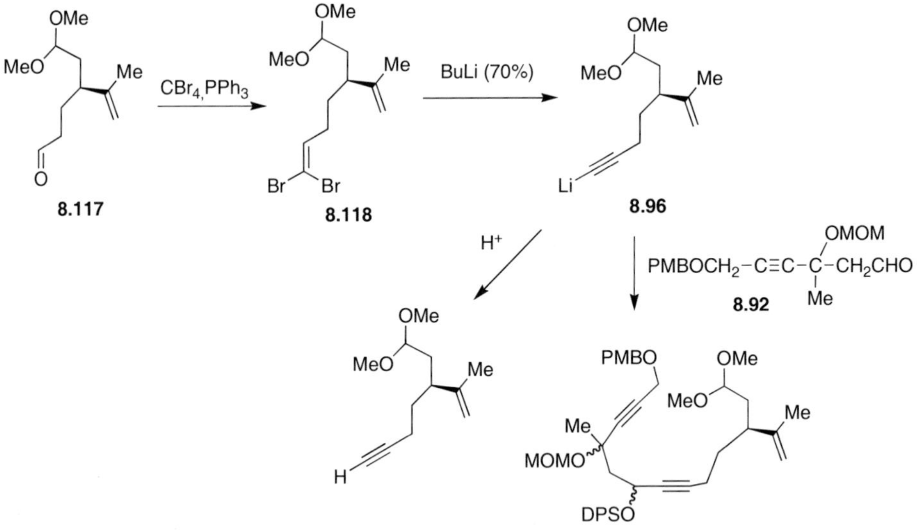

8.117   **8.118**   **8.96**

$PMBOCH_2-C\equiv C-C-CH_2CHO$

8.92

8.119   **8.97**

Difficulties to attach fused heterocyclic rings to existent macrocyclic system seem to be a very general problem. The difficulties encountered during roseophilin tricyclic core synthesis discussed below are another example. In this case the macrocycle cyclophanic structure was constructed in a relatively easy way by using a metathesis ring closure step. However, the building of the fused pyrrole ring in a very advanced intermediate turned to be problematic.

8.6 Roseophilin Tricyclic Core: The Formal Total Synthesis of Roseophilin [20]

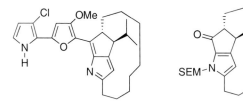

8.120 roseophilin   8.121 roseophilin tricyclic core

Target relevance

Roseophilin **8.120** was isolated as a deeply colored hydrochloride salt from the culture broth of *Streptomyces griseoviridis* [21]. It showed significant cytotoxicity against K562 human erythroid leukemia cells and KB human epidermoid carcinoma cells.

Synthetic plan

The approach to roseophilin involves the final connection of the heterocyclic side chain **8.122** to the tricyclic core **8.121**. This tricyclic core was the target of this synthesis, since other authors have reported the joining of fragments **8.121** and **8.122**. The formation of the pyrrole moiety will be effected from the allyl alcohol **8.123**. The isopropyl group will be introduced on diene **8.124** using the macrocyclic bridge to control the stereochemistry. Diene **8.124** may in turn derive from the enyne metathesis of **8.125** (Scheme 8.38).

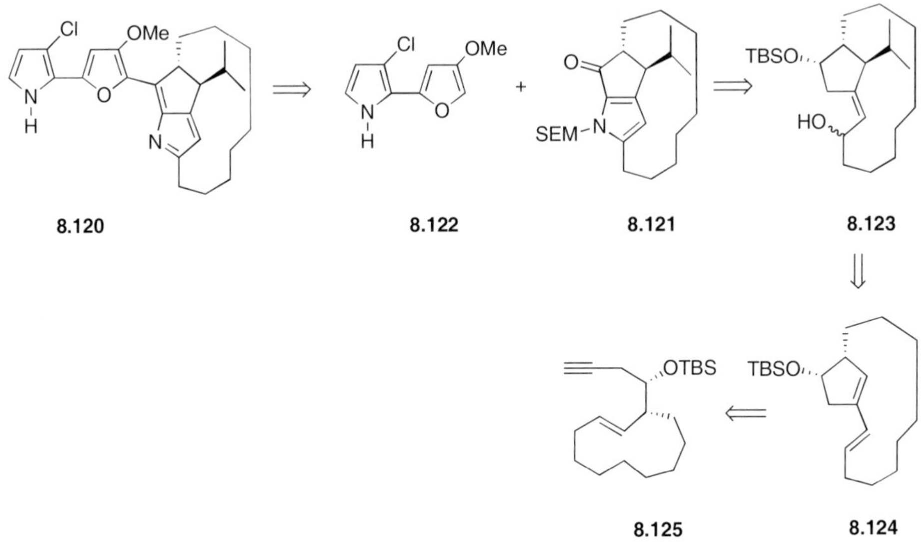

Scheme 8.38

Step 1. Ring closure metathesis of 8.125 and incorporation of the isopropyl moiety (Scheme 8.39):

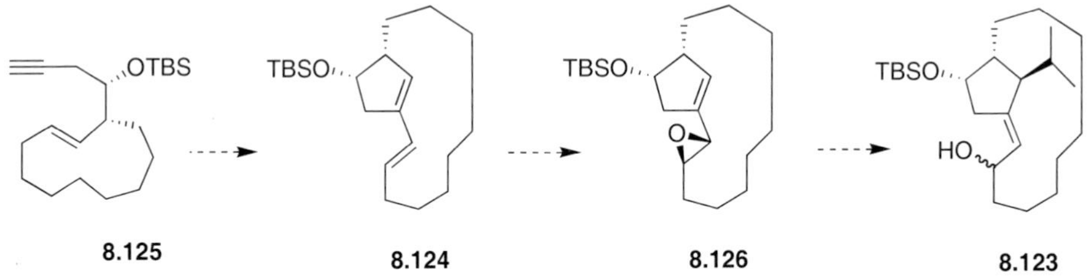

Scheme 8.39

Compound **8.125** was the metathesis substrate and was cyclized both with PtCl$_2$ or (*p*-cymene)RuCl$_2$ dimer in the presence of tri-*o*-tolyl phosphite in essentially quantitative yields to afford diene **8.124**. The next step requires a regio- and diastereocontrolled addition of a hydroxyl and an isopropyl group across the diene fragment. This was devised to occur by addition of a cuprate from the less hindered face of epoxide **8.126**. The chemoselective epoxidation of the less reactive double bond of diene **8.124** was achieved by reaction with NBS in wet THF. These reaction conditions gave the corresponding mixture of bromohydrins, which were highly unstable. This instability required *in situ* treatment with base to form the mixture of the two regioisomeric epoxides **8.126** and **8.127**. This last epoxide was the minor product (9:1 ratio). Epoxide **8.126** was in turn a 3:1 mixture of diastereomeric epoxides **8.126a** and **8.126b**.

It is known that cuprate additions to vinyl epoxides like **8.126** occur *syn* to the epoxide fragment [22]. Therefore, the isopropyl group was introduced on the same face of the cyclopentene of both diastereomeric epoxides **8.126** by reaction with isopropyl-magnesium bromide in the presence of catalytic amounts of CuBr. These reactions lead to the stereoisomeric (at the double bond) alkenes **8.128** and **8.129**, which were oxidized to the corresponding ketones **8.130** and **8.131**. Only **8.130** has the right stereochemistry to produce roseophilin (Scheme 8.40).

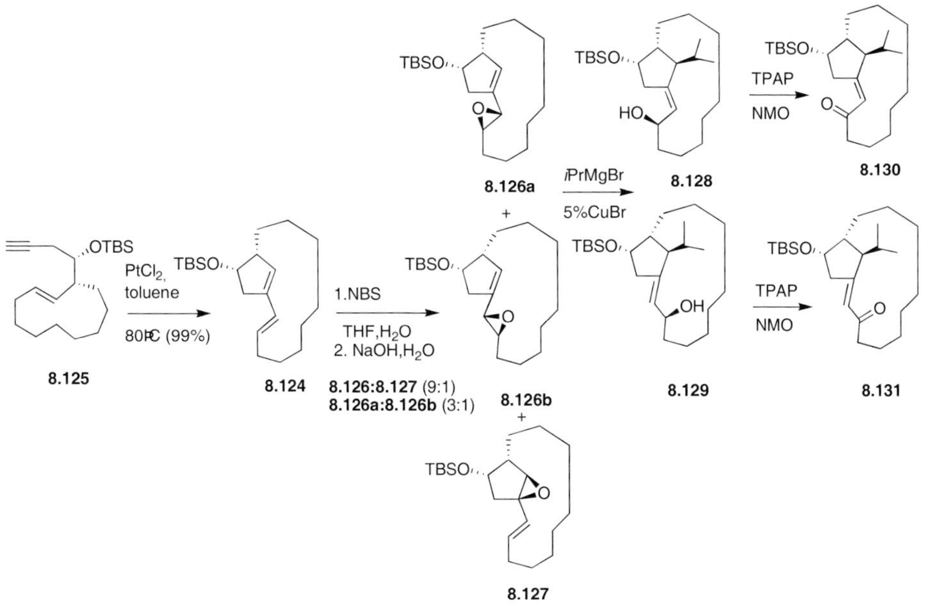

Scheme 8.40

Step 2. Construction of pyrrole nucleus and completion of the roseophilin core: In order to finish the synthesis of the roseophilin core only the construction of the pyrrole nucleus remains. There are several methods of achieving this task. The attractive route that uses the introduction of the needed amino-group via an intramolecular insertion of an acylnitrene, was considered first. The removal of the TBS protecting group resulted in alcohol **8.132**, which dehydrated easily to form enone **8.133** during derivatization. To avoid this undesirable dehydratation, the double bond of **8.130** was hydrogenated and the ketone **8.134** was transformed into the azidoformate derivative **8.135** after removal of the TBS-protecting group by treatment with $Cl_3COCO_2CCl_3$ and reaction with lithium azide. Neither photolysis of hydrazide **8.135**, nor thermolysis, afforded any of the expected oxazoline **8.136** (Scheme 8.41).

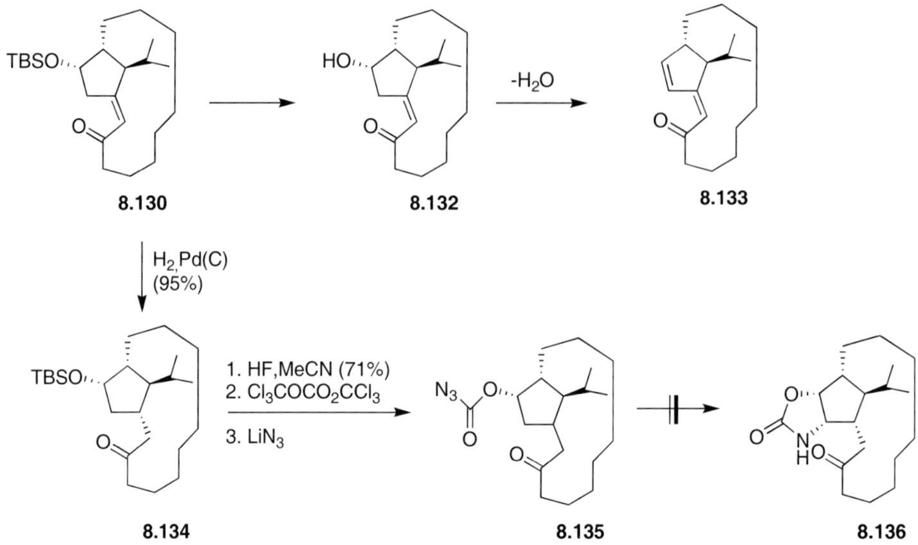

Scheme 8.41

Second attempt: The mixture of alcohols **8.128** and **8.129** was easily dehydratable with methanesulfonyl chloride to form diene **8.137** that could be, in turn, used in the preparation of dihydroisoxazines **8.139** by oxidation of the intermediate hydroxamic acids **8.138**. The idea was to convert compound **8.139** to diketone **8.140** through a fragmentation reaction of the heterocyclic ring. Diketone **8.140** would then be converted to the desired pyrrole **8.141** through intramolecular ring closure (Scheme 8.42). The problem emerged when compound **8.139** spontaneously cyclo-reverted to the corresponding diene **8.137**. This thwarted the entry to the fused pyrrole ring. The synthesis of **8.141** was next pursued by hydrogenation of **8.139** leading to alcohol **8.142** that, upon oxidation, formed ketone **8.143**. All attempts to effect the heterocycle ring closure to convert ketone **8.143** into pyrrole **8.141** met with no success.

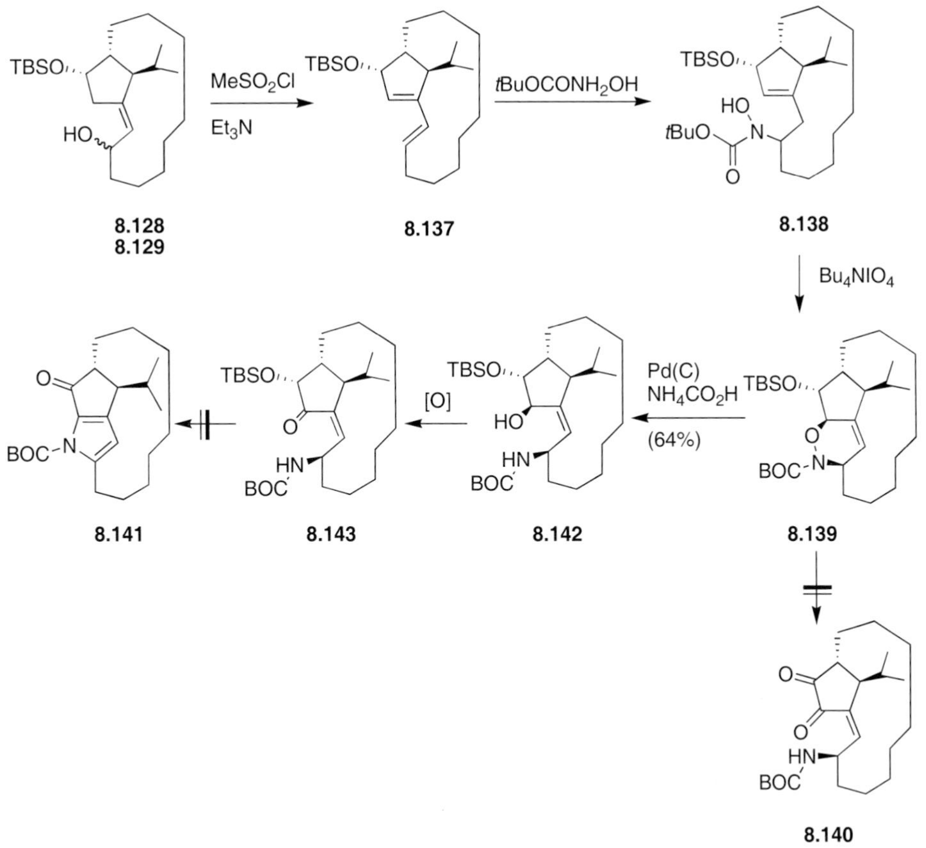

Scheme 8.42

Third and final, successful, attempt: the venerable Paal–Knorr ring closure: The Paal–Knorr pyrrole ring synthesis [23] uses the condensation of an amine and a 1,4-diketone. Therefore, to construct the roseophilin pyrrole nucleus a diketone like **8.144** is needed. Treatment of the mixture of enones **8.128** and **8.129** with DBU formed the deconjugated ketone **8.145**. Borane treatment of **8.145** resulted in the diastereomeric mixture of alcohols **8.146** after H_2O_2 work-up. Oxidation of the diol with catalytic TPAP formed diketone **8.144** as a single product. The best conditions for achieving pyrrole formation were treatment of **8.144** with ammonium acetate in the presence of catalytic amounts of CSA, followed by capturing *in situ* the parent pyrrole as its SEM-derivative **8.145**. Finally, TBS-protecting group removal following by oxidation of the alcohol **8.147** gave the target **8.121** that has been converted by Fürstner and Weintritt [24] in roseophilin **8.120**. Therefore, the formal synthesis of this compound was achieved (Scheme 8.43).

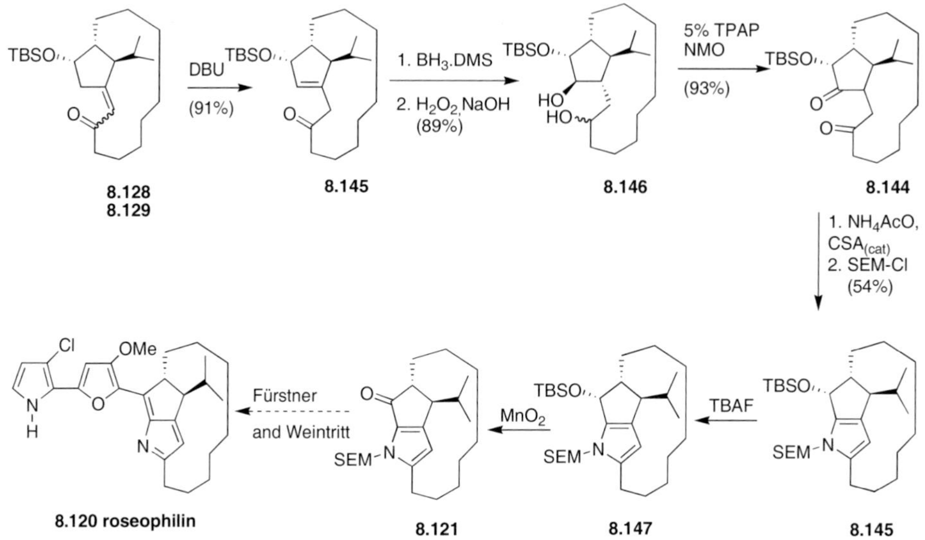

Scheme 8.43

Evaluation

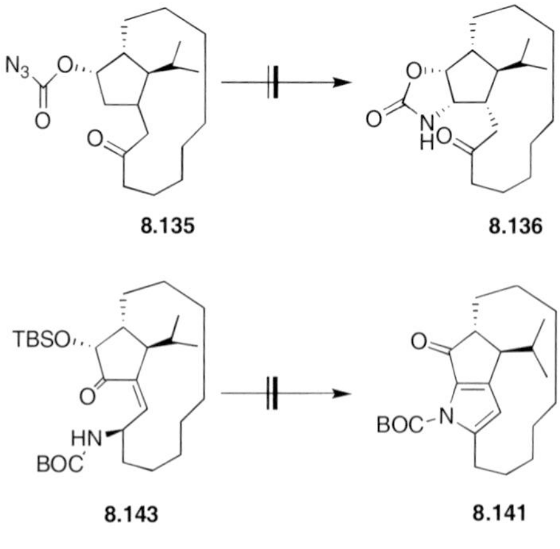

The synthesis of roseophilin exemplifies the difficulties encountered during the apparently slightly complex task of building a simple heterocyclic ring in a preformed macrocyclic substrate. This example, together with the above synthesis of *ent*-rubifolide **8.84**, demonstrates that these ring closures may determine the fate of a synthetic scheme. This assertion may be contrary to the general belief that it is the macrocycle part of a fused macrocyclic policycle, which is the most difficult part of a synthesis. The truth is that, as always, this part of the synthesis has attracted much more attention than the "simpler" building of furane or pyrrole rings, because the question of aromatic five and six-membered ring synthesis was considered solved since the 19th century. Evidently, this is far from the truth.

The last example in this chapter, shows clearly that, in a highly convergent synthesis where, at the end, two fragments are to be joined to yield the skeleton of the final product, orchestration of the sequence, in which the fragments are joined to effect a macrocyclization, may become the cornerstone of the synthesis.

8.7   (–)-Papuamine [25]

8.148 (–)-papuamine

Target relevance

Papuamine **8.148** and its epimer haliconadiamine **8.149** display significant inhibitory activity against the growth of *Candida albicans, Bacillus subtilis,* and *Staphylococcus aureus.* Papuamine also shows antifungal activity against *Trichophyton mentagrophytes.* Papuamine is the major metabolite of the bright red encrusting sponge *Haliclona sp,* growing near South Lion Head Island, Papua New Guinea [26].

8.148 (–)-papuamine   8.149 haliclonadiamine

Synthetic planning

The key point of the synthetic planning is the homocoupling of an enantiomerically pure perhydroindane system **8.150**, to first generate a tetracyclic *E,E*-1,3-diene **8.151**, followed by introduction of a three-carbon unit between the nitrogens to generate the 13-membered macrocyclic ring. This approach effects the ring closure by the north (diamine moiety) of the molecule and takes advantage of the C$_2$ symmetry axis present in (–)-papuamine.

Implementation of this planning requires the vinyl synthon **8.150**, derived from alkyne **8.152** by *syn*-addition of HX across the triple bond (Scheme 8.44). Alkene **8.150** would be the substrate for the homocoupling leading to **8.151**. Alkynyl amine **8.152** would be, in turn, derived from the intramolecular cyclization of an imino-allenyl silane **8.153**. Imino-allenyl silane **8.153** would be derived from lactone (+)-**8.155** through the successive incorporation of the triple bond to form the alkyne diol **8.154**. Further manipulation of the triple bond and the alcohol groups of **8.154** would form the required allene and imine moieties of **8.153**, respectively (Scheme 8.44).

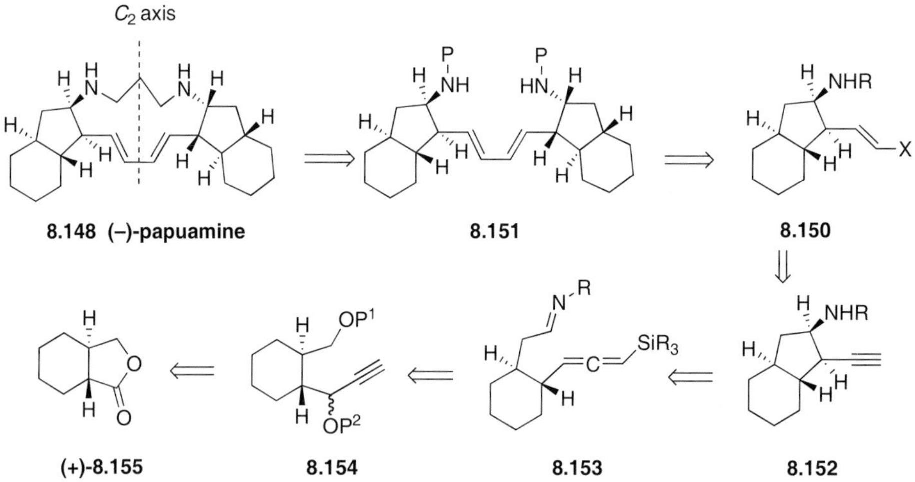

8.148 (–)-papuamine   **8.151**   **8.150**

(+)-**8.155**   **8.154**   **8.153**   **8.152**

Scheme 8.44

Predictable problems

The transformation imino-allenyl silane 8.153 → bicyclic amine 8.152, had been demonstrated only in intermolecular reactions [27]. The question was whether the intramolecular transformation would be as efficient, both in yield and in stereoselectivity as the intermolecular reaction.

Synthesis

Step 1. Synthesis of formyl-allenyl silane 8.155 (Scheme 8.45):

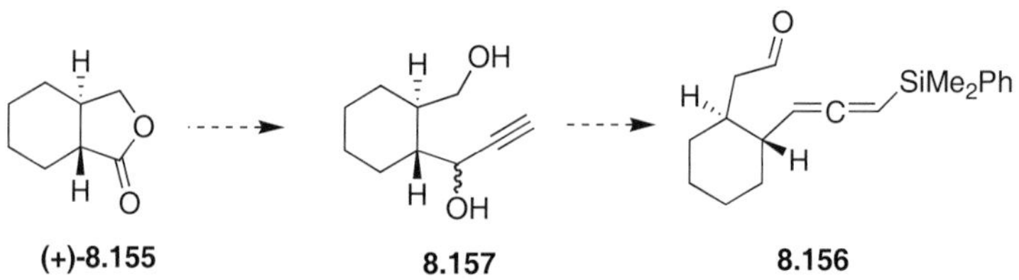

(+)-**8.155**   **8.157**   **8.156**

Scheme 8.45

The preparation of the key intermediate **8.156** started with DIBALH reduction of lactone (+)-**8.155** to the corresponding lactol, followed by addition of ethynylmagnesium bromide to form the mixture of epimeric alcohols **8.157**. It is not evident at this point that the formation of an epimeric mixture of alcohols **8.157** may be of relevance in the development of the synthesis, especially in the cyclization of the imino-allenyl silane derived from **8.156**. This assumption turned out to be wrong (see below). To continue with the synthesis, both alcohols **8.157** were separated. The less-hindered primary alcohol was selectively tritylated, while the propargilic secondary alcohol was acetylated to form epimers **8.158**. The allene moiety was introduced using the *anti* S$_N$2′ addition of a silylcuprate **8.159** (the Fleming–Terret cuprate) [28] to alkynes **8.158**. Allenyl silanes **8.160a** and **8.160b** were formed stereospecifically. Nitriles **8.161a** and **8.161b** were obtained from **8.160a-b** by detritylation followed by mesylation of the primary alcohol and cyanide displacement. The cyanide group was then cleanly reduced by DIBALH to yield the desired aldehydes **8.156a** and **8.156b** (Scheme 8.46).

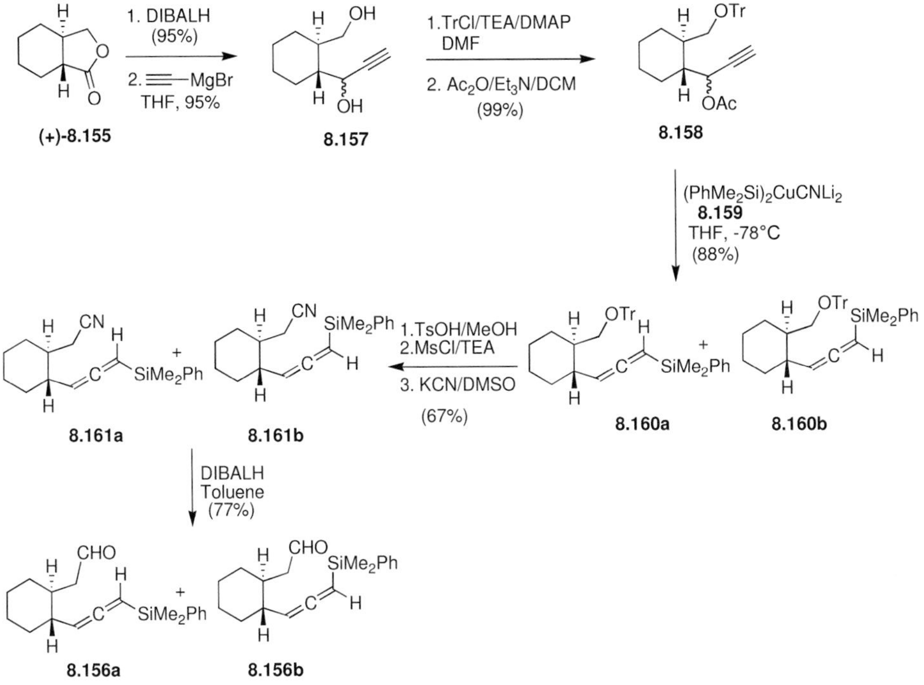

Scheme 8.46

Step 2. Cyclization of imino-allenyl silane 8.162 → bicyclic amine 8.163 (Scheme 8.47):

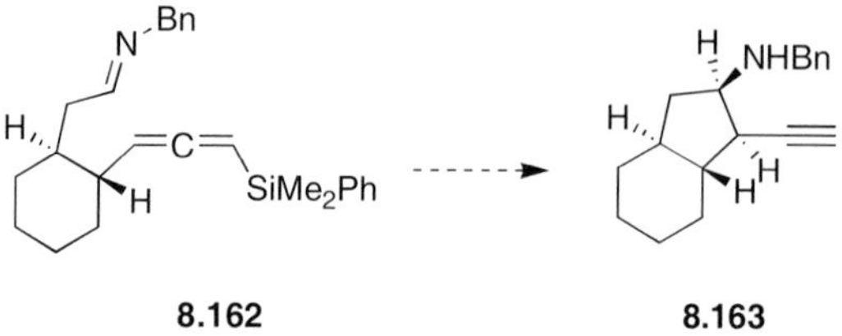

8.162   **8.163**

Scheme 8.47

It was presumed that the Danheiser cyclization [27] of either benzylimines **8.162a** or **8.162b** would yield the desired bicyclic amine **8.163**. However, the reaction was stereoespecific with compound **8.162a** yielding product **8.163** having the desired stereochemistry, while diastereomer **8.162b** yielded isomer **8.164**, which has the wrong stereocheistry. It should be noted that reaction products **8.163** and **8.164** were neither terminal alkynes nor chloro vinylsilanes. These would be the expected products if the cyclization of **8.162** involved an intermediate vinyl cation (as in the Danheiser process). In fact, the reaction occurs more probably through a concerted imino-ene reaction [29] involving the transition states depicted in Scheme 8.48.

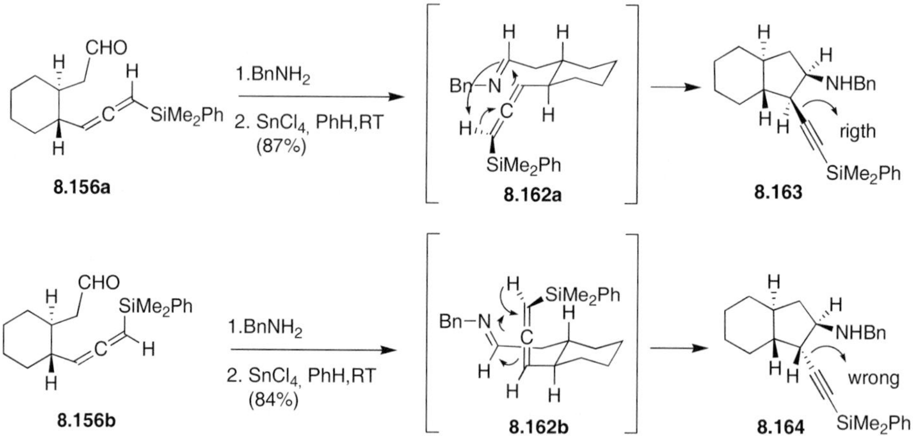

8.156a   **8.162a**   **8.163**

8.156b   **8.162b**   **8.164**

Scheme 8.48

From a synthetic point of view, this means dividing the amount of material in an advanced stage of the synthesis. However, this is a rather useful means of increasing chemical knowledge and it was therefore thoroughly investigated.

Step 3. Homocoupling of perhydroindane system 8.166 and ring closure (Scheme 8.49):

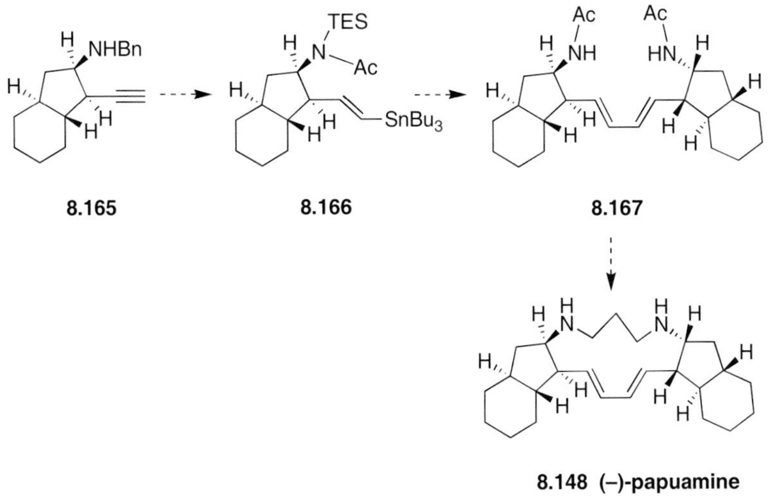

8.165   **8.166**   **8.167**

8.148 (–)-papuamine

Scheme 8.49

To complete the synthesis of (–)-papuamine **8.14** it is necessary to accomplish the following steps:

- Stereoselective formation of *E*-vinyl perhydroindane **8.166**.
- Homocoupling of olefin **8.166** to form the macrocyclization substrate **8.167**.
- Macrocycle ring closure to obtain the papuamine **8.148** skeleton.

These final stages of the synthesis were not effected using amine **8.165** but the β-tosylethylamine derivative as a benzylamine substitute [30]. This change would avoid potential interferences derived from the reducing conditions needed to eliminate the benzyl group and the instaurations present in the compound. The imino ene reaction was repeated from the imine derived from aldehyde **8.156a** and β-tosylethylamine, to yield **8.168** in 62% yield. The amine was aceylated and converted to the *E*-vinyl stannane **8.166** by Pd-catalyzed addition of TBTH to the terminal alkyne [31]. Cu(II) homocoupling [32] of **8.166** yielded the desired *E,E*-diene **8.169** in acceptable yields. Base elimination of the TSE groups formed the bisacetamide **8.167**, which was designed to form the skeleton of papuamine by double alkylation. However, the diverse experiments made to accomplish this target, including treatment with 1,3-dihalopropanes and the change of the acetate groups by trifluoracetate groups in order to increase the nucleophicity of the nitrogens, met with no success (Scheme 8.50). Product **8.168**, the immediate precursor of papuamide, could not be accessed. This is a major and unpredicted drawback to the synthetic scheme.

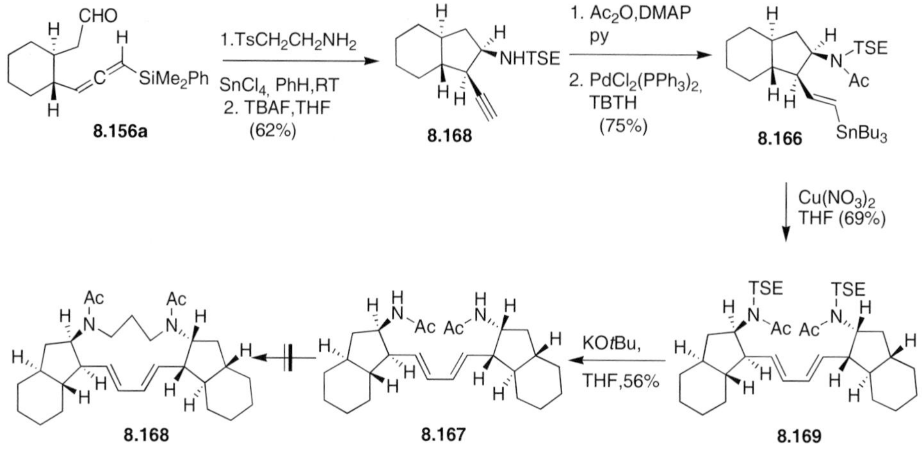

Scheme 8.50

Step 4. Change in the synthetic plan (Scheme 8.51): The failure of diamide **8.167** to yield **8.168** by ring closure of the northern hemisphere of papuamine **8.148**, made it necessary to re-evaluate the synthetic plan. This time the southern hemisphere of papuamine would be closed first by using a double imino ene reaction from aldehyde **8.156a** to form **8.169**. The aldehyde **8.156a** is the connecting point of the new synthetic scheme with the original plan. The closure of the southern hemisphere will be effected on the intermediate **8.170** derived from dialkyne **8.171**.

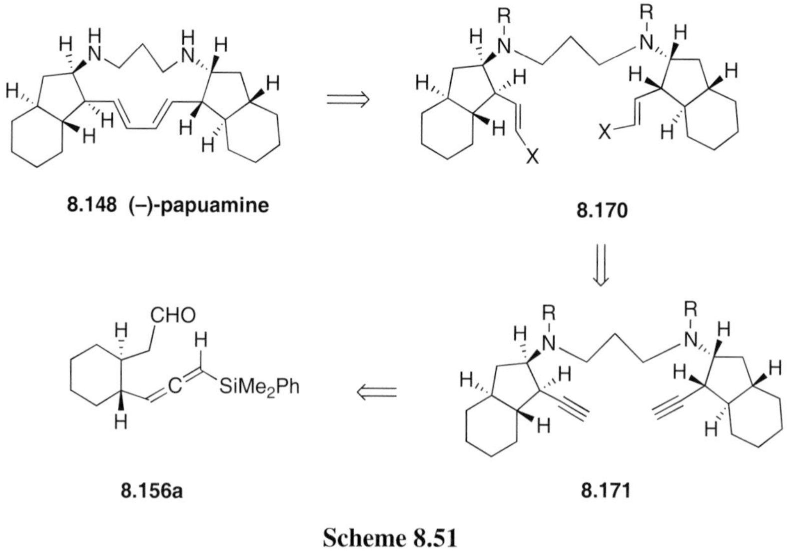

Scheme 8.51

*Step 5. Double imino-ene reaction and southern ring closure. Formation of (−)-papuamine **8.148**:* The key to the success of this new route to (−)-papuamine **8.148** is the ability to perform the double imino-ene reaction on the diimine formed from aldehyde **8.156a** and 1,3-diaminopropane. Heating of **8.156a** and 0.5 equiv. of 1,3-diamino-propane in toluene, smoothly formed the bis-acetylene **8.172** in a respectable 70% yield and as a single stereoisomer. The highly concerted nature of this reaction, which should occur through transition state **8.173**, is responsible for this exceptional result. Removing the silyl groups and hydrostannylation under radical conditions, formed the bis-stannyl derivative **8.174**. This compound experiences Pd-catalyzed ring closure to yield (−)-papuamine **8.148** in 48% yield (Scheme 8.52).

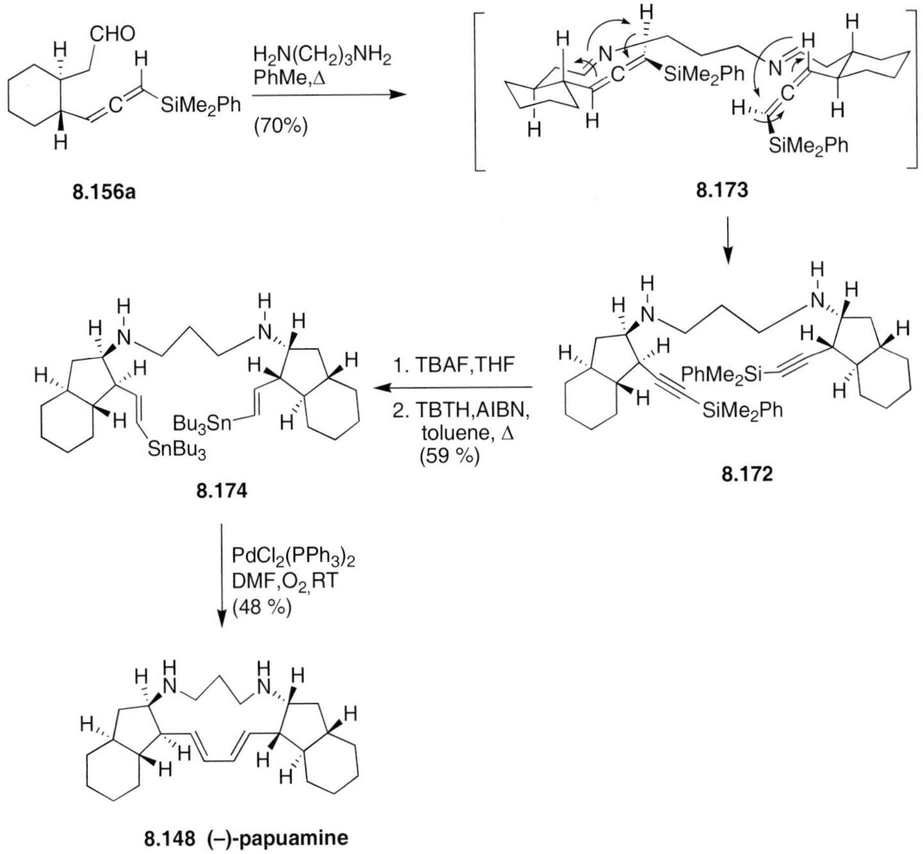

Scheme 8.52

Evaluation

- A major change in strategy was made during the synthesis of (–)-papuamine
  8.148. By comparing the original synthetic plan with the route that finally led to
  (–)-papuamine (Scheme 8.55) it is clear that the orchestration of the sequence in
  which fragments are joined may prove to be the cornerstone of the synthesis. The
  unpredictable refusal of **8.167** (or any of its amide derivatives) to experience the
  ring closure by a double S_N, led to a re-evaluation of the synthetic planning
  inverting the sequence to effect the ring closure. It is not clear why intermediate
  8.167 fails to yield the closed-ring compound while **8.173** leads to (–)-papuamine
  8.148 in good yields (Scheme 8.53). One can reason that a conformational change
  to the *s-cis*-conformer has to occur prior to the closure of the ring in compounds
  8.151, while this conformational change is not necessary in **8.170**. The energy
  associated with this change is 6.5 kcal mol^{-1}. However, the *N-N* distance on *s-
  trans*-**8.167** is 5.1 Å shorter than the 6.3 Å on *s-cis*-**8.167**. Moreover, *s-trans*-
  papuamine is 1.6 kcal mol^{-1} more stable than *s-cis*-papuamine [33]. The
  perplexing conclusion, which is drawn is that cyclization should in theory occur,
  but does not.

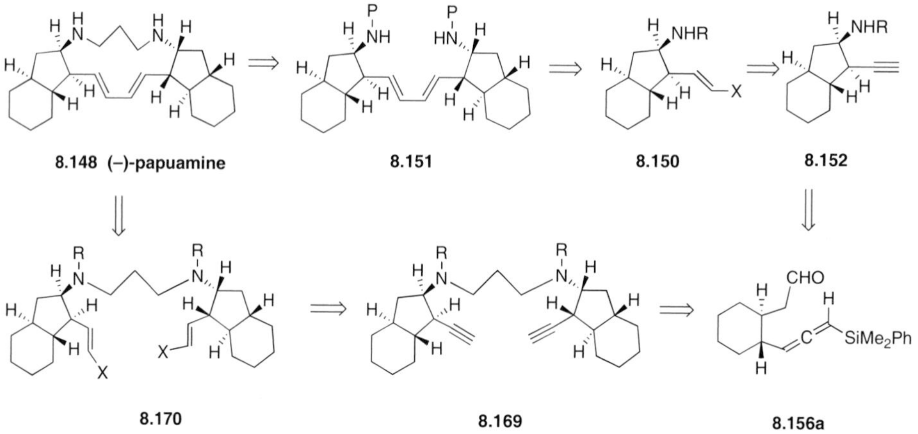

8.148 (–)-papuamine   8.151   8.150   8.152

8.170   8.169   8.156a

Scheme 8.53

- The expected problem of effecting the intramolecular version of the Danheiser
  cyclization of imino-allenyl silanes was not encountered. Furthermore, the
  reaction led to the discovery of a new stereoespecific imino-ene cyclization,
  which resulted in an impressive novel synthetic methodology. Nevertheless, it
  caused a practical synthetic problem, since it was expected that both diastereomeric allenes, derived from **8.156a** and **8.156b**, would cyclize to the
  single isomer of the bicyclic amine **8.163**. The stereoespecificity of the imino-ene
  cyclization causes a major advance step in synthetic methodology, but it meant
  that half of the material in an advanced intermediate was of no further use in the
  synthesis of (–)-papuamine, **8.148**.

References

1. Kita, Y.; Tohma, H.; Inagaki, M.; Hatanaka, K.; Yakura, T. *J. Am. Chem. Soc.* **1992**, *114*, 2175.
2. a) Perry, N. B.; Blunt, J. W.; McCombs, J. D.; Munro, M. H. G. *J. Org. Chem.* **1986**, *51*, 547; b) Blunt, J. W.; Calder, V. L.; Fenwick, G. D.; Lake, R. J.; McCombs, J. D.; Munro, M. G. H.; Perry, N. B. *J. Nat. Prod.* **1987**, *50*, 290; c) Perry, N. B.; Blunt, J. W.; Munro, M. H. G.; Higa, T.; Sakai, R. *J. Org. Chem.* **1988**, *53*, 4127; d) Perry, N. B.; Blunt, J. W.; Munro, M. H. G. *Tetrahedron* **1988**, *44*, 1727.
3. Kita, Y.; Tohma, H.; Inagaki, M.; Hatanaka, K.; Kikuchi, K.; Yakura, T. *Tetrahedron Lett.* **1989**, *30*, 1119.
4. Tohma. H.; Harayama, Y.; Hashizume, M.; Iwata, M.; Kiyono, Y.; Egi, M.; Kita, Y. *J. Am. Chem. Soc.* **2003**, *125*, 11235.
5. Inoue, M.; Carson, M. W.; Frontier, A. J.; Danishefsky, S. J. *J. Am. Chem. Soc.* **2001**, *123*, 1878.
6. Patil, A. D.; Freyer, A. J.; Kilmer, L.; Offen, P.; Carte, B.; Jurewiez, A. J.; Johnson, R. K. *Tetrahedron*, **1997**, *53*, 5047.
7. Tebbe, F. N.; Marshall, G. W.; Reddy, G. S. *J. Am. Chem. Soc.* **1978**, *100*, 3611.
8. a) Jones, G.; Stanforth, S. P *Org. React.* **1997**, *49*, 1.
9. a) Petasis, N. A.; Lu, S. -P.; Bzowej, E. I.; Fu, D.-K.; Staszewski, J. P.; Akritopoulou-Zanze, I.; Patane, M. A.; Hu, Y. -H. *Pure & Appl. Chem.*, **1996**, *68*, 667; b) Petasis, N. A.; Hu, Y. -H. *J. Org. Chem.*, **1997**, *62*, 782; c) Petasis, N. A.; Hu, Y.-H. *Current Organic Chemistry*, **1997**, *1*, 249.
10. Ghosh, A. K.; Wang, Y.; Kim, J. T. *J. Org. Chem.* **2001**, *66*, 8973.
11. Quinoa, E.; Kakou, Y.; Crews, P. *J. Org. Chem.* **1988**, *53*, 3642.
12. Julia, M.; Paris, J. M. *Tetrahedron Lett.* **1973**, *14*, 4833.
13. Fürstner, A. *Angew. Chem. Int. Ed.* **2000**, *39*, 3012.
14. a) Roush, W. R.; Spada, A. P. *Tetrahedron Lett.* **1983**, *24*, 3693; b) Roush, W. R.; Blizzard, T. A. *J. Org. Chem.* **1994**, *59*, 7549.
15. Corey, E. J.; Fuchs, P. L. *Tetrahedron Lett.* **1973**, *14*, 376.
16. Marshall, J. A.; Sehon, C. A. *J. Org. Chem.* **1997**, *62*, 4313.
17. Williams, D.; Andersen, R. J.; VanDuyne, G. D.; Clardy, J. *J. Org. Chem.* **1987**, *52*, 332.
18. Marshall, J. A.; DuBay, W. J. *J. Org. Chem.* **1994**, *59*, 1703.
19. Marshall, J. A.; Wang, X.-J. *J. Org. Chem.* **1992**, *57*, 3387.
20. Trost, B. M.; Doherty, G. A. *J. Am. Chem. Soc.* **2000**, *122*, 3801.
21. Hayakawa, Y.; Kawakami, K.; Seto, H.; Furihata, K. *Tetrahedron Lett.* **1992**, *33*, 2701.
22. Marshall, J. A.; Audia, V. H. *J. Org. Chem.* **1987**, *52*, 1106.
23. a) Paal, G. *Chem. Ber.* **1884**, *17*, 2756; b) Knorr, L. *Chem. Ber.* **1884**, *17*, 2863.
24. Fürstner, A.; Weintritt, H. *J. Am. Chem. Soc.* **1998**, *120*, 2817.
25. Borzilleri, R. M.; Weinreb, S. M.; Parvez, M. *J. Am. Chem. Soc.* **1995**, *117*, 10905.
26. Baker, B. J.; Scheuer, P. J.; Shoolery, J. N. *J. Am. Chem. Soc.* **1988**, *110*, 965.

27. a) Danheiser, R. L.; Carini, D. J. *J. Org. Chem.* **1980**, *45*, 3925; b) Danheiser, R. L.; Carini, D. J.; Kawasigroch, C. A. *J. Org. Chem.* **1986**, *51*, 3870; c) Danhaeiser, R. L.; Kawasigroch, C. A.; Tsai, Y.-M. *J. Am. Chem. Soc.* **1985**, *107*, 7233; d) Danheiser, R. L.; Stoner, E. J.; Koyama, H.; Yamshita, D. S.; Klade, C. A. *J. Am. Chem. Soc.* **1989**, *111*, 4407; e) Becker, D. A.; Danheiser, R. L. *J. Am. Chem. Soc.* **1989**, *111*, 389.

28. a) Fleming, I.; Terret, N. K. *J. Organomet. Chem.* **1984**, *264*, 99; b) Fleming, I.; Takakasi, K.; Thomas, A. P. *J. Chem. Soc., Perkin Trans. I* **1987**, 2269.

29. Boezilleri, R. M.; Weinreb, S. M. *Synthesis* **1995**, 347.

30. DiPietro, D.; Borzilleri, R. M.; Weinreb, S. M. *J. Org. Chem.* **1994**, *59*, 5956.

31. Zhang, H. X.; Guibe, F.; Balavoine, G. *J. Org. Chem.* **1990**, *55*, 1857.

32. Ghosal, S.; Luke, G. P.; Kyler, K. S. *J. Org. Chem.* **1987**, *52*, 4296.

33. Sierra, M. A.; de la Torre, M. C. *Angew. Chem. Int. Ed.* **2000**, *39*, 1538.

Subject Index

Dead Ends and Detours: Direct Ways to Successful Total Synthesis
Miguel A. Sierra and María C. de la Torre
Copyright © 2004 WILEY-VCH Verlag GmbH & Co. KGaA, Weinheim
ISBN: 3-527-30644-7

DATE DUE